75

Symmetry and its Applications in Science

Symmetry and its Applications in Science

A. D. Boardman, D. E. O'Connor, and
P. A. Young

University of Salford

A HALSTED PRESS BOOK

John Wiley and Sons
New York

Published in the U.S.A.
by **Halsted Press**
A Division of **John Wiley and Sons, Inc., New York**

ISBN 0 470 08412 X
LC 72-13908

PRINTED AND BOUND IN GREAT BRITAIN

Table of Contents

Preface

This book was motivated by a teach-in conducted by the authors at the University of Salford. The majority of the audience, comprising students and staff, had no previous knowledge of group theory so it was necessary to go back to first principles. It seemed to us, then, that there was no reason why, at the elementary level, the teaching of how to exploit the inherent symmetry of physical phenomena should not be more widespread. This view is supported by the fact that this sort of material is increasingly finding its way into British undergraduate syllabi. A major problem seems to be that although there are a number of excellent books on group theory and its applications they all seem to be, in one respect or another, just too difficult for an undergraduate to assimilate in the time allowed.

We have therefore produced a preliminary text which we hope will give the student the ability and the enthusiasm to tackle the more advanced works. Indeed we have tried, as far as possible, to present the subject in such a way that it is unencumbered by rigorous proofs; a feature all too familiar and all too forbidding to the novice reading any advanced text. All the necessary mathematical background is contained in Appendix A; however some previous contact with vectors and matrices would be an advantage. Also since many applications involve quantum mechanics we have assumed the reader to have a basic knowledge of this subject.

The first three chapters of the book contain the basic group theory needed to understand the applications found in chapters 4 to 9. In these basic chapters we do not put the emphasis on rigour and we have omitted those aspects which we feel can be left out by a reader making a first acquaintance with the subject. The chapters containing the applications can, to a large extent, be read independently of each other. For this reason each chapter contains a preamble of background material and, in the main, it is not necessary to refer to the other chapters on applications. We should emphasize that the chapters on applications concentrate on the role of symmetry and are not designed to teach the subject *ab initio*.

The book is aimed at science students particularly those reading physics, chemistry, and electrical engineering. However since the authors are physicists there is some bias towards physics. As a tentative guide to the use of the book we offer the following schemes of study.

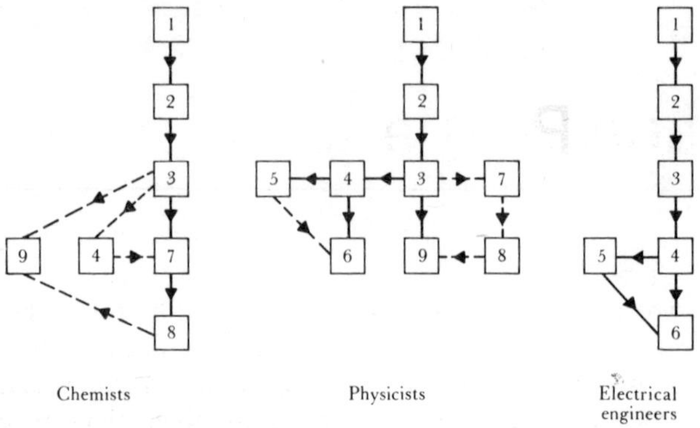

The chapters are represented by labelled boxes and the suggested schemes of study are denoted by full lines. Dotted lines are possible alternatives.

We would like to thank Professor John Leech of the Physics Department in the University of Waterloo, Ontario, Canada, for reading some of the draft. We also acknowledge that the basic material of chapter 8 was provided by Dr M. A. Slifkin of the Physics Department in the University of Salford. Finally, we must thank our typists Mrs M. Fairhurst and Mrs V. E. Hague for working so quickly and so efficiently.

A. D. Boardman
D. E. O'Connor
P. A. Young

Introduction

The notion of symmetry is such common currency that most people instinctively feel that they understand what it means. Indeed, this is reasonable because many objects such as buildings and so on can be seen, at a glance, to possess external features which are related to each other in a regular manner. For instance a church tower with a clock on each wall, will look much the same when viewed from several angles; furthermore a stationary observer would observe that the clock-faces of each side of a rectangular tower would re-appear every quarter of a turn if only the tower could be rotated about a vertical axis. This sort of property is really what is meant by the symmetry of an object.

Now many objects in nature possess this kind of symmetry; for example a crystal or a leaf which may be picked up on a country walk. On the other hand many man-made things possess regular shapes simply because human beings prefer them this way. It seems that a certain love of regularity is built into human nature with the result that a regular object or design is, for most of us, more pleasing than one which is randomly shaped or organized.

This concern with symmetry is not perhaps an accidental one because even atoms, the fundamental building bricks of matter, possess symmetry to a high degree, in fact they are basically spherically symmetric. In addition the fundamental particles making up the nucleus can be classified using a scheme founded on symmetry. Molecules and crystals also possess symmetry and the latter, especially, have an external form which is a manifestation of their regular internal arrangement of atoms.

From a scientific point of view it is necessary to produce a more specific definition of symmetry which really means that a mathematical formulation is required. Fortunately, as will soon be appreciated, there is a branch of mathematics, now over one hundred years old, which is ideally suited to this task.

In order to illustrate the latter remark let us imagine a cube being rotated about an axis perpendicular to one of its faces. If this is done then the cube will appear to be in the same position after every quarter turn (i.e., a rotation of $\pi/2$). Thus the cube, during the rotation assumes a position which is physically indistinguishable from its starting position. A rotation such as the one just mentioned is what is called a *symmetry operation* and obviously there are many such operations which can be performed on the cube (in fact there are forty-eight operations). Hence there really exists a *set* of operations. These sets have rather special mathematical properties and are called *groups*. A knowledge of group theory is therefore indispensable to a proper exploitation of symmetry in solving scientific problems.

The ideas of group theory were first established, in approximately 1840, by Cauchy, but the importance of them to the teaching of mathematics is only now being fully realized. By teaching that the integers for example form a group the underlying logical structure of mathematics can be exposed in an aesthetic and satisfying manner. The satisfaction in doing this rests, to some extent, on the same grounds as the pleasure we derive from architectural symmetry or the symmetry of a flower; but more important, from a technical point of view, is the justification of seeing how an examination of the logical structure can provide us with an insight into the meaning of mathematics. This insight tends to be lost in the more conventional approach in which the structure is tackled from the outside in a piece-meal fashion.

Can the same be said for physics? It would be presumptuous to be too categorical here; but we (the authors) feel that a knowledge of group theory, which provides the ability to completely exploit the symmetry of a given physical situation, provides not only an efficient means of arriving at a solution to a problem but also a clarity of vision which enables the scientist to see the subject in better perspective and even to discover things which are difficult to find by other means. Therefore the exploitation of the symmetry of a given situation should always precede the real task of providing detailed quantitative results.

There are, indeed, many concepts in physics which are underpinned by symmetry arguments but we may not realize that this is so. For example, the atomic quantum numbers and the reason an atom emits a certain type of spectrum have their origin in the fact that the atom possesses spherical symmetry. In fact, the quantum numbers arise in such a logical and natural manner by using group theory that the group-theoretical techniques give a meaning to quantum mechanics which might otherwise be overlooked.

Symmetry, then, can be exploited in many situations, many of which are entirely classical. For example, a consideration of the reflection of light from the surface of a homogeneous dielectric gives the well-known result that, if the incident light is polarized in the plane of incidence its reflected and transmitted components will also be polarized in the same plane. This result can be derived, *at the outset*, without solving any equations, simply by imposing the constraint that the plane of incidence is a plane of symmetry. Indeed if the plane of symmetry is destroyed by the introduction of a magnetic field this result will no longer hold; as, for example, in the case of light reflected from magnetic materials. Further examples concern the laws of conservation of total energy and momentum. The law of conservation of energy is a consequence of the invariance of a system with respect to time translations and the conservation of momentum is a consequence of the invariance of the system with respect to spatial translations.

Another advantage of learning group theory involves computing. Modern computers obviously take much of the labour out of numerical calculations but it is equally apparent that we can learn a great deal about the physics of a problem if, prior to the numerical work, as much analytical progress as possible is made. Such progress can often be made only by exploiting the symmetry of the problem and indeed there are many results which, by using group theory, can be established *exactly*. Also, by preparing the ground in this way, we are led to a more *efficient* use of the computer and to an understanding which enables us to interpret the final results more readily.

1 Basic group theory

1.0 Introduction

A group, to the layman, can be anything from the source of music at the discotheque to merely a collection of objects with something in common. A mathematician considers a group to be something very much more specific. To him a group is a collection of entities which are related one to another in a particular way. A familiar example which we may not have previously recognized as a group is the set of all integers together with the connecting operation of addition. These statements are, of course, not very meaningful as yet; but we hope to make the meaning clear as the chapter progresses.

In this chapter we define the concept of a group and a number of terms related to groups which will be required in order to understand the subsequent chapters. We also discuss symmetry operators, because they act as group elements, and because they are most important in the physical applications of group theory. Finally we discuss some of the groups which are required to provide a description of the total symmetry of physical situations.

1.1 Sets

The entities which form a group are called GROUP ELEMENTS and the collection of group elements is known as a SET. A set of entities is only a collection because no relationship between them is implied. The number of elements (entities) in the set is called the ORDER of the set and is given the symbol g. If we take a fraction of the elements in a set we call this fraction a SUBSET. Obviously we can divide any set into a number of subsets. Examples of sets are the set of all integers; the set of negative integers; the set of students reading Social Science at Oxbridge. Clearly the set of negative integers is a subset of the set of all integers.

Set theory is a subject in its own right as can be appreciated by reference to most books on modern mathematics. However, for the purpose of using this book it is sufficient merely to know what is meant by a set and a subset.

1.2 Set relationships

A group requires that the elements in a set are related to each other. This means that if two elements A and B, say, of a set are selected (for convenience, the group elements will be designated as $A, B, C, \ldots$, etc.) then the relationship between them defines a third element C. Mathematically this statement can be written as

$$AB = C \tag{1.1}$$

We must consider AB as an *ordered* pair because BA is not necessarily equal to C. Equation (1.1) can be 'read' A times B equals C but this does not imply arithmetic multiplication.

Examples of this equation as applied to the set of positive integers are

$$4 + 2 = 6 \text{ (arithmetic addition)} \tag{1.2}$$

$$4 \times 2 = 8 \text{ (arithmetic multiplication)} \tag{1.3}$$

More complicated relationships can be devised in which the operation connecting A and B is not simply addition or multiplication.

1.3 Definition of a group

For a set of elements, with a relationship between them, to form a group four rules must be fulfilled. These rules are

(1) Closure

For every ordered pair of elements A and B in the set there is a unique element C which is also a member of the set. That is,

$$AB = C \tag{1.4}$$

where A, B, and C are all elements of the set.

(2) Associativity

If three or more elements are related, the resulting element must be independent of the way in which two adjacent elements are paired. Thus, if

$$A(BC) = D \tag{1.5}$$

then

$$(AB)C = D \tag{1.6}$$

so that if

$$BC = F \tag{1.7}$$

and

$$AB = G \tag{1.8}$$

then

$$AF = GC = D \tag{1.9}$$

If the elements of the set obey this rule then the set is said to be associative.

(3) Identity element

The set must contain a unique element labelled E such that for all elements of the set

$$EA = AE = A \tag{1.10}$$

The element E which satisfies eq. (1.10) is called the IDENTITY or sometimes the UNIT ELEMENT.

(4) Inverse element

For every element A in the set there must exist *in the set* a unique element labelled A^{-1} such that

$$AA^{-1} = A^{-1}A = E \tag{1.11}$$

The element A^{-1} which satisfies eq. (1.11) is called the INVERSE ELEMENT of A. We should now observe that if

$$AB = BA = E \tag{1.12}$$

then

$$B = A^{-1} \tag{1.13}$$

and

$$A = B^{-1} \tag{1.14}$$

The set of elements with a relationship between them, such that the four rules given above are fulfilled, is called a GROUP.

The number of elements in the group is called the ORDER of the group and will be designated by the symbol g.

1.4 Commutations and multiplications

The order in which two elements are related is of the utmost importance because in general

$$AB \neq BA \tag{1.15}$$

If it does turn out that

$$AB = BA \tag{1.16}$$

then A and B are said to COMMUTE.

We can now see by using the definition of a group that the identity element commutes with all elements of the group (eq. (1.10)) and every element commutes with its inverse (eq. (1.11)).

A group in which every element commutes with every other element is called an ABELIAN GROUP [named after the Norwegian mathematician Abel (1802–29)].

In general elements do not commute so that we must exercise some caution when manipulating equations containing group elements. It is therefore recommended that an equation be changed by one of the two following processes:

(1) Pre-multiplication

In this process we pre-multiply both sides of the equation by the same element. For example if

$$AB = C \tag{1.17}$$

then pre-multiplying by the general element D gives us

$$DAB = DC \tag{1.18}$$

It is interesting to note that if $D = A^{-1}$ then

$$A^{-1}AB = A^{-1}C \tag{1.19}$$

which immediately reduces to

$$B = A^{-1}C \tag{1.20}$$

The use of $A^{-1}A = E$ is probably one of the most frequently used devices in group theory.

(2) Post-multiplication

Here we post-multiply both sides of the equation by some element F, say, to give

$$ABF = CF \tag{1.21}$$

which, assuming that $F = B^{-1}$, gives

$$ABB^{-1} = CB^{-1} \tag{1.22}$$

Thus giving

$$A = CB^{-1} \tag{1.23}$$

Division or cancellation does not exist *per se*. However, instead of division we now have multiplication by inverse elements. For example if

$$AB = CB \tag{1.24}$$

then by post-multiplying by B^{-1} one sees that

$$A = C \tag{1.25}$$

It appears therefore as if we have divided by B. On the other hand if we start with

$$AB = BC \tag{1.26}$$

then one has either

$$A = BCB^{-1} \tag{1.27}$$

or

$$C = B^{-1}AB \tag{1.28}$$

1.5 Multiplication table

Because of the property of closure of a group we can construct a MULTIPLICATION TABLE for the group which will have g rows and g columns. This table requires a knowledge of the relationship between the elements and allows us to find C in eq. (1.1) for *any* ordered pair AB.

As a trivial example let us consider a group of order 2 which has two elements $+1$ and -1 together with the relation of arithmetic multiplication.

The multiplication table, which can easily be constructed, is given in table 1.1

Table 1.1

	1	−1
1	1	−1
−1	−1	1

This table can be symbolized by writing $E = 1$ and $B = -1$. It then becomes table 1.2.

Table 1.2

	E	B
E	EE	EB
B	BE	BB

We should note that the entries in table 1.2 are obtained by pre-multiplying the elements along the top of the table by the elements down the side. All multiplication tables are constructed in this way and they exhibit the fundamental properties of the group. If two groups have the same multiplication table they are said to be ISOMORPHIC, i.e., identical from a mathematical point of view.

1.6 Subgroups and cyclic groups

If we take a subset of the group then its elements may or may not form a group. If the subset happens to obey the four rules for a group (see section 1.3) then the subset itself is a group and is called a SUBGROUP.

Apart from subgroups another interesting group is a cyclic group. The latter group can be constructed by successive multiplication of any element such as C from any group $A, B, C, \ldots$ with itself. Thus, the first multiplication may give

$$CC = B \tag{1.29}$$

and the second

$$CCC = CB = F \tag{1.30}$$

In this way we can attach a meaning to the element C^n which is the element obtained by n successive multiplications of C with itself. If the group containing C is finite then a particular value of n exists such that $C^n = E$. Thus the set of elements $C, C^2, C^3, \ldots, C^n = E$ fulfills all the conditions for a group and is known as a CYCLIC GROUP. Note that this group is ABELIAN because all elements commute with each other. This can be seen by examining the associativity of eq. (1.30). Since we can do successive multiplication of any element in a group and a group has the property of closure it must always contain one or more cyclic subgroups.

1.7 Similarity transformation, conjugation, and classes

Elements of a group can be related to one another by a transformation which is called a SIMILARITY TRANSFORMATION. This transformation is defined (for an *arbitrary* element R) by

$$R^{-1}AR = B \tag{1.31}$$

where A is transformed into B. The action of R is mitigated by the action of R^{-1} leaving B with *similar* nature to A. We should point out here that it is merely conventional to write R^{-1} on the left of A. We could of course define a similarity transformation as $RAR^{-1} = C$.

If A, B, and the element R in eq. (1.31) are all elements of the *same group*, then B is said to be CONJUGATE to A. We can now show that if B is conjugate to A then A must be conjugate to B. This can be seen, for any element R, by noting that if we perform the similarity transformation

$$R^{-1}AR = B \tag{1.32}$$

and re-arrange the result to

$$AR = RB \tag{1.33}$$

then we obtain

$$A = RBR^{-1} = (R^{-1})^{-1}BR^{-1} \tag{1.34}$$

showing that A is indeed conjugate to B.

We can also show that if B is conjugate to A, and C is conjugate to A, then B is conjugate to C. This can be shown from the similarity transformations

$$R^{-1}AR = B \tag{1.35}$$

$$S^{-1}AS = C \tag{1.36}$$

Thus

$$A = RBR^{-1} = SCS^{-1} \tag{1.37}$$

which shows that

$$B = R^{-1}SCS^{-1}R \tag{1.38}$$

Thus if we let

$$R^{-1}S = T^{-1} \tag{1.39}$$

then

$$S^{-1}R = T \tag{1.40}$$

so that

$$B = T^{-1}CT \tag{1.41}$$

which proves that B is conjugate to C.

We now come to the concept of a CLASS which is defined as a subset of the group, consisting of all the elements of the group which are conjugate to each other. The class containing A is found by performing the similarity transformation of eq. (1.31) where R ranges through all the elements of the group.

Any element F, say, which commutes with all the elements R of the group satisfies, for *every* R,

$$R^{-1}FR = FR^{-1}R = R^{-1}RF = F \tag{1.42}$$

Thus starting with the element F we cannot produce another element. F therefore, because it commutes with all the elements of the group, stands in a class by itself. Examples of F are the identity element E and the elements of a cyclic group.

Classes are mutually exclusive so that every element belongs to one and only one class; a fact which follows from the preceding analysis. A group can therefore be divided up into classes.

It should be noted that for elements A, B to be conjugate, the element R in eq. (1.31) must belong to the group of A and B. Therefore if A and B also belong to a subgroup of the group they will not be conjugate with respect to the subgroup unless R also belongs to the subgroup. Consequently if two elements belong to the same class of a group they need not belong to the same class in a subgroup.

1.8 Symmetry operations

Obviously all physical situations exhibit, in the most general sense, spatial symmetry. Some show a high degree of symmetry, as for example atoms and crystals. The concepts of symmetry are not however confined to microscopic phenomena or crystallography, they can in fact be applied with equal facility to electrical networks and to problems involving general engineering structures.

We now introduce the idea of a SYMMETRY OPERATION by considering the movement of an object in such a way that it leaves behind a trace of itself. If after this movement the new orientation of the object is superimposed on its trace we cannot then detect that the object has moved. This kind of movement is called a symmetry operation. As a familiar illustration let us consider the rotation of a sphere about an axis through its centre. Such a rotation will clearly leave the sphere superimposed on its trace showing that any, arbitrary, rotation about any axis through the centre is a symmetry operation. Another example is a rotation of 120°, about a perpendicular axis, through the centre of an equilateral triangle. This operation also leaves the object superimposed on its trace. However, a rotation of 100° about this axis is clearly not a symmetry operation.

We are now going to discuss, in some detail, the list of symmetry operations which act in an ordinary Euclidean space, i.e., those operations which act in such a way that a system is left in the same or an equivalent position. These operations which we label as C_n, σ_h, σ_v, σ_d, S_n, and i are defined as follows.

C_n: This is an operation which effects a rotation through an angle $2\pi/n$ about an axis, fixed in space, where n is an integer. In addition we can have C_n^k, which is C_n raised to the power k, obviously a rotation through an angle $2\pi k/n$ about the same axis. C_n^n is a rotation through 2π and is the identity operation since a rotation through 2π leaves the object as if it had *not* been moved at all. C_n^n is of course the same as C_1. n is known as the multiplicity of the axis, and the latter is called an *n-fold axis.* If a system has more than one axis of symmetry then the axis with the highest value of n is called the PRINCIPAL AXIS. It should be pointed out that some systems do not have a unique principal axis as will be seen later. The effect of the rotation C_n can be seen in Fig. 1.1(a), in which the point B is the new position of point A; the axis of rotation being perpendicular to the plane of the page.

σ_d or σ_v: If we consider a vertical plane, fixed in space, containing the principal axis (this axis is always considered to be vertical) then the operation which effects a reflection in this plane is defined as σ_v. The new position B, of the point A, after such an operation σ_v is shown in Fig. 1.1(b). B is simply the mirror image of A. σ_d is a form of σ_v and is discussed in section 1.9(9).

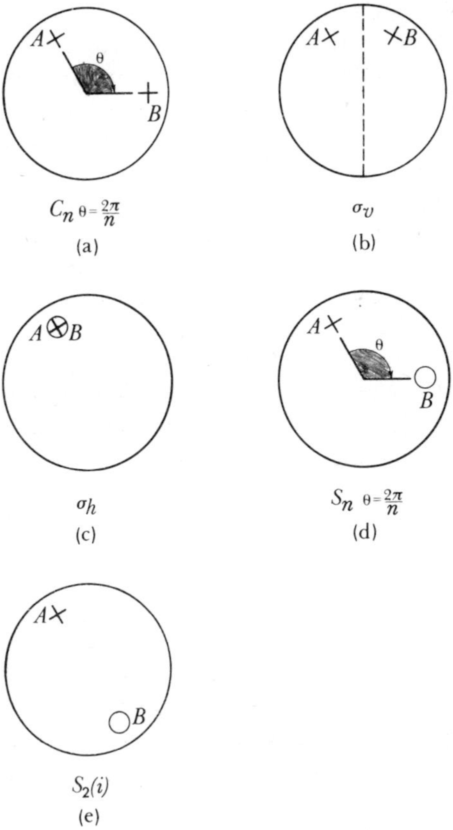

Fig. 1.1 *The basic symmetry operations.* × *denotes a point above the horizontal plane.* ○ *denotes a point below the plane* – – – – – – *denotes a vertical mirror plane. B is the new position of A after the symmetry operation.*

σ_h: This is an operation which produces a reflection in a horizontal plane perpendicular to the principal axis. The new position B, of a point A situated above the plane, is in the mirror image position below the plane, as shown in Fig. 1.1(c). We should observe that there is only one horizontal plane whereas there can be many vertical planes.

S_n: This operation produces an IMPROPER ROTATION which is defined as a rotation about a principal axis fixed in space, followed by a reflection σ_h or vice versa. The full operation can be written sequentially as

$$S_n = C_n \sigma_h = \sigma_h C_n \tag{1.43}$$

The new position of A (i.e., B) after the operation S_n, can be seen in Fig. 1.1(d).

$S_2 = i$: A most interesting operation which also produces an improper rotation is called the INVERSION operation and is designated by the symbol i or S_2. The action of i on the Cartesian coordinates (x, y, z) is to transform them to $(-x, -y, -z)$. The inversion operation can only be performed if there exists a unique point O in the system such that every other point can be joined through O by means of a straight line to an equivalent point. O is called a centre of symmetry. The new position B, of the point A, after inversion is shown in Fig. 1.1(e).

The action of the above collection of operations always leaves at least one point in the system unmoved. For example, any point on a rotation axis will remain unmoved, as will any point on a reflection plane.

In the case of a periodic lattice (cf. chapter 4) we must introduce a set of translational operations, **t**, which we add to the list of symmetry elements where **t** is the vector representing the translational displacement. This operation will move every point in the system and in this respect is fundamentally different from the rotation and reflection operations.

The translational operation could be combined with the rotation and reflection operations and these types of operations will be discussed in chapter 4.

1.9 Point groups

Symmetry operations can be combined to produce other operations as for example the combination of C_2 with σ_h to produce i (eq. (1.43)). Indeed relationships (cf. (1.2)) exist between the symmetry operations so that we can select appropriate sets which behave like the *elements* of a group (cf. (1.3)).

In this connection we should now emphasize that, in any product of operations, AB say, we must 'read' this as the operation B *followed* by the operation A.

If we exclude translational operations, for the time being, we can then construct POINT GROUPS containing elements consisting entirely of rotation and reflection operations which leave at least one point in the system unmoved. These point groups are described below and are denoted by the Schoenflies symbols. Actually there are alternative notations and a comparison is listed in Appendix B.

The first four groups we consider contain only rotations (in future we will, for convenience, drop the word 'operations' in favour of a looser terminology such as rotations or reflections)
These are

(1) C_n: Although this symbol has been used before it must in this context be regarded as the cyclic group obtained by successive multiplication of the *element* C_n. It is therefore used in the description of a system possessing an axis of symmetry (cf. Fig. 1.2(a)) and possesses the *n elements*, C_n, $C_n^2, C_n^3, \ldots, C_n^n = E$. An example of such a group is C_3 which has three elements namely C_3, C_3^2, and E which are rotations of 120°, 240°, and 0°. Note that a rotation of 240° is equivalent to a rotation of −120°.

(2) D_n: This group concerns a system possessing one n-fold axis called the principal axis and n 2-fold axes symmetrically placed in a plane perpendicular to it (cf. Fig. 1.2(b)). The n-fold axis provides the n elements of the cyclic group C_n (thus C_n is a subgroup of D_n). The group also contains one C_2 element provided by every perpendicular 2-fold axis where we do not count $C_2^2 = E$ because E only occurs once in a set of group elements and it has already appeared in the n elements of C_n. The group D_n therefore contains a total of $2n$ elements. Actually the group D_n contains all the rotations which are symmetry operations of a regular polygon; the group D_3 for example, describes the symmetry of an equilateral triangle and has elements

$$E, C_3, C_3^2, C_2^{(1)}, C_2^{(2)}, C_2^{(3)}$$

where the superscripts in parentheses differentiate between the three 2-fold axes.

(3) T: This group is the group of symmetry rotations of the regular tetrahedron which has four 3-fold axes and three 2-fold axes as shown in Fig. 1.2(c) and Fig. 1.2(d). The four 3-fold axes pass through each apex and the centre of the opposite face. The three 2-fold axes go through the centres of opposite edges. Each three-fold axis contributes two elements C_3 and C_3^2; each two-fold axis contributes the element C_2; and of course there is the element E giving a total of 12 elements. This group differs from the two previous groups in so far as the system does not possess a unique principal axis. In fact there are four 3-fold axes which are equivalent to each other.

(4) O: This is the group of symmetry rotations of the regular octahedron. An octahedron can be formed from a cube by slicing off its corners in a regular fashion as shown in Fig. 1.2(e). A regular octahedron is therefore invariant under all of the symmetry operations of a cube. The cube shown in Fig. 1.2(f) has three 4-fold axes passing through the centres of opposite faces, four 3-fold axes joining diagonal apices and six 2-fold axes joining

the centres of opposite parallel edges. Each four-fold axis contributes to the group O three elements C_4, $C_4^2 = C_2$, and C_4^3; each three-fold axis contributes two elements C_3 and C_3^2; and each two-fold axis contributes one element C_2. Thus, including E, the group O contains 24 elements. We can see that the four 3-fold axes of T are equivalent to the four 3-fold axes of O and the three 2-fold axes of T are contained in the six 2-fold axes of O. Therefore all the elements of T must be contained in the elements of O. Thus T is a subgroup of O.

We will now go on to consider point groups which contain reflections and improper rotations. These groups are

(5) S_n (where n is an even number): This is a cyclic group and describes the symmetry of a system possessing one n-fold improper rotation axis and therefore possesses the *elements* $S_n, S_n^2, \ldots, S_n^n = E$. Note that n *must* be even so that the group can contain $E = S_n^n$. Otherwise S_n^n would

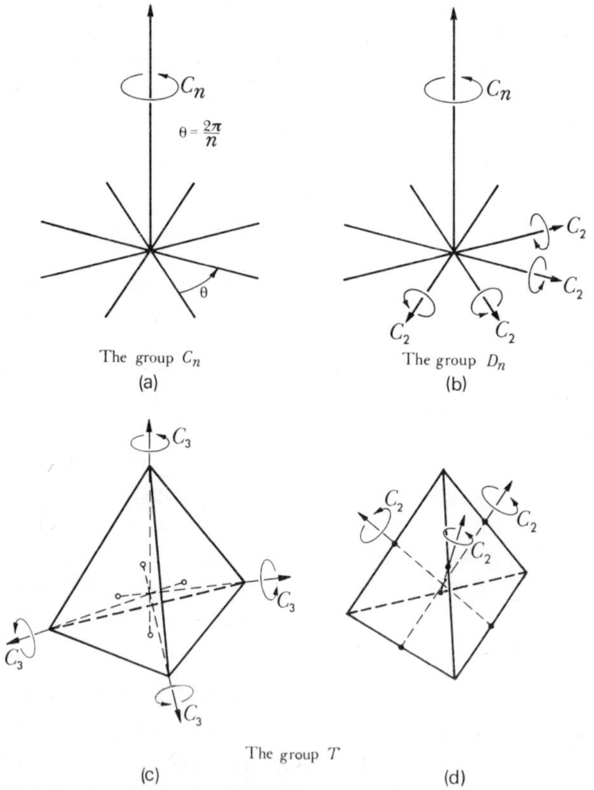

Fig. 1.2 *(a) and (b) illustrate respectively the operations of the group C_n and D_n. (c) and (d) illustrate the operations of the group T. (e) and (f) illustrate the operations of the group O.*

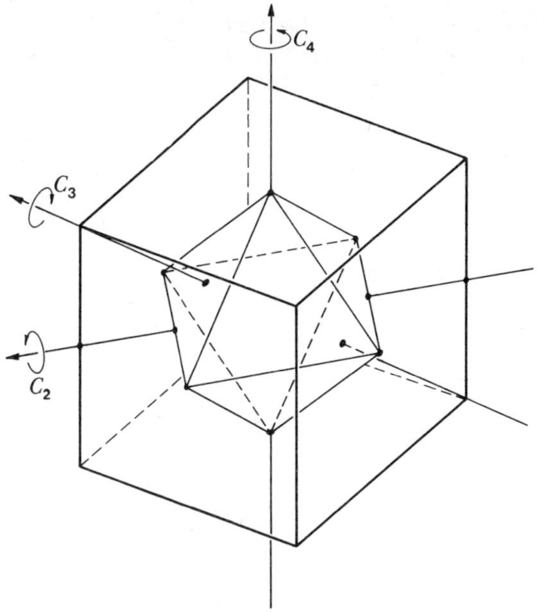

The group O
(e)

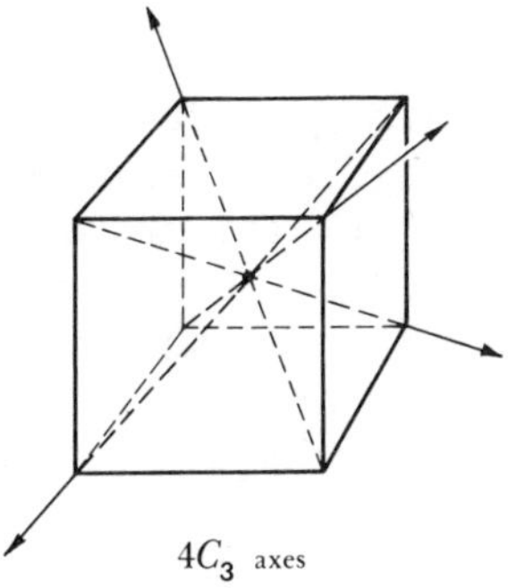

$4C_3$ axes

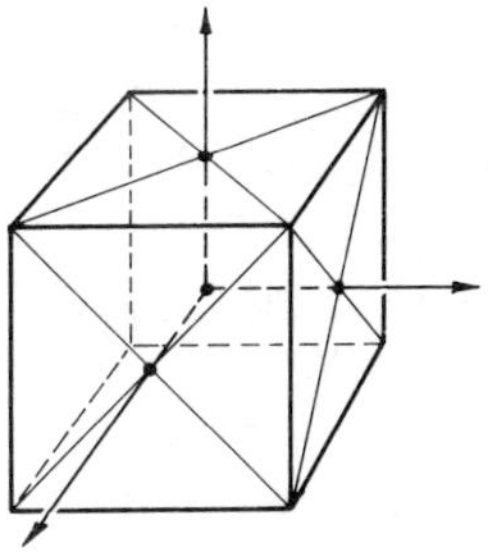

$3C_4$ axes

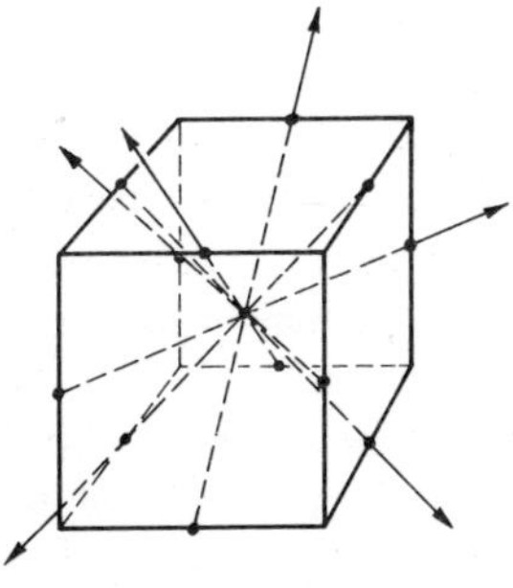

$6C_2$ axes

The group O
(f)

simply be σ_h. If $n/2$ is an odd number then the element $S_n^{n/2}$ becomes $S_2 = i$ the inversion element, where we have used the fact that $C_n^{n/2} = C_2$.

(6) C_{nh}: This is the group of the symmetry elements of a system possessing a single n-fold rotation axis which is called the principal axis and a reflection plane which is perpendicular to the principal axis. It contains the group C_n as a subgroup together with the n elements produced by multiplying the n elements of C_n with σ_h. The group therefore contains $2n$ elements and is Abelian because σ_h always commutes with C_n. The group C_{1h} has only two elements E and σ_h; and we can see that the group C_{nh} is the group produced by multiplying all the elements of the group C_n by the elements of C_{1h}. A group formed in this way is called a DIRECT PRODUCT GROUP. This type of group is given a more detailed consideration in chapter 3, and is written symbolically as

$$C_{nh} = C_n \otimes C_{1h}$$

Finally, if n is even the group contains the inversion element in which case it contains S_n as a subgroup.

(7) C_{nv}: This is the group of the symmetry operations of a regular pyramid which has one n-fold principal axis and n reflection planes passing through this axis. These n vertical reflection planes are symmetrically arranged so that there is an angle π/n between adjacent planes. The group contains the n elements of C_n as a subgroup and $n\sigma_v$ elements, i.e., a total of $2n$ elements.

(8) D_{nh}: This is the group of the symmetry operations of a regular prism and is in fact another direct product group, i.e., the direct product of D_n and the group C_{1h}; the latter possessing the elements E and σ_h. In a similar fashion to the production of C_{nh}, D_{nh} is obtained by multiplying all the elements of D_n by the two elements of C_{1h}; hence it contains $2n$ elements. Once again if n is even the group contains the inversion element. Symbolically we can write D_{nh} as $D_n \otimes C_{1h}$ which if n is even becomes $D_n \otimes S_2$.

(9) D_{nd}: This group contains the n elements of the group D_n as a subgroup and contains symmetry operations produced by vertical reflection planes, *bisecting* (the suffix d refers to the fact that these planes lie diagonally between the C_2 axes) the angles between the n horizontal two-fold axes of the group. These planes produce n elements σ_d and also convert the principal axis (the n-fold axis) into a $2n$-fold improper rotation axis; thus introducing n new elements which are the odd components of the cyclic group S_{2n}, i.e., $S_{2n}, S_{2n}^3, \ldots, S_{2n}^{2n-1}$. The even terms of this cyclic group are in fact already contained in the group D_n. The total number of elements in D_{nd} is $4n$, i.e., the $2n$ elements of D_n plus n elements from reflection planes σ_d and n elements from S_{2n}. If n is odd then D_{nd} is the direct product group. $D_n \otimes S_2$.

(10) T_h: The group T_h is the direct product group $T \otimes S_2$ which obviously contains T as a subgroup plus the 12 new elements obtained by multiplying all the elements of T by S_2. T_h therefore contains 24 elements.

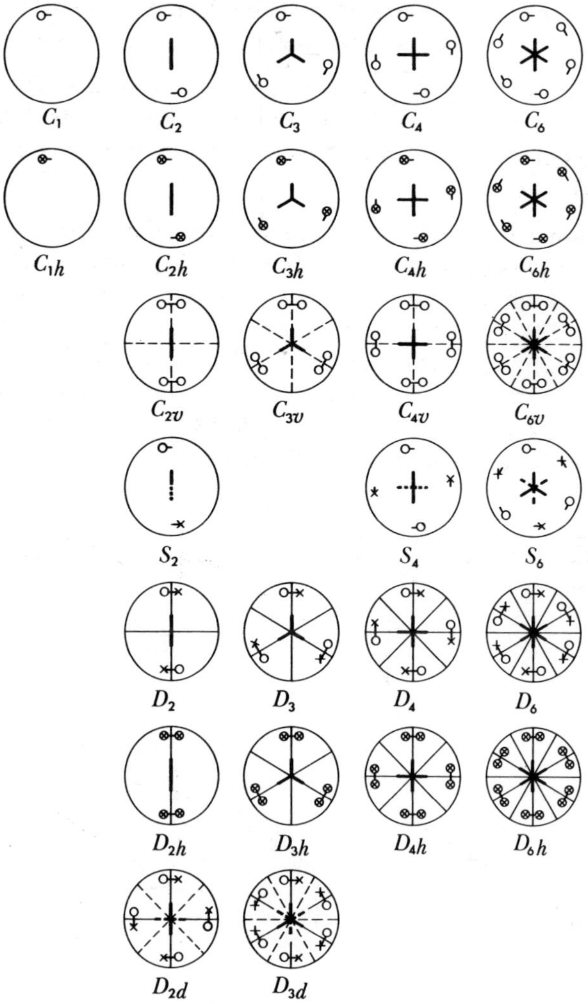

Fig. 1.3 *Projection diagrams illustrating the finite point groups which do not possess more than one principal axis of order greater than two.* ○ *denotes a point below the horizontal plane and* × *denotes a point above that plane. The tails on* ○ *and* × *indicate that the points are displaced from a diameter of the circle: this has been done to remove vertical reflection planes where necessary.* ⊗ *denotes* × *directly above* ○ *indicating the presence of a horizontal reflection plane.* – – – – – – *represents a vertical reflection plane. The principal rotation axis, which is vertical and passes through the centre of the circle, is represented by a figure such as* ⅄ *denoting its multiplicity.* ——— *represents a two-fold rotation axis in the horizontal plane. Only groups with* $n \leqslant 6$ *are shown.*

(11) T_d: This is the group of symmetry operations of a regular tetrahedron and has a subgroup T. Thus T_d consists of T together with six reflection planes and three 4-fold improper rotation axes. T_d therefore contains 24 elements comprising the 12 elements of T plus the 6 σ_v elements (one from each reflection plane) and two elements S_4 and S_4^3 arising from each of the three 4-fold improper rotation axes.

(12) O_h: This is the group of symmetry operations of a cube and is the direct product group of O with S_2 possessing the 24 elements of the group O plus the 24 elements obtained by multiplying each of the elements of O by i, i.e., $O_h = O \otimes S_2$.

There are a number of point groups with infinite-fold rotation axes which have physical significance. These are $C_{\infty v}$ which is, for example, the symmetry group of a heteronuclear diatomic molecule such as HCl or CO; and $D_{\infty h}$ which is, for example, the symmetry group of a homonuclear diatomic molecule such as H_2, N_2, and O_2; and R_3 which is the group of all rotations of a sphere which has an infinite number of infinite-fold rotation axes all passing through the centre of the sphere; and finally $R_3 \otimes S_2$ which is the full group of the sphere.

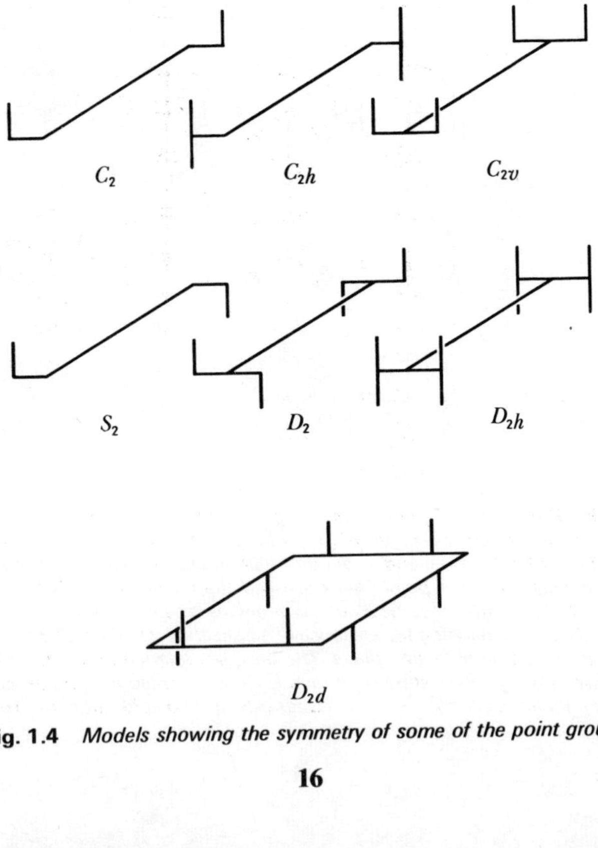

Fig. 1.4 *Models showing the symmetry of some of the point groups.*

All the finite point groups can be illustrated using diagrams such as those shown in Fig. 1.3. These diagrams are to be interpreted as follows. The plane of the paper is the horizontal plane so that the principal axis is perpendicular to it. Any arbitrary point in a three-dimensional figure is represented by a mark (projection) on this plane; a cross indicating a point above the plane and a circle representing a point below the plane. Solid lines represent two-fold rotation axes in the horizontal plane and dotted lines represent vertical mirror planes. By performing all the operations of the group we generate a set of points which are said to be equivalent points. This set of points characterizes the group. In addition we have added, at the centre of the diagram, a figure possessing a rotational symmetry corresponding to the principal axis.

Using Fig. 1.3 we can construct three-dimensional models which have the symmetry of a particular group. We construct a model with g points (g is the order of the group) such that each point is an equivalent point. Figure 1.4 shows some possible models for some of the groups listed in Fig. 1.3.

1.10 Illustration of fundamental concepts

In order to illustrate the concepts developed in the earlier sections of this chapter we will now consider in detail the group D_3 which is the group of the rotations of the equilateral triangle shown in Fig. 1.5(a). The six operations of the group are

$E =$ the identity.
$A = C_2$ a rotation of π about a *fixed* axis initially through a.
$B = C_2$ a rotation of π about a *fixed* axis initially through b.
$C = C_2$ a rotation of π about a *fixed* axis initially through c.
$D = C_3$ a rotation (taken for convenience to be anticlockwise) of $2\pi/3$ about the perpendicular axis to bring the apex a to position c.
$F = C_3^2$ a clockwise rotation of $2\pi/3$ which brings the apex a to position b.

The triangle is of course transformed to a new position which is physically indistinguishable from its old position. The labels a, b, c on the vertices are introduced merely as an aid to help us to follow the operations through.

Each basic operation on the triangle is shown in Fig. 1.5(a) and the action of some products of operations is shown in Fig. 1.5(b). The latter diagram enables us to find the single operation which is equivalent to the product of any two operations and hence the multiplication table (cf. table 1.3). D_3 is obviously not Abelian because $AB = F$ and $BA = D$ showing that $AB \neq BA$. Also from table 1.3 we determine the inverse of each element by noting $DF = E$ giving $F^{-1} = D$ and $D^{-1} = F$ and the rest of the elements are their own inverses.

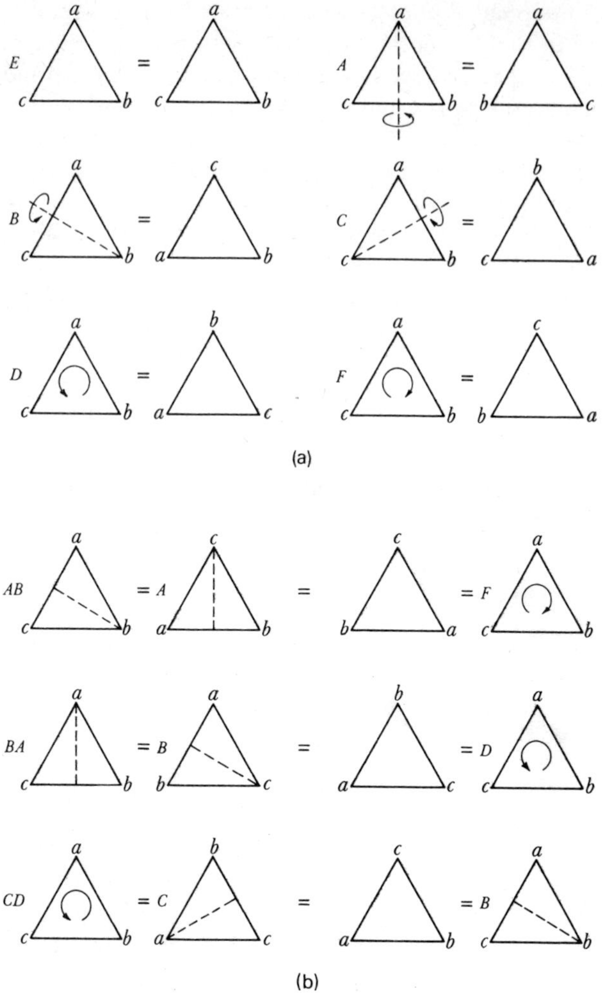

Fig. 1.5 *(a) Symmetry operations of the group D_3. (b) Some products of the symmetry operations of D_3.*

Table 1.3 Multiplication table of the group D_3

	E	A	B	C	D	F
E	E	A	B	C	D	F
A	A	E	F	D	C	B
B	B	D	E	F	A	C
C	C	F	D	E	B	A
D	D	B	C	A	F	E
F	F	C	A	B	E	D

The classes of the group (cf. section 1.7) can be now found by performing a similarity transformation on each element of D_3, using eq. (1.31) with all the elements of the group.

For example using the element A and performing all the possible similarity transformations we obtain

$$
\begin{aligned}
E^{-1}AE &= A \\
A^{-1}AA &= A \\
B^{-1}AB &= C \\
C^{-1}AC &= B \\
D^{-1}AD &= B \\
F^{-1}AF &= C
\end{aligned}
\tag{1.44}
$$

We therefore conclude that A, B, and C form a class. There is in fact no point in performing the same calculation for B and C because classes are mutually exclusive, i.e., B and C must necessarily be in the same class as A. Considering the element D we obtain

$$
\begin{aligned}
E^{-1}DE &= D \\
A^{-1}DA &= F \\
B^{-1}DB &= F \\
C^{-1}DC &= F \\
D^{-1}DD &= D \\
F^{-1}DF &= D
\end{aligned}
\tag{1.45}
$$

which shows that D and F also form a class. Since E is obviously in a class by itself, the group D has therefore three classes, namely,

$$
\begin{aligned}
\mathscr{C}_1 &= E \\
\mathscr{C}_2 &= D, F \\
\mathscr{C}_3 &= A, B, C
\end{aligned}
\tag{1.46}
$$

We should observe that all the elements in a class are of a similar nature. Thus D and F are rotations of $2\pi/3$, one clockwise and the other anti-clockwise whereas A, B, and C are rotations of π about different axes. D and F are in the same class in the case of D_3 because of the existence of C_2 rotations which bring them into equivalence. In the group C_{3h}, for example, rotations of $\pm 2\pi/3$ are in different classes.

Finally we can show that there are five cyclic subgroups (cf. section 1.6) which are

$$\begin{aligned} G_1 &= E \\ G_2 &= E, A \\ G_3 &= E, B \\ G_4 &= E, C \\ G_5 &= E, D, F \end{aligned} \tag{1.47}$$

This example completes a survey of basic group theory in which we have established the concepts necessary to an understanding of the next chapter and considered in some detail the important point groups. It is necessary now to proceed to a new aspect of group theory namely that of group representations.

1.11 Problems

1.1 The connecting operators (a) addition, (b) subtraction, (c) multiplication, (d) division, (e) scalar product, (f) vector product could act in the sets, (1) all integers, (2) all real numbers, (3) all vectors in two dimensions, (4) all vectors in three dimensions to produce groups. Determine the appropriate combinations of operator and set and explain which of the group rules (cf. section (1.3) are not satisfied in the cases where a group is not formed.

1.2 Using figures, such as those in Fig. 1.5, establish the multiplication table 1.3.

1.3 Use the multiplication table of the group D_3 of table 1.3 to find the product of the elements $A\ B\ C\ D\ F$ and check that the result is independent of the way in which you combine the elements in pairs thus demonstrating associativity.

1.4 Determine the group of rotations which are the symmetry operations of a square. By using a method similar to that used in problem 1.2 determine the multiplication table. Hence determine the group class structure and its subgroups.

1.5 Show that the introduction of horizontal or vertical reflection planes to the groups S_n does not give rise to any new groups outside those already discussed in section 1.9. In particular show that

$$\text{`}S_{4v}\text{'} = D_{2d}, \qquad \text{`}S_{4h}\text{'} = C_{4h}$$

1.6 Consider the three equi-triangular prisms of the figure (opposite) in which (a) is entirely white while (b) and (c) consist of white and black parts. Determine the point groups to which the three objects belong. If an operator exists, which will turn black to white and vice versa, show that one can construct another symmetry group for each of the objects (b) and (c). These latter groups are called black and white groups and are of great importance in magnetism.

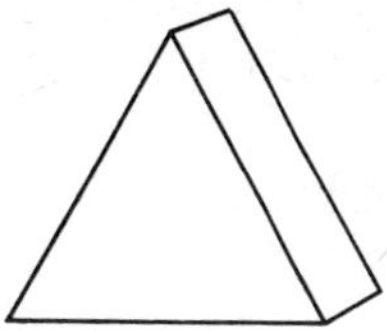

(a)

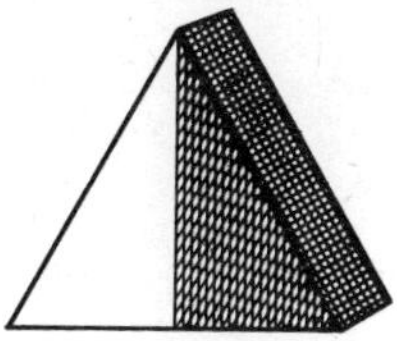

(b)

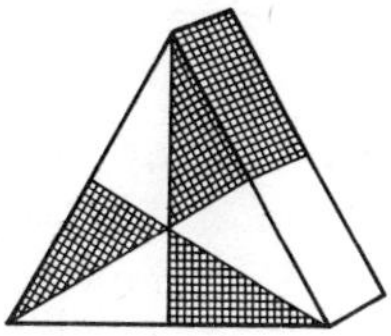

(c)

2 Group representations

2.0 Mapping

Up until now only rather abstract quantities like symmetry operations have been considered. However, if we wish to apply group theory to physical situations it is necessary to introduce a means of quantifying the theory. We do this by representing the symmetry operations by matrices, whose elements can be *real or complex*.

Obviously an arbitrary matrix will not be a satisfactory representation of a symmetry operation; the matrix must be, in some way, compatible with it. The sets of such compatible matrices will eventually enable us to draw certain quantitative conclusions about a given physical situation as for example the degeneracy of an eigenstate of a hydrogen atom. However, before we consider the relationship of matrices to symmetry operations we will first develop the concept of *mapping*, because it is by this process that we represent symmetry operations by matrices.

If we have a set of elements it is possible to produce another set of elements in such a way that there is some relationship between each element in one set and an element in the other. For example, consider a set of five men, Albert, Bill, Colin, David, and Edward; then we can construct the set of numbers which correspond to their ages, i.e.,

$$
\text{Set } A \text{ of men} \quad
\begin{array}{l}
\text{Albert} \;\nearrow\; 28 \\
\text{Bill} \;\nearrow\; 33 \\
\text{Colin} \;/\rightarrow\; 36 \\
\text{David} \\
\text{Edward} \nearrow\; 40
\end{array}
\quad \text{Set } B \text{ of numbers}
$$

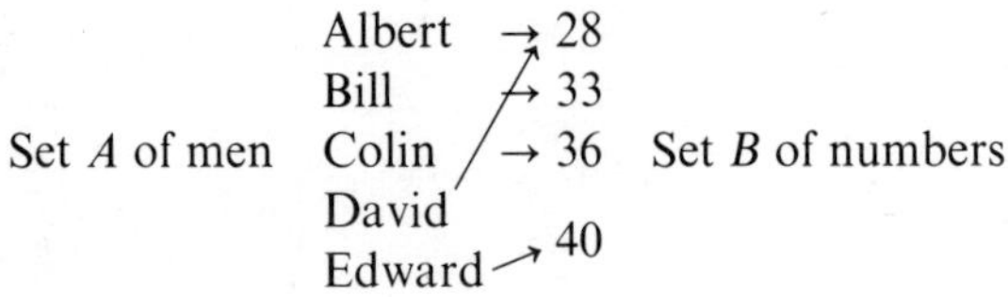

We now say that the set A has been *mapped* onto the set B, where each *object* (a man) in set A has a unique *image* (a number) in set B.

The direction of the arrow goes from object to image and in this example the arrow cannot be reversed as there is no unique relationship between the set of numbers and the men. Obviously the arrows relating a set of husbands to the set of their wives could be reversed, for example

Albert ↔ Angela
Bill ↔ Barbara
Colin ↔ Catherine
David ↔ Dorothy
Edward ↔ Ellen

A mapping in which the arrows are reversible, i.e., where either set can be considered as objects is called a *reversible mapping*, or alternatively a *one-to-one correspondence*. This type of mapping is the basis of counting. For example the fingers and thumbs form a handy reference set on to which we can map the set of objects to be counted. We can also have a *two-to-one* and indeed a *many-to-one correspondence*. The mapping of a set of parents on to the set of their eldest children is a two-to-one correspondence, i.e.,

Albert ↘
Angela ↗ Agnes
Bill ↘
Barbara ↗ Bob
Colin ↘
Catherine ↗ Chris

The essence of group theory lies in the mapping of the elements in the group (e.g., symmetry operations, etc.) on to a representation group (i.e., square matrices). However in order to be useful the mapping must be a one-to-one or a many-to-one correspondence. Furthermore it is most important that the elements in the representation set form a group, possessing the same multiplication table as the original group so that the representation group will have all the properties (mathematically speaking) of the original group.

If the correspondence is one-to-one the representation is said to be *isomorphic* or a *true representation.* On the other hand if the correspondence is many-to-one the representation is said to be *homomorphic*.

2.1 Representation of the group D_3

We will now illustrate this rather abstract introduction to representations, with the aid of the group D_3. In table 2.1 we list the symmetry elements (shown in Fig. 1.5(a)) of the group D_3 together with square matrices $\mathbf{\Gamma}(R)$ which are used to represent them. The use of the superscript in parentheses indicates the axis of rotation and we do not at this stage give any justification for the choice of $\mathbf{\Gamma}(R)$. The six different representations are each denoted by $\mathbf{\Gamma}^i(R)$ and the superscripts are used to differentiate between them. (Note that each element is given a letter A, B, C, etc., denoting the abstract group.)

Table 2.1 Six representations of the group D_3

D_3	E	$C_2^{(a)}$	$C_2^{(b)}$	$C_2^{(c)}$	C_3	C_3^2
Abstract notation	E	A	B	C	D	F
$\mathbf{\Gamma}^1(R)$	1	1	1	1	1	1
$\mathbf{\Gamma}^2(R)$	1	-1	-1	-1	1	1
$\mathbf{\Gamma}^3(R)$	$\begin{bmatrix} 1 & 0 \\ 0 & 1 \end{bmatrix}$	$\begin{bmatrix} 1 & 0 \\ 0 & -1 \end{bmatrix}$	$\begin{bmatrix} -\frac{1}{2} & -\frac{\sqrt{3}}{2} \\ -\frac{\sqrt{3}}{2} & \frac{1}{2} \end{bmatrix}$	$\begin{bmatrix} -\frac{1}{2} & \frac{\sqrt{3}}{2} \\ \frac{\sqrt{3}}{2} & \frac{1}{2} \end{bmatrix}$	$\begin{bmatrix} -\frac{1}{2} & \frac{\sqrt{3}}{2} \\ -\frac{\sqrt{3}}{2} & -\frac{1}{2} \end{bmatrix}$	$\begin{bmatrix} -\frac{1}{2} & -\frac{\sqrt{3}}{2} \\ \frac{\sqrt{3}}{2} & -\frac{1}{2} \end{bmatrix}$
$\mathbf{\Gamma}^4(R)$	$\begin{bmatrix} 1 & 0 \\ 0 & 1 \end{bmatrix}$	$\begin{bmatrix} 0 & -1 \\ -1 & 0 \end{bmatrix}$	$\begin{bmatrix} -\frac{\sqrt{3}}{2} & \frac{1}{2} \\ \frac{1}{2} & \frac{\sqrt{3}}{2} \end{bmatrix}$	$\begin{bmatrix} \frac{\sqrt{3}}{2} & \frac{1}{2} \\ \frac{1}{2} & -\frac{\sqrt{3}}{2} \end{bmatrix}$	$\begin{bmatrix} -\frac{1}{2} & \frac{\sqrt{3}}{2} \\ -\frac{\sqrt{3}}{2} & -\frac{1}{2} \end{bmatrix}$	$\begin{bmatrix} -\frac{1}{2} & -\frac{\sqrt{3}}{2} \\ \frac{\sqrt{3}}{2} & -\frac{1}{2} \end{bmatrix}$
$\mathbf{\Gamma}^5(R)$	$\begin{bmatrix} 1 & 0 & 0 \\ 0 & 1 & 0 \\ 0 & 0 & 1 \end{bmatrix}$	$\begin{bmatrix} 1 & 0 & 0 \\ 0 & -1 & 0 \\ 0 & 0 & 1 \end{bmatrix}$	$\begin{bmatrix} -\frac{1}{2} & -\frac{\sqrt{3}}{2} & 0 \\ -\frac{\sqrt{3}}{2} & \frac{1}{2} & 0 \\ 0 & 0 & 1 \end{bmatrix}$	$\begin{bmatrix} -\frac{1}{2} & \frac{\sqrt{3}}{2} & 0 \\ \frac{\sqrt{3}}{2} & \frac{1}{2} & 0 \\ 0 & 0 & 1 \end{bmatrix}$	$\begin{bmatrix} -\frac{1}{2} & \frac{\sqrt{3}}{2} & 0 \\ -\frac{\sqrt{3}}{2} & -\frac{1}{2} & 0 \\ 0 & 0 & 1 \end{bmatrix}$	$\begin{bmatrix} -\frac{1}{2} & -\frac{\sqrt{3}}{2} & 0 \\ \frac{\sqrt{3}}{2} & -\frac{1}{2} & 0 \\ 0 & 0 & 1 \end{bmatrix}$
$\mathbf{\Gamma}^6(R)$	$\begin{bmatrix} 1 & 0 & 0 \\ 0 & 1 & 0 \\ 0 & 0 & 1 \end{bmatrix}$	$\begin{bmatrix} 1 & 0 & 0 \\ 0 & 0 & 1 \\ 0 & 1 & 0 \end{bmatrix}$	$\begin{bmatrix} 0 & 0 & 1 \\ 0 & 1 & 0 \\ 1 & 0 & 0 \end{bmatrix}$	$\begin{bmatrix} 0 & 1 & 0 \\ 1 & 0 & 0 \\ 0 & 0 & 1 \end{bmatrix}$	$\begin{bmatrix} 0 & 1 & 0 \\ 0 & 0 & 1 \\ 1 & 0 & 0 \end{bmatrix}$	$\begin{bmatrix} 0 & 0 & 1 \\ 1 & 0 & 0 \\ 0 & 1 & 0 \end{bmatrix}$

The most important fact, about these representation matrices, as we emphasized in section 2.0, is that they all obey the same multiplication table (table 1.3) so that as can be readily checked, using eq. (A.13),

$$
\begin{aligned}
AB &= F \\
C_2^{(a)}C_2^{(b)} &= C_3^2 \\
\boldsymbol{\Gamma}^1(A)\boldsymbol{\Gamma}^1(B) &= \boldsymbol{\Gamma}^1(F) \\
\boldsymbol{\Gamma}^2(A)\boldsymbol{\Gamma}^2(B) &= \boldsymbol{\Gamma}^2(F) \\
\boldsymbol{\Gamma}^3(A)\boldsymbol{\Gamma}^3(B) &= \boldsymbol{\Gamma}^3(F) \\
\boldsymbol{\Gamma}^4(A)\boldsymbol{\Gamma}^4(B) &= \boldsymbol{\Gamma}^4(F) \\
\boldsymbol{\Gamma}^5(A)\boldsymbol{\Gamma}^5(B) &= \boldsymbol{\Gamma}^5(F) \\
\boldsymbol{\Gamma}^6(A)\boldsymbol{\Gamma}^6(B) &= \boldsymbol{\Gamma}^6(F)
\end{aligned}
\tag{2.1}
$$

It is clear that $\boldsymbol{\Gamma}^1(R)$ and $\boldsymbol{\Gamma}^2(R)$ are homomorphic representations, i.e., all the group elements are mapped on to one, and two matrices respectively; (numbers are considered as one-dimensional matrices cf. (A.1)) while $\boldsymbol{\Gamma}^3(R)$, $\boldsymbol{\Gamma}^4(R)$, $\boldsymbol{\Gamma}^5(R)$ and $\boldsymbol{\Gamma}^6(R)$ are isomorphic.

An important feature of representations is the number of rows or columns in the matrices—this number is called the *dimension* of the representation and is denoted by l. $\boldsymbol{\Gamma}^1(R)$ and $\boldsymbol{\Gamma}^2(R)$ are one-dimensional, $\boldsymbol{\Gamma}^3(R)$ and $\boldsymbol{\Gamma}^4(R)$ are two-dimensional while $\boldsymbol{\Gamma}^5(R)$ and $\boldsymbol{\Gamma}^6(R)$ are three-dimensional representations.

There is no obvious limit to the number of possible representations of a group. In fact we will see later there is an infinite number of possibilities. However, the situation is not as hopeless as it might at first appear because there exists a basic set of representations from which *all* other representations can be produced. These fundamental representations are called *inequivalent irreducible representations.*

2.2 Equivalent and reducible representations

It is necessary before we consider irreducible representations to examine the nature of equivalent and reducible representations.

Two representations are said to be equivalent if the two matrices representing any element R of the group are related by the equation

$$\boldsymbol{\Gamma}'(R) = \mathbf{T}^{-1}\boldsymbol{\Gamma}(R)\mathbf{T} \tag{2.2}$$

where $\mathbf{T}$ is any non-singular square matrix (cf. section A.5) with the same dimension as $\boldsymbol{\Gamma}(R)$. $\boldsymbol{\Gamma}(R)$ is said to be related to $\boldsymbol{\Gamma}(R)$ by a similarity transformation (cf. section A.13). Note that equivalent representations have the same dimensions. In order to see that $\boldsymbol{\Gamma}'(R)$ is indeed a representation of the group, it is only necessary to show that if

$$\mathbf{\Gamma}(A)\mathbf{\Gamma}(B) = \mathbf{\Gamma}(AB) \tag{2.3}$$

then

$$\mathbf{\Gamma}'(A)\mathbf{\Gamma}'(B) = \mathbf{\Gamma}'(AB) \tag{2.4}$$

This follows immediately because

$$\begin{aligned}\mathbf{\Gamma}'(A)\mathbf{\Gamma}'(B) &= \mathbf{T}^{-1}\mathbf{\Gamma}(A)\mathbf{T}\mathbf{T}^{-1}\mathbf{\Gamma}(B)\mathbf{T} \\ &= \mathbf{T}^{-1}\mathbf{\Gamma}(A)\mathbf{\Gamma}(B)\mathbf{T} \\ &= \mathbf{\Gamma}'(AB)\end{aligned} \tag{2.5}$$

As an example of equivalent representations let us now consider $\mathbf{\Gamma}^3(R)$ and $\mathbf{\Gamma}^4(R)$ of table 2.1. If we select the transformation matrix

$$\mathbf{T} = \begin{bmatrix} \frac{1}{\sqrt{2}} & -\frac{1}{\sqrt{2}} \\ \frac{1}{\sqrt{2}} & \frac{1}{\sqrt{2}} \end{bmatrix}$$

then $\mathbf{\Gamma}^3(A)$ can be transformed to

$$\mathbf{\Gamma}^4(A) = \begin{bmatrix} \frac{1}{\sqrt{2}} & \frac{1}{\sqrt{2}} \\ -\frac{1}{\sqrt{2}} & \frac{1}{\sqrt{2}} \end{bmatrix} \mathbf{\Gamma}^3(A) \begin{bmatrix} \frac{1}{\sqrt{2}} & -\frac{1}{\sqrt{2}} \\ \frac{1}{\sqrt{2}} & \frac{1}{\sqrt{2}} \end{bmatrix} \tag{2.6}$$

likewise the similarity transform of $\mathbf{\Gamma}^3(B)$ is

$$\mathbf{\Gamma}^4(B) = \begin{bmatrix} \frac{1}{\sqrt{2}} & \frac{1}{\sqrt{2}} \\ -\frac{1}{\sqrt{2}} & \frac{1}{\sqrt{2}} \end{bmatrix} \mathbf{\Gamma}^3(B) \begin{bmatrix} \frac{1}{\sqrt{2}} & -\frac{1}{\sqrt{2}} \\ \frac{1}{\sqrt{2}} & \frac{1}{\sqrt{2}} \end{bmatrix} \tag{2.7}$$

The other elements of $\mathbf{\Gamma}^3(R)$ can be transformed in a similar manner. $\mathbf{\Gamma}^4(R)$ is therefore an *equivalent* representation to $\mathbf{\Gamma}^3(R)$; however there is no matrix $\mathbf{T}$ which will transform $\mathbf{\Gamma}^1(R)$ into $\mathbf{\Gamma}^2(R)$. Thus $\mathbf{\Gamma}^1(R)$ and $\mathbf{\Gamma}^2(R)$ are said to be *inequivalent.*

We are now going to show that representations of certain dimensions (size of matrix) may often be expressed in terms of representations of lower dimensions.

Mathematically, the way this is done, is to express the representation in *block form* by a similarity transformation. In block form a matrix is a direct sum (cf. section A.4(d)) of a number of matrices of lower dimensions. These matrices of lower dimensions are, in fact, other representations of the group. For example $\mathbf{\Gamma}^5(R)$ of table 2.1 is a direct sum of $\mathbf{\Gamma}^3(R)$ and $\mathbf{\Gamma}^1(R)$, so that we can see that $\mathbf{\Gamma}^5(R)$ has the matrix $\mathbf{\Gamma}^3(R)$ in the upper left-hand corner and $\mathbf{\Gamma}^1(R)$ in the lower right-hand corner; the remaining elements being zero. A representation when cast into block form is said to be reduced provided the matrices in the blocks cannot themselves be reduced into block form. If a matrix representation cannot be cast into block form by a similarity transformation then it is said to be *irreducible*. We emphasize that casting a representation into block form is *not* the same thing as diagonalizing a matrix. Thus a reduced representation is in block form where each block is the matrix of an irreducible representation. Finite groups possess a finite number of irreducible representations; and *all* other representations can be produced either by building up a reducible representation from irreducible ones or by producing *equivalent* representations by similarity transformations.

A reduced representation $\mathbf{\Gamma}^{\text{red}}(R)$ is composed of two or more irreducible representations, e.g.,

$$\mathbf{\Gamma}^{\text{red}}(R) = \left[\begin{array}{c|c} \mathbf{\Gamma}^i(R) & 0 \\ \hline 0 & \mathbf{\Gamma}^j(R) \end{array}\right] \tag{2.8}$$

where R is any element of the group. Clearly, $\mathbf{\Gamma}^{\text{red}}(R)$ is a square matrix of dimension $l_i + l_j$.

In order to show that direct sum matrices form a representation of the group it is necessary to show that

$$\mathbf{\Gamma}^{\text{red}}(A)\mathbf{\Gamma}^{\text{red}}(B) = \mathbf{\Gamma}^{\text{red}}(AB) \tag{2.9}$$

This can be shown by inspection as follows:

$$\begin{aligned} \left[\begin{array}{c|c} \mathbf{\Gamma}^i(A) & 0 \\ \hline 0 & \mathbf{\Gamma}^j(A) \end{array}\right] \left[\begin{array}{c|c} \mathbf{\Gamma}^i(B) & 0 \\ \hline 0 & \mathbf{\Gamma}^j(B) \end{array}\right] &= \left[\begin{array}{c|c} \mathbf{\Gamma}^i(A)\mathbf{\Gamma}^i(B) & 0 \\ \hline 0 & \mathbf{\Gamma}^j(A)\mathbf{\Gamma}^j(B) \end{array}\right] \\ &= \left[\begin{array}{c|c} \mathbf{\Gamma}^i(AB) & 0 \\ \hline 0 & \mathbf{\Gamma}^j(AB) \end{array}\right] \end{aligned} \tag{2.10}$$

the first step follows from the rule for matrix multiplication; while the second step follows from the group properties. Equation (2.10) can be verified for D_3 by using two matrices of $\mathbf{\Gamma}^5(R)$ where the matrices are already in block form so that a similarity transformation is unnecessary. As we have already stated $\mathbf{\Gamma}^5(R)$ is the direct sum of $\mathbf{\Gamma}^3(R)$ and $\mathbf{\Gamma}^1(R)$ and can be written as

$$\mathbf{\Gamma}^5(R) = \mathbf{\Gamma}^3(R) \oplus \mathbf{\Gamma}^1(R) \tag{2.11}$$

where the symbol $\oplus$ indicates a direct sum of matrices. Equation (2.11) is a particular case of the rule

$$\mathbf{\Gamma}^{\text{red}}(R) = \sum_i a_i \mathbf{\Gamma}^i(R) \tag{2.12}$$

where a_i is a coefficient which determines the numbers of times $\mathbf{\Gamma}^i(R)$ appears in the direct sum matrix.

In general reducible matrices are not in block form and must be brought into block form by a similarity transformation which means that the block form is an *equivalent* representation. In fact we must find a *single* similarity transformation which operates on *each* matrix of the representation (e.g., all the $\mathbf{\Gamma}^6(R)$ matrices representing E, A, B, C, D, and F) in such a way that *all* are cast into block form with the *same* block structure. These reservations are necessary in order to preserve the group properties. For example the matrix $\mathbf{T}$ where

$$\mathbf{T} = \tfrac{1}{6}\begin{bmatrix} -2 & 0 & 2 \\ 1 & -\sqrt{3} & 2 \\ 1 & \sqrt{3} & 2 \end{bmatrix} \tag{2.13}$$

and

$$\mathbf{T}^{-1} = \begin{bmatrix} -2 & 1 & 1 \\ 0 & -\sqrt{3} & \sqrt{3} \\ 1 & 1 & 1 \end{bmatrix} \tag{2.14}$$

will transform $\mathbf{\Gamma}^6(R)$ into $\mathbf{\Gamma}^5(R)$ by a similarity transformation.

For example considering $\mathbf{\Gamma}^6(F)$ then

$$\tfrac{1}{6}\begin{bmatrix} -2 & 1 & 1 \\ 0 & -\sqrt{3} & \sqrt{3} \\ 1 & 1 & 1 \end{bmatrix}\begin{bmatrix} 0 & 0 & 1 \\ 1 & 0 & 0 \\ 0 & 1 & 0 \end{bmatrix}\begin{bmatrix} -2 & 0 & 2 \\ 1 & -\sqrt{3} & 2 \\ 1 & \sqrt{3} & 2 \end{bmatrix}$$
$$= \left[\begin{array}{cc:c} -\frac{1}{2} & -\frac{\sqrt{3}}{2} & 0 \\ \frac{\sqrt{3}}{2} & -\frac{1}{2} & 0 \\ \hdashline 0 & 0 & 1 \end{array}\right] \tag{2.15}$$

i.e.,

$$\mathbf{T}^{-1}\mathbf{\Gamma}^6(F)\mathbf{T} = \mathbf{\Gamma}^5(F) \tag{2.16}$$

Similarly it can be shown that

$$\begin{aligned}
\mathbf{T}^{-1}\mathbf{\Gamma}^6(A)\mathbf{T} &= \mathbf{\Gamma}^5(A) \\
\mathbf{T}^{-1}\mathbf{\Gamma}^6(B)\mathbf{T} &= \mathbf{\Gamma}^5(B) \\
\mathbf{T}^{-1}\mathbf{\Gamma}^6(C)\mathbf{T} &= \mathbf{\Gamma}^5(C) \\
\mathbf{T}^{-1}\mathbf{\Gamma}^6(D)\mathbf{T} &= \mathbf{\Gamma}^5(D)
\end{aligned} \tag{2.17}$$

We therefore conclude that $\mathbf{\Gamma}^6(R)$ is equivalent to $\mathbf{\Gamma}^5(R)$; however since $\mathbf{\Gamma}^5(R)$ is in block form we also conclude that $\mathbf{\Gamma}^6(R)$ is *reducible.*

Therefore $\mathbf{\Gamma}^{6\,\mathrm{red}}(R)$ in reduced form can be written as

$$\mathbf{\Gamma}^{6\,\mathrm{red}}(R) = \mathbf{\Gamma}^3(R) \oplus \mathbf{\Gamma}^1(R) \tag{2.18}$$

Finally we should point out that we can only bring matrices to block form provided we adopt unitary matrix (cf. section A.4(j)) representations of the group operations. Although the proof is a little involved it can be shown that any matrix representation can be turned into unitary form. We will therefore assume for the rest of our discussions that all representations are indeed unitary. This assumption in fact effects some simplification in the presentation of the results we will be using later on.

2.3 Characters

Before discussing the properties of the inequivalent irreducible representations of a group we wish to introduce an important property of a matrix namely its *trace* which is defined in eq. (A.3) as the sum of its diagonal elements. The trace of a matrix therefore is simply a number. In group theory the usual symbol for the trace of a matrix is the Greek letter $\chi(R)$ where R is the element of the group. The trace of the matrix corresponding to the element R in the ith representation is therefore

$$\chi^i(R) = \sum_{\kappa} \Gamma^i_{\kappa\kappa}(R) \tag{2.19}$$

where we are now using the suffix notation (cf. section A.1).

The reason why the trace of a matrix is so important in the theory of group representations is because the trace is unaltered by a similarity transformation (cf. section A.19). This statement can be summarized mathematically as

$$\chi(T^{-1}RT) = \chi(R) \tag{2.20}$$

for all operations T. We are now led to two important results: the first being that the traces of all matrices representing elements in a class are identically equal, the second is that the set of traces of the matrices of a group representation are the same for all equivalent representations.

We can therefore, by examining the traces of the representations determine those which are *equivalent* and those which are *inequivalent*. All equivalent representations are defined by a *single* set of traces. Thus if we have two sets of traces, say, which are not identical they are then said to characterize two inequivalent representations. From the representations of table 2.1 we can construct table 2.2 which is a table of traces which

Table 2.2 Characters of the representations of table 2.1

	E	A	B	C	D	F
$\Gamma^1(R)$	1	1	1	1	1	1
$\Gamma^2(R)$	1	-1	-1	-1	1	1
$\Gamma^3(R)$	2	0	0	0	-1	-1
$\Gamma^4(R)$	2	0	0	0	-1	-1
$\Gamma^5(R)$	3	1	1	1	0	0
$\Gamma^6(R)$	3	1	1	1	0	0

characterize the representations. It is now immediately clear from table 2.2 that $\mathbf{\Gamma}^3(R)$ and $\mathbf{\Gamma}^4(R)$ are equivalent. Similarly $\mathbf{\Gamma}^5(R)$ and $\mathbf{\Gamma}^6(R)$ are also equivalent. The traces under A, B, and C are the *same* in each representation as they are for D and F. This derives from the fact that (A, B, and C) and (D and F) each form classes (cf. section 1.7).

The trace of the matrix representation of R is called the *character* of R in that representation so that table 2.2 can be referred to as a table of characters, where $\Gamma^i(R)$ is simply being used as a label.

The trace of a one-dimensional matrix is the matrix itself so that in $\mathbf{\Gamma}^1(R)$ and $\mathbf{\Gamma}^2(R)$ the row of characters is also the row of the original matrices. The matrix representation of E is a unit matrix so that its trace is obviously equal to the dimension of the representation.

2.4 Orthogonality theorem

The inequivalent irreducible unitary representations (i.e., those representations which have different sets of traces and which cannot be reduced into block form) have special properties the essence of which is contained in the *orthogonality theorem*. This theorem involves the matrix elements of the matrix representations. We take a matrix element from the same position in each matrix in one representation and we imagine these to be the components of a vector. The number of components is obviously equal to g, the number of elements in the group. The vector therefore can be considered to exist in a g-dimensional space. We can clearly construct

Table 2.3 Selected representations of the group D_3

	E	A	B	C	D	F
$\mathbf{\Gamma}^1(R)$	1	1	1	1	1	1
$\mathbf{\Gamma}^2(R)$	1	−1	−1	−1	1	1
$\mathbf{\Gamma}^3(R)$	$\begin{bmatrix} 1 & 0 \\ 0 & 1 \end{bmatrix}$	$\begin{bmatrix} 1 & 0 \\ 0 & -1 \end{bmatrix}$	$\begin{bmatrix} -\frac{1}{2} & -\frac{\sqrt{3}}{2} \\ -\frac{\sqrt{3}}{2} & \frac{1}{2} \end{bmatrix}$	$\begin{bmatrix} -\frac{1}{2} & \frac{\sqrt{3}}{2} \\ \frac{\sqrt{3}}{2} & \frac{1}{2} \end{bmatrix}$	$\begin{bmatrix} -\frac{1}{2} & \frac{\sqrt{3}}{2} \\ -\frac{\sqrt{3}}{2} & -\frac{1}{2} \end{bmatrix}$	$\begin{bmatrix} -\frac{1}{2} & -\frac{\sqrt{3}}{2} \\ \frac{\sqrt{3}}{2} & -\frac{1}{2} \end{bmatrix}$

such vectors for all inequivalent irreducible representations of the group. The orthogonality theorem (which we give without proof) states that all the vectors formed in this way are orthogonal to each other, i.e.,

$$\sum_R \Gamma^{i*}_{\mu\nu}(R)\Gamma^{j}_{\mu'\nu'}(R) = \frac{g}{l_i}\delta_{ij}\delta_{\mu\mu'}\delta_{\nu\nu'} \tag{2.21}$$

where i and j denote the representation, μ and μ' denote the rows of the matrix elements ν and ν' denote the columns of the matrix elements, g is the order of the group, and l_i is the dimensionality of the ith representation. $\Gamma^i_{\mu\nu}(R)$ is the matrix element of the μth row and the νth column of the matrix representing the group element R in the ith representation, and * denotes its complex conjugate ($\mathbf{\Gamma}^{i*}(R)$ is not necessarily equivalent to $\mathbf{\Gamma}^{i}(R)$).

Consider the representations $\mathbf{\Gamma}^1(R)$, $\mathbf{\Gamma}^2(R)$, and $\mathbf{\Gamma}^3(R)$ in table 2.3. These representations produce six row vectors as shown in table 2.4.

Table 2.4 The six mutually orthogonal vectors obtained from the representations $\mathbf{\Gamma}^1(R)$, $\mathbf{\Gamma}^2(R)$ and $\mathbf{\Gamma}^3(R)$

Vector	L^2	$\frac{g}{l_i}$	i	μ	ν
$[1,\ 1,\ 1,\ 1,\ 1,\ 1]$	6	6	1	1	1
$[1,\ -1,\ -1,\ -1,\ 1,\ 1]$	6	6	2	1	1
$\left[1,\ 1,\ -\frac{1}{2},\ -\frac{1}{2},\ -\frac{1}{2},\ -\frac{1}{2}\right]$	3	3	3	1	1
$\left[0,\ 0,\ -\frac{\sqrt{3}}{2},\ \frac{\sqrt{3}}{2},\ \frac{\sqrt{3}}{2},\ -\frac{\sqrt{3}}{2}\right]$	3	3	3	1	2
$\left[0,\ 0,\ -\frac{\sqrt{3}}{2},\ \frac{\sqrt{3}}{2},\ -\frac{\sqrt{3}}{2},\ \frac{\sqrt{3}}{2}\right]$	3	3	3	2	1
$\left[1,\ -1,\ \frac{1}{2},\ \frac{1}{2},\ -\frac{1}{2},\ -\frac{1}{2}\right]$	3	3	3	2	2

The square of the length L of the vector is obtained by summing the square of each element of the vector. Thus L^2 is by virtue of the orthogonality theorem, given by g/l_i (by putting $i = j$, $\mu = \mu'$, $\nu = \nu'$ in eq. (2.21).

The fact that each of these six vectors is orthogonal to each other can be readily verified. Since the theorem only relates to inequivalent irreducible representations we can conclude that $\mathbf{\Gamma}^1(R)$, $\mathbf{\Gamma}^2(R)$, and $\mathbf{\Gamma}^3(R)$ are three such representations. We could enquire whether there are any more inequivalent irreducible representations of the group D_3. The answer to this question can be obtained by a further consideration of the vectors. In a three-dimensional space we cannot have more than three mutually orthogonal vectors. It follows therefore, intuitively, that in a six-dimensional space there cannot be more than six mutually orthogonal vectors. In a group representation each group element contributes one dimension so that the group D_3 which has six elements has six-dimensional vectors. $\mathbf{\Gamma}^1(R)$, $\mathbf{\Gamma}^2(R)$, and $\mathbf{\Gamma}^3(R)$ yield six mutually orthogonal vectors; hence there cannot be any more inequivalent irreducible representations.

Each l_i-dimensional inequivalent irreducible representation contributes l_i^2 vectors. The total number of vectors contributed by all the inequivalent irreducible representations must clearly not exceed the dimension g of the space that they span. Therefore at least we have

$$\sum_{i=1}^{n} l_i^2 \leqslant g \tag{2.22}$$

where n is the number of irreducible representations. In fact it is possible to show that

$$\sum_{i=1}^{n} l_i^2 = g \tag{2.23}$$

2.5 Orthogonality of characters

If in eq. (2.21) we now set $\mu = \nu$ and $\mu' = \nu'$ we obtain

$$\sum_R \Gamma^{i*}_{\mu\mu}(R)\Gamma^{j}_{\mu'\mu'}(R) = \frac{g}{l_i}\delta_{ij}\delta_{\mu\mu'} \tag{2.24}$$

which if we sum over μ and μ' becomes

$$\sum_{R\mu\mu'} \Gamma^{i*}_{\mu\mu}(R)\Gamma^{j}_{\mu'\mu'}(R) = \frac{g}{l_i}\delta_{ij}\sum_{\mu\mu'}\delta_{\mu\mu'} \tag{2.25}$$

However, eq. (2.25) now contains the characters (traces) of the representations of R, i.e.,

$$\sum_{\mu} \Gamma^{i}_{\mu\mu}(R) = \chi^i(R); \qquad \sum_{\mu'} \Gamma^{j}_{\mu'\mu'}(R) = \chi^j(R) \tag{2.26}$$

Therefore eq. (2.25) now becomes

$$\sum_R \chi^{i*}(R)\chi^j(R) = \frac{g}{l_i}\delta_{ij}l_i \tag{2.27}$$

so that we obtain the following important property of the characters of irreducible representations

$$\sum_R \chi^{i*}(R)\chi^j(R) = g\delta_{ij} \tag{2.28}$$

This equation states that the vectors formed by the g characters of the representation form a set of mutually orthogonal vectors.

For example the characters of $\mathbf{\Gamma}^1(R)$, $\mathbf{\Gamma}^2(R)$, and $\mathbf{\Gamma}^3(R)$ of D_3 give rise to the three vectors

[1,	1,	1,	1,	1,	1]	from $\mathbf{\Gamma}^1(R)$
[1,	−1,	−1,	−1,	1,	1]	from $\mathbf{\Gamma}^2(R)$
[2,	0,	0,	0,	−1,	−1]	from $\mathbf{\Gamma}^3(R)$

These three vectors can be verified to be mutually orthogonal.

Since the characters of the representation of each element in the same class are equal, eq. (2.28) can be summed over the classes instead of over individual group elements. Thus

$$\sum_k \chi^{i*}(\mathscr{C}_k)\chi^j(\mathscr{C}_k)N_k = g\delta_{ij} \tag{2.29}$$

where N_k is the number of elements in the class $\mathscr{C}_k$. Equation (2.29) can also be written as

$$\sum_k \{\chi^{i*}(\mathscr{C}_k)\sqrt{N_k}\}\{\chi^j(\mathscr{C}_k)\sqrt{N_k}\} = g\delta_{ij} \tag{2.30}$$

Therefore the characters of a class multiplied by the square root of the number of elements in the class, form a set of orthogonal vectors. The dimension of these vectors is equal to *number of classes in the group.* $\mathbf{\Gamma}^1(R)$, $\mathbf{\Gamma}^2(R)$, and $\mathbf{\Gamma}^3(R)$ of the group D_3 give three such vectors since it contains three classes; these vectors are

Class:	E	ABC	DF	
	[1,	$\sqrt{3}$,	$\sqrt{2}$]	from $\mathbf{\Gamma}^1(R)$
	[1,	$-\sqrt{3}$,	$\sqrt{2}$]	from $\mathbf{\Gamma}^2(R)$
	[2,	0,	$-\sqrt{2}$]	from $\mathbf{\Gamma}^3(R)$

We now draw the important conclusion from these vectors that the maximum number of irreducible representations is less than or equal to the number of classes in the group. Indeed, although we do not give a proof, it can be shown that

The number of irreducible representations = *number of classes* (2.31)

2.6 Character tables

At first sight it might seem that a table of irreducible representations is the way to characterize a group. However, by referring to table 2.1 we can see that such a table is *not unique.* For example we might use the first three representations but the use of $\Gamma^4(R)$ instead of $\Gamma^3(R)$ would provide an entirely equivalent table. On the other hand since the traces of equivalent irreducible representations are identical a table of characters is a *unique* way to characterize a group. Tables of characters are, as we shall see in later chapters, very useful in the solution of many physical problems.

A character table takes the form of table 2.5 where at the head of each column is a symbol $C(r)$ to denote the nature of the elements in the class, (e.g., E, C_2, S_4, etc.) together with N_r giving the number of the elements in the class.

Table 2.5 General form of a character table

	$N_1C(1)$	$N_2C(2)$	$N_3C(3)$	...	$N_rC(r)$
$\Gamma^1(R)$	$\chi^1(C(1))$	$\chi^1(C(2))$	...	...	$\chi^1(C(r))$
$\Gamma^2(R)$	$\chi^2(C(1))$	$\chi^2(C(2))$	...	...	...
$\Gamma^3(R)$	...	...	...	...	...
	...	...	...	...	...
$\Gamma^r(R)$	$\chi^r(C(1))$	...	...	...	$\chi^r(C(r))$

$\Gamma^1(R)$ is always the homomorphic representation where all the elements are represented by the same one-dimensional unit matrix, i.e., the number 1. Also the column under E contains numbers equal to the dimensions of the representations.

From table 2.2 we can see that the character table for the group D_3 is as shown in table 2.6.

Table 2.6 Character table of D_3 with labels (cf. section 2.7)

	D_3	E	$3C_2$	$2C_3$
A_1	$\Gamma^1(R)$	1	1	1
A_2	$\Gamma^2(R)$	1	-1	1
E	$\Gamma^3(R)$	2	0	-1

The character tables of the groups discussed in chapter 1 are given in Appendix C many of which can be obtained by the application of the following rules:

(1) Number of inequivalent irreducible representations = number of classes

(2) $\sum_{i=1}^{n} l_i^2 = g$; n = number of irreducible representations

(3) $\sum_R \chi^{i*}(R)\chi^j(R) = g\delta_{ij}$

$$(4) \quad \sum_{i=1}^{n} \chi^{i*}(\mathscr{C}_k)\chi^{i}(\mathscr{C}_{k'}) = \frac{g}{N_k}\delta_{kk'}$$

where eq. (4), which we also give without proof, states that the columns of a character table form mutually orthogonal vectors.

As an example, let us now consider the group D_4 which has eight elements namely E, C_4, C_4^2, C_4^3, and four C_2 rotations about axes perpendicular to the C_4 axis (cf. Fig. 2.1). The group has five classes which are:

$$\begin{aligned} \mathscr{C}_1 &= E \\ \mathscr{C}_2 &= C_4, C_4^3 \\ \mathscr{C}_3 &= C_4^2 \\ \mathscr{C}_4 &= C_2^{(1)}, C_2^{(2)} \\ \mathscr{C}_5 &= C_2^{(3)} C_2^{(4)} \end{aligned} \qquad (2.32)$$

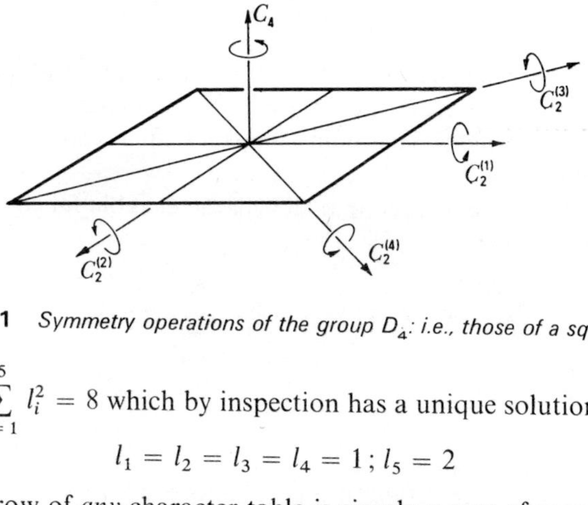

Fig. 2.1 *Symmetry operations of the group D_4: i.e., those of a square.*

Therefore $\sum_{i=1}^{5} l_i^2 = 8$ which by inspection has a unique solution:

$$l_1 = l_2 = l_3 = l_4 = 1; l_5 = 2 \qquad (2.33)$$

The first row of *any* character table is simply a row of ones, i.e., there will always be an obvious representation of the elements of a group, by one-dimensional unit matrices (homomorphic representation). Thus we can account for the first row of table 2.7. Equation (2.33) establishes

Table 2.7 Character table of D_4 with labels (cf. section 2.7)

	D_4	E	$2C_4$	C_4^2	$2C_2'$	$2C_2''$
A_1	$\Gamma^1(R)$	1	1	1	1	1
A_2	$\Gamma^2(R)$	1	1	1	−1	−1
B_1	$\Gamma^3(R)$	1	−1	1	1	−1
B_2	$\Gamma^4(R)$	1	−1	1	−1	1
E	$\Gamma^5(R)$	2	0	−2	0	0

the first column because as we stated previously the representations of E are diagonal unit matrices whose traces are equal to their dimensions.

If we now remember that the characters of one-dimensional representations are also the matrices which obey the multiplication table, then the square of the character under C_4 must be equal to the character under C_4^2. Thus the characters of the representations $\mathbf{\Gamma}^1(R)$ to $\mathbf{\Gamma}^4(R)$ of C_4^2 must be $+1$ and the characters of C_4 in these representations must be $+1$ or -1.

We can fill in the remainder of the characters of $\mathbf{\Gamma}^2(R)$ to $\mathbf{\Gamma}^4(R)$ by using rule 3 which states the rows must be mutually orthogonal (remembering to take into account the number of elements in each class when performing the summation).

The characters of $\mathbf{\Gamma}^5(R)$ can now be obtained by using rule 4, e.g., using the column under $2C_4$ we obtain putting $k = k'$

$$1 + 1 + 1 + 1 + \{\chi^5(C_4)\}^2 = \tfrac{8}{2} \tag{2.34}$$

The character $\chi^5(C_4)$ is therefore zero.

The rest of the characters of $\mathbf{\Gamma}^5(R)$ can be determined in a similar way.

2.7 Labelling of characters

Each representation is commonly given a label which is intended to provide a systematic guide to the representations. These labels are specified by the following rules:

(1) One-dimensional representations are labelled: A if the character of the elements C_n^k about the principal rotation axis are $+1$ for all k; B if the characters of C_n^k are $(-1)^k$ for all k.

(2) If a group has more than one A or B representation they are given subscripts 1 and 2 according to whether the character is $+1$ or -1 in the column representing

(a) a rotation or improper rotation about an axis other than the principal axis. For example in the groups D_n a representation is given a subscript 1 if the character under C_2 about the x axis is $+1$ and 2 if it is -1.

(b) reflection in a plane. For example in the groups C_{nv} a representation is given a subscript 1 or 2 if the character under σ_v is $+1$ or -1, respectively.

(3) Two-dimensional representations are labelled E.

(4) Three-dimensional representations are usually labelled T.

(5) If the group contains i (which is in a class by itself) the label is given a suffix:

g if character under i is $+1$
u if character under i is -1

(6) If the group contains σ_h the label is dashed:
a single dash if character under σ_h is $+1$
a double dash if character under σ_h is -1

All the character tables in Appendix C are labelled according to these rules; however as an illustration we reproduce the character tables of D_3 and D_4, completely labelled in tables 2.6 and 2.7.

2.8 Reduction of reducible representations

In many physical applications of group theory it is necessary to determine the irreducible components of a reducible representation. This reduction can be simply carried out with a knowledge of the character table. We begin with eq. (2.12) namely

$$\boldsymbol{\Gamma}^{\text{red}}(R) = \sum_i a_i \boldsymbol{\Gamma}^i(R)$$

where a_i is the number of times $\boldsymbol{\Gamma}^i(R)$ appears in the reducible representation. Hence, since each $\boldsymbol{\Gamma}^i(R)$ is situated on the diagonal of $\boldsymbol{\Gamma}^{\text{red}}(R)$ (block form!) the trace of $\boldsymbol{\Gamma}^{\text{red}}(R)$ is equal to the sum of the traces of the irreducible matrices, i.e.,

$$\chi^{\text{red}}(R) = \sum_i a_i \chi^i(R) \tag{2.35}$$

If we now multiply this equation by $\chi^{j*}(R)$ and sum over R we obtain

$$\begin{aligned} \sum_R \chi^{j*}(R)\chi^{\text{red}}(R) &= \sum_{iR} a_i \chi^{j*}(R)\chi^i(R) \\ &= \sum_i a_i g \delta_{ij} \end{aligned} \tag{2.36}$$

Therefore we finally obtain

$$a_j = \frac{1}{g}\sum_R \chi^{j*}(R)\chi^{\text{red}}(R) \tag{2.37}$$

Thus from eq. (2.37) we can determine how many times each irreducible representation appears in a given reducible representation. For example $\boldsymbol{\Gamma}^6(R)$ of table 2.1 is a reducible representation whose characters are (using $\Gamma^6(R)$, as opposed to $\boldsymbol{\Gamma}^6(R)$, to *label* the representation)

	E	$3C_2$	$2C_3$
$\Gamma^6(R)$	3	1	0

so that using the character table of D_3 (i.e., table 2.6) and eq. (2.37) we obtain

$$\left.\begin{aligned} a_1 &= \tfrac{1}{6}\{(1 \times 3) + 3(1 \times 1) + 2(1 \times 0)\} = 1 \\ a_2 &= \tfrac{1}{6}\{(1 \times 3) + 3(-1 \times 1) + 2(1 \times 0)\} = 0 \\ a_3 &= \tfrac{1}{6}\{(2 \times 3) + 3(0 \times 1) + 2(-1 \times 0)\} = 1 \end{aligned}\right\} \tag{2.38}$$

We therefore conclude that

$$\boldsymbol{\Gamma}^{6\,\mathrm{red}}(R) = \boldsymbol{\Gamma}^1(R) \oplus \boldsymbol{\Gamma}^3(R) \tag{2.39}$$

2.9 Problems

2.1 Show that the matrix

$$\mathbf{T} = \begin{bmatrix} 1/\sqrt{2} & -1/\sqrt{2} \\ 1/\sqrt{2} & 1/\sqrt{2} \end{bmatrix}$$

will transform all the matrices $\boldsymbol{\Gamma}^3(R)$ to $\boldsymbol{\Gamma}^4(R)$ by a similarity transformation.

2.2 Write down the six row vectors obtainable from $\boldsymbol{\Gamma}^1(R)$, $\boldsymbol{\Gamma}^2(R)$, and $\boldsymbol{\Gamma}^4(R)$ of table 2.1, and show that they are mutually orthogonal. Hence show that these matrices form a set of inequivalent irreducible representations of the group D_3.

2.3

$$\begin{array}{ccccccc} & E & A & B & C & D & F \\ \boldsymbol{\Gamma}(R): & \begin{bmatrix} 1 & 0 & 0 \\ 0 & 1 & 0 \\ 0 & 0 & 1 \end{bmatrix} & \begin{bmatrix} -2 & 1 & 1 \\ -\frac{3}{2} & \frac{3}{2} & \frac{1}{2} \\ -\frac{3}{2} & \frac{1}{2} & \frac{3}{2} \end{bmatrix} & \begin{bmatrix} -2 & 1 & 1 \\ -\frac{3}{2} & \frac{3}{2} & \frac{1}{2} \\ -\frac{3}{2} & \frac{1}{2} & \frac{3}{2} \end{bmatrix} & \begin{bmatrix} -2 & 1 & 1 \\ -\frac{3}{2} & \frac{3}{2} & \frac{1}{2} \\ -\frac{3}{2} & \frac{1}{2} & \frac{3}{2} \end{bmatrix} & \begin{bmatrix} 1 & 0 & 0 \\ 0 & 1 & 0 \\ 0 & 0 & 1 \end{bmatrix} & \begin{bmatrix} 1 & 0 & 0 \\ 0 & 1 & 0 \\ 0 & 0 & 1 \end{bmatrix} \end{array}$$

Show that matrices $\boldsymbol{\Gamma}(R)$ given above form a representation of the group D_3. Prove that it is a reducible representation and find its component irreducible representations.

2.4 By using the character table of the group T_d, given in Appendix C, write down the vectors of eq. (2.30) and show that they are mutually orthogonal.

2.5 The group C_{4v} is a subgroup of O_h and therefore the irreducible representation matrices of O_h corresponding to the elements of C_{4v} will form a reducible representation of C_{4v}. By using the character tables of C_{4v} and O_h find the irreducible components of the reducible representations obtained from the irreducible representations T_{1g} and T_{2u}.

2.6 Show that the character table of C_{4h} can be constructed by considering the direct product $C_4 \otimes S_2$.

3 Basis vectors, basis functions, and quantum mechanics

3.0 Basis vectors

In this chapter the origin of the matrices which represent group operations is going to be considered. In order to see where these matrices come from it is necessary to consider in more detail the action of the symmetry operators, where we shall henceforward consider the *operations* described in chapter 1 as being effected via an *operator* acting on the system.

Symmetry operators commonly used in physics, for example rotation and reflection operators, can formally be regarded as acting on vectors to produce other vectors in the same space. A vector in three-dimensional space, for example, may be defined in terms of some co-ordinate system determined by three mutually perpendicular unit vectors which are called *basis vectors*. The effect of a rotation operator R, say, can be considered either as a rotation of the vector in a fixed co-ordinate system or as the rotation of the co-ordinate system in a reverse sense. Both operations change the components of the vector. In order to be specific let us write the basis vectors as $\mathbf{e}_1, \mathbf{e}_2, \mathbf{e}_3$; than an arbitrary vector $\mathbf{a}$ is given by

$$\mathbf{a} = x_1\mathbf{e}_1 + x_2\mathbf{e}_2 + x_3\mathbf{e}_3 \tag{3.1}$$

where x_1, x_2, and x_3 are the components of $\mathbf{a}$. Equation (3.1) can obviously be written as

$$\mathbf{a} = \sum x_i\mathbf{e}_i \tag{3.2}$$

The action of R *on the basis* produces a *new basis* $\mathbf{e}'_i$ where

$$\mathbf{e}'_i = R\mathbf{e}_i \tag{3.3}$$

For convenience, from now on, we will write the n basis vectors $\mathbf{e}_1, \mathbf{e}_2, \mathbf{e}_3, \ldots, \mathbf{e}_n$ of an n-dimensional space as a row vector which we represent by the symbol

$$(\tilde{\mathbf{e}}) = (\mathbf{e}_1, \mathbf{e}_2, \mathbf{e}_3, \ldots, \mathbf{e}_n) \tag{3.4}$$

or as a column vector which we will write as

$$(\mathbf{e}) = \begin{bmatrix} \mathbf{e}_1 \\ \mathbf{e}_2 \\ \mathbf{e}_3 \\ \cdot \\ \cdot \\ \cdot \\ \mathbf{e}_n \end{bmatrix} \tag{3.5}$$

As an illustration, a diagram of a rotation in a two-dimensional space is shown in Fig. 3.1, where $\mathbf{e}'_1$ and $\mathbf{e}'_2$ are new vectors resulting from a

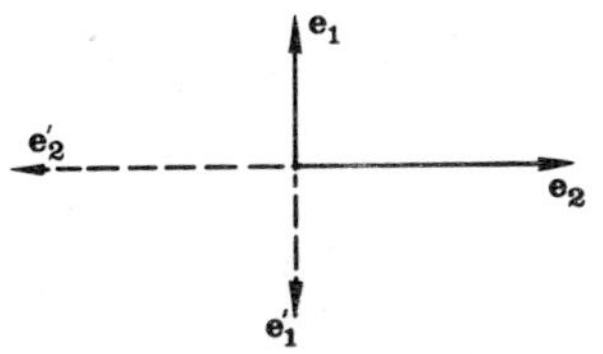

Fig. 3.1 *Rotation of a two-dimensional basis,* $\mathbf{e}_1$, $\mathbf{e}_2$, *through an angle* π *about an axis perpendicular to their plane.*

rotation through π about an axis perpendicular to the plane containing $\mathbf{e}_1$ and $\mathbf{e}_2$. It is clear from Fig. 3.1 that

$$\begin{aligned} R\mathbf{e}_1 &= \mathbf{e}'_1 = -\mathbf{e}_1 \\ R\mathbf{e}_2 &= \mathbf{e}'_2 = -\mathbf{e}_2 \end{aligned} \tag{3.6}$$

Equation (3.6) can therefore be written in the matrix form

$$R(\tilde{\mathbf{e}}) = (\tilde{\mathbf{e}}) \begin{bmatrix} -1 & 0 \\ 0 & -1 \end{bmatrix} = (\tilde{\mathbf{e}})\mathbf{\Gamma}(R) \tag{3.7}$$

where

$$\mathbf{\Gamma}(R) = \begin{bmatrix} -1 & 0 \\ 0 & -1 \end{bmatrix}$$

is said to be the matrix which *represents* R using the basis vectors $(\tilde{\mathbf{e}})$.

For *any* complete set of basis vectors, i.e., the complete set of vectors defining the space we can write

$$R(\tilde{\mathbf{e}}) = (\tilde{\mathbf{e}})\mathbf{\Gamma}(R) \tag{3.8}$$

where eq. (3.8) is the basic equation defining the *matrix representation* of R. If the set of vectors is orthonormal then the representation $\mathbf{\Gamma}(R)$ will be unitary and if possible we will always choose such a set. The dimension of the matrix R is equal to the number of basis vectors of the space.

The choice of the *row* $(\tilde{\mathbf{e}})$ to define $\mathbf{\Gamma}(R)$ is dictated by the fact that $\mathbf{\Gamma}(R)$ must be a representation of the group of R (that is the group of which R is a member), i.e., $\mathbf{\Gamma}(R)$ must obey the same multiplication table as does R. The need to employ row vectors is necessitated by the fact that matrices multiply from left to right; a matrix product AB means B multiplied by A from the left. This directionality is also implied in the multiplication of group elements. The fact that column vectors $(\mathbf{e})$ are unsatisfactory for forming representations can be readily shown by considering the representation of a product of two group elements.

The matrices of $\mathbf{\Gamma}^3(R)$ in table 2.1 are the representation matrices of the group operators with respect to a particular two-dimensional basis and Fig. 3.2 shows these basis vectors in relation to the triangle which locates the symmetry operators of the group D_3. The operator B, for example, of D_3 is, as shown in Fig. 3.2, a rotation of π about the axis b.

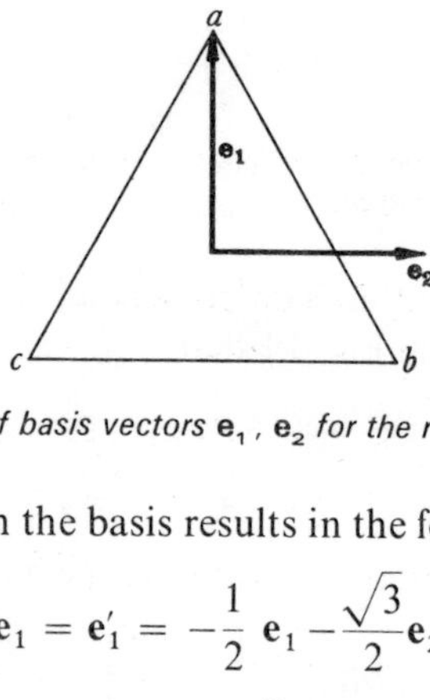

Fig. 3.2 *Location of basis vectors* $\mathbf{e}_1$, $\mathbf{e}_2$ *for the representation* $\mathbf{\Gamma}^3(R)$.

This operation acting on the basis results in the following new vectors

$$
\begin{aligned}
B\mathbf{e}_1 &= \mathbf{e}_1' = -\frac{1}{2}\,\mathbf{e}_1 - \frac{\sqrt{3}}{2}\mathbf{e}_2 \\
B\mathbf{e}_2 &= \mathbf{e}_2' = -\frac{\sqrt{3}}{2}\mathbf{e}_1 + \frac{1}{2}\mathbf{e}_2
\end{aligned}
\tag{3.9}
$$

Equation (3.9) is, in matrix form,

$$
B(\tilde{\mathbf{e}}) = (\tilde{\mathbf{e}}) \begin{bmatrix} -\dfrac{1}{2} & -\dfrac{\sqrt{3}}{2} \\ -\dfrac{\sqrt{3}}{2} & \dfrac{1}{2} \end{bmatrix} = (\tilde{\mathbf{e}})\mathbf{\Gamma}(B) \tag{3.10}
$$

where

$$
\mathbf{\Gamma}(B) = \begin{bmatrix} -\dfrac{1}{2} & -\dfrac{\sqrt{3}}{2} \\ -\dfrac{\sqrt{3}}{2} & \dfrac{1}{2} \end{bmatrix} \tag{3.11}
$$

and is $\mathbf{\Gamma}^3(B)$ of table 2.1.

We now need to show that the matrices defined by eq. (3.8) satisfy the multiplication table of the group of R. This can be done by establishing that is $RS = Q$ in the group of R then $\mathbf{\Gamma}(R)\mathbf{\Gamma}(S) = \mathbf{\Gamma}(Q)$ in the group of $\mathbf{\Gamma}(R)$. We can see this from the following

$$\begin{aligned} R(\tilde{\mathbf{e}}) &= (\tilde{\mathbf{e}})\mathbf{\Gamma}(R) \\ S(\tilde{\mathbf{e}}) &= (\tilde{\mathbf{e}})\mathbf{\Gamma}(S) \\ Q(\tilde{\mathbf{e}}) &= (\tilde{\mathbf{e}})\mathbf{\Gamma}(Q) \end{aligned} \tag{3.12}$$

$$Q(\tilde{\mathbf{e}}) = RS(\tilde{\mathbf{e}}) = R(\tilde{\mathbf{e}})\mathbf{\Gamma}(S) = (\tilde{\mathbf{e}})\mathbf{\Gamma}(R)\mathbf{\Gamma}(S) \tag{3.13}$$

Showing that $\mathbf{\Gamma}(Q) = \mathbf{\Gamma}(R)\mathbf{\Gamma}(S)$ so that the multiplication table of the group of R is indeed satisfied by $\mathbf{\Gamma}(R)$.

If $\mathbf{e}_3$ is chosen to be perpendicular to the plane of $\mathbf{e}_1$ and $\mathbf{e}_2$ in Fig. 3.2, we find that the effect of the operators of D_3 is to produce vectors $\mathbf{e}_3$ or $-\mathbf{e}_3$ so that $\mathbf{e}_3$ will form the basis of a one-dimensional representation of the group. This representation is $\mathbf{\Gamma}^2(R)$ of table 2.1. Obviously the three vectors $\mathbf{e}_1$, $\mathbf{e}_2$, and $\mathbf{e}_3$ will form the basis of a three-dimensional representation of the group; however this representation is reducible. If there are l basis vectors they will produce an l-dimensional representation.

It is clear then that the matrices $\mathbf{\Gamma}(R)$ depend on both the operator R and the basis $(\tilde{\mathbf{e}})$. Let us therefore see what happens to $\mathbf{\Gamma}(R)$ if we choose a new basis $(\tilde{\mathbf{e}}')$ which is a linear combination of the old basis vectors (a simple rotation of axes will give us such a combination).

The new basis is related to the old basis by

$$\begin{aligned} (\tilde{\mathbf{e}}') &= (\tilde{\mathbf{e}})\mathbf{T} \\ (\tilde{\mathbf{e}}) &= (\tilde{\mathbf{e}}')\mathbf{T}^{-1} \end{aligned} \tag{3.14}$$

where $\mathbf{T}$ is a non-singular matrix. The operation of R on these two bases gives us *two* representations of R determined from

$$R(\tilde{\mathbf{e}}) = (\tilde{\mathbf{e}})\mathbf{\Gamma}(R) \tag{3.15}$$

$$R(\tilde{\mathbf{e}}') = (\tilde{\mathbf{e}}')\mathbf{\Gamma}'(R) \tag{3.16}$$

where eq. (3.16) can now, with the aid of eq. (3.14), be written as

$$R(\tilde{\mathbf{e}}') = R(\tilde{\mathbf{e}})\mathbf{T} = (\tilde{\mathbf{e}})\mathbf{\Gamma}(R)\mathbf{T} = (\tilde{\mathbf{e}}')\mathbf{T}^{-1}\mathbf{\Gamma}(R)\mathbf{T} \tag{3.17}$$

We therefore deduce from eqs. (3.16) and (3.17) that

$$\mathbf{\Gamma}'(R) = \mathbf{T}^{-1}\mathbf{\Gamma}(R)\mathbf{T} \tag{3.18}$$

which shows that the new representation is *equivalent* (cf. section 2.2) to the old representation.

In this section we have obtained two very important results, namely that we can obtain the representation of the group by choosing a basis

$(\tilde{\mathbf{e}})$; and we can obtain equivalent representations by a linear transformation of the basis vectors.

3.1 Operators which act on functions

The description of many physical problems is in terms of functions of vectors, rather than the vectors themselves; for example the eigenfunctions which are used in quantum mechanics. It is therefore of some considerable importance to define operators P_R which act on functions and obey the multiplication table of the operators R. In the previous section we used basis vectors to obtain representations of the group of R. Here we must use *basis functions* to obtain representations of the group of P_R. Thus, as previously we were transforming basis *vectors* now we are interested in transforming basis *functions*. The relevance of basis functions in quantum mechanics is that the eigenfunctions form basis functions of the irreducible representations of the group of the Hamiltonian (to be discussed later in the chapter).

A function which depends upon co-ordinates, that is a function which is determined by the components of a vector, will change if an operator acts on the co-ordinates.

As an example let us consider

$$f(x, y) = \frac{x^2}{a^2} + \frac{y^2}{b^2} \tag{3.19}$$

where $f(x, y)$ can be evaluated if x and y, i.e., the components of the vector $|x\rangle$ are known. Thus we can alternatively write $f(x, y)$ as $f(|x\rangle)$. If an operator changes x to $-y$ and y to x (i.e., a rotation of 90° about the z axis) then $f(x, y)$ goes to $f(-y, x)$ which is equal to $y^2/a^2 + x^2/b^2$. The latter is therefore a *new* function in the *old* co-ordinate system.

The effect of *a change of basis*, i.e., $R(\tilde{\mathbf{e}}) = (\tilde{\mathbf{e}})\mathbf{\Gamma}(R)$ on the components of the vector $|x\rangle$ due to an operation R (discussed in Appendix A.11) is to change x to x' and y to y' so that (cf. eq. (A.132)) in *two dimensions*

$$\begin{bmatrix} x' \\ y' \end{bmatrix} = R\begin{bmatrix} x \\ y \end{bmatrix} = \mathbf{\Gamma}^{-1}(R)\begin{bmatrix} x \\ y \end{bmatrix} \tag{3.20}$$

where $\mathbf{\Gamma}(R)$ is the matrix representation of R in the original basis. The column vector $\begin{bmatrix} x \\ y \end{bmatrix}$ should be regarded as the *representation* of an arbitrary vector in the basis $(\tilde{\mathbf{e}})$. Thus we can define in general an operator P_R which *acts on an arbitrary function* $f(|x\rangle)$ to produce

$$P_R f(|x\rangle) = f(\mathbf{\Gamma}^{-1}(R)|x\rangle) \tag{3.21}$$

where R is a group element.

Clearly for every operator R of a group we can define an operator P_R and these operators, as will be shown, form a group isomorphic with the group of R. The proof of this statement requires us to show that

$$P_R P_S f(|x\rangle) = P_{RS} f(|x\rangle) \tag{3.22}$$

We can begin the proof by writing

$$P_S f(|x\rangle) = f(\mathbf{\Gamma}^{-1}(S)|x\rangle) = g(|x\rangle) \tag{3.23}$$

so that

$$P_R P_S f(|x\rangle) = P_R g(|x\rangle) = g(\mathbf{\Gamma}^{-1}(R)|x\rangle) = f(\mathbf{\Gamma}^{-1}(S)\mathbf{\Gamma}^{-1}(R)|x\rangle) \tag{3.24}$$

Hence

$$P_R P_S f(|x\rangle) = f(\mathbf{\Gamma}^{-1}(RS)|x\rangle) = P_{RS} f(|x\rangle) \tag{3.25}$$

which shows that the group of P_R possesses the same multiplication table as the group of R.

Now for every group of operators P_R a space exists which is spanned by a complete set of functions ϕ_i such that the action of any operator P_R on one of the functions is to produce a new function which is a linear combination of the original set, i.e.,

$$P_R \phi_i = \sum_{j=1}^{l} \phi_j \Gamma_{ji}^p(R) \tag{3.26}$$

where l is the number of functions in the set of ϕ_i and $\Gamma_{ji}^p(R)$ are now simply the coefficients (numbers) in the expansion.

The action of P_R on all the functions can be written as the matrix equation

$$P_R(\tilde{\boldsymbol{\phi}}) = (\tilde{\boldsymbol{\phi}})\mathbf{\Gamma}^p(R) \tag{3.27}$$

where $(\tilde{\boldsymbol{\phi}}) \equiv (\phi_1, \phi_2, \ldots)$ is a row vector in function space (entirely analogous to $(\tilde{\mathbf{e}})$ introduced earlier) and $\mathbf{\Gamma}^p(R)$ is the matrix which *represents* P_R using the *basis functions* ϕ_i (N.B. the use of ϕ_i^* produces $\mathbf{\Gamma}^{p*}(R)$).

We emphasize that eq. (3.27) is exactly analogous to eq. (3.8) which defines the matrices representing the operator R. Because of this we can see that the matrices $\mathbf{\Gamma}^p(R)$ defined by eq. (3.27) will satisfy the same multiplication table as the group of P_R and also the group of R. If we select a new set of basis functions ϕ_i which happen to be linear combinations of the old functions ϕ_i we have, in matrix form,

$$(\tilde{\boldsymbol{\phi}}') = (\tilde{\boldsymbol{\phi}})\mathbf{T} \tag{3.28}$$

These new basis functions now give rise to a new representation which is equivalent to the old one, i.e.,

$$P_R(\tilde{\boldsymbol{\phi}}') = (\tilde{\boldsymbol{\phi}}')\mathbf{\Gamma}^{p'}(R) = (\tilde{\boldsymbol{\phi}})\mathbf{T}^{-1}\mathbf{\Gamma}^p(R)\mathbf{T} \tag{3.29}$$

As an example of the use of basis functions let us consider, once again, the group D_3 which consists of the operators shown in table 2.1 and possesses the multiplication table given in table 1.3. The basis vectors $\mathbf{e}_1$ and $\mathbf{e}_2$ (x axis and y axis) shown in Fig. 3.2 form a basis for the representation $\boldsymbol{\Gamma}^3(R)$ of this group so that the matrices of $\boldsymbol{\Gamma}^3(R)$ in table 2.1 can be used to transform the components of a vector in the x–y plane according to eq. (3.20)

We proceed by considering a function $f(x) = f_1$ and perform the operations P_R and in this way we obtain

$$\begin{aligned}
P_E f_1 &= f(x) = f_1 \\
P_A f_1 &= f(x) = f_1 \\
P_B f_1 &= f\left(-\frac{1}{2}x - \frac{\sqrt{3}}{2}y\right) = f_2 \\
P_C f_1 &= f\left(-\frac{1}{2}x + \frac{\sqrt{3}}{2}y\right) = f_3 \\
P_D f_1 &= f\left(-\frac{1}{2}x - \frac{\sqrt{3}}{3}y\right) = f_2 \\
P_F f_1 &= f\left(-\frac{1}{2}x + \frac{\sqrt{3}}{2}y\right) = f_3
\end{aligned} \tag{3.30}$$

Equations (3.30) are actually produced by considering the operation of P_R on $f(x)$ to produce a function $f(x')$ where $f(x')$ has been obtained from eq. (3.21) by the application of $\boldsymbol{\Gamma}^{-1}(R)$ on the vector $|x\rangle$. Thus the action of P_B on $f(x)$ produces $f(x')$; where x' is a component of

$$\begin{bmatrix} x' \\ y' \end{bmatrix} = \{\boldsymbol{\Gamma}^3(B)\}^{-1} \begin{bmatrix} x \\ y \end{bmatrix} = \begin{bmatrix} -\frac{1}{2} & -\frac{\sqrt{3}}{2} \\ -\frac{\sqrt{3}}{2} & \frac{1}{2} \end{bmatrix} \begin{bmatrix} x \\ y \end{bmatrix} \tag{3.31}$$

giving

$$x' = -\frac{x}{2} - \frac{\sqrt{3}}{2}y \tag{3.32}$$

i.e.,

$$f(x') = f\left(-\frac{x}{2} - \frac{\sqrt{3}}{2}y\right) = f_2 \tag{3.33}$$

We have now obtained two new functions f_2 and f_3 and we must now consider the operations P_R on these. For example

$$P_E f_2 = f_2$$
$$P_A f_2 = P_A(P_B f_1) = P_F f_1 = f_3$$

Proceeding in this way we can, similarly, obtain equations for the action of all the operations of the group D_3 on f_2 and f_3.

Ultimately we find that the functions f_1, f_2, and f_3 form a *closed* set and hence form a *basis* of the group of P_R.

Using eq. (3.27) we can therefore write, symbolically (dropping the superscript p),

$$\begin{aligned}
P_E(\tilde{\mathbf{f}}) &= (\tilde{\mathbf{f}})\begin{bmatrix} 1 & 0 & 0 \\ 0 & 1 & 0 \\ 0 & 0 & 1 \end{bmatrix} = (\tilde{\mathbf{f}})\boldsymbol{\Gamma}(E) \\
P_A(\tilde{\mathbf{f}}) &= (\tilde{\mathbf{f}})\begin{bmatrix} 1 & 0 & 0 \\ 0 & 0 & 1 \\ 0 & 1 & 0 \end{bmatrix} = (\tilde{\mathbf{f}})\boldsymbol{\Gamma}(A) \\
P_B(\tilde{\mathbf{f}}) &= (\tilde{\mathbf{f}})\begin{bmatrix} 0 & 1 & 0 \\ 1 & 0 & 0 \\ 0 & 0 & 1 \end{bmatrix} = (\tilde{\mathbf{f}})\boldsymbol{\Gamma}(B) \\
P_C(\tilde{\mathbf{f}}) &= (\tilde{\mathbf{f}})\begin{bmatrix} 0 & 0 & 1 \\ 0 & 1 & 0 \\ 1 & 0 & 0 \end{bmatrix} = (\tilde{\mathbf{f}})\boldsymbol{\Gamma}(C) \\
P_D(\tilde{\mathbf{f}}) &= (\tilde{\mathbf{f}})\begin{bmatrix} 0 & 0 & 1 \\ 1 & 0 & 0 \\ 0 & 1 & 0 \end{bmatrix} = (\tilde{\mathbf{f}})\boldsymbol{\Gamma}(D) \\
P_F(\tilde{\mathbf{f}}) &= (\tilde{\mathbf{f}})\begin{bmatrix} 0 & 1 & 0 \\ 0 & 0 & 1 \\ 1 & 0 & 0 \end{bmatrix} = (\tilde{\mathbf{f}})\boldsymbol{\Gamma}(F)
\end{aligned} \qquad (3.34)$$

Thus with these three functions we have obtained a three-dimensional representation of the group of P_R which is identical to $\boldsymbol{\Gamma}^6(R)$ of table 2.1.

The character of this representation is

	E	A	B	C	D	F
$\chi(R)$	3	1	1	1	0	0

so that eq. (2.37) and the character table of D_3, given in table 2.6, shows that the representation $\mathbf{\Gamma}^6(R)$ is reducible, i.e., $\mathbf{\Gamma}^6(R)$ in block form is $\mathbf{\Gamma}^{6\,\mathrm{red}}(R) = \mathbf{\Gamma}^1(R) \oplus \mathbf{\Gamma}^3(R)$.

3.2 Classification of functions

All the irreducible representations have their own sets of basis functions so that basis functions can therefore be classified according to the representation for which they form a basis. We can devise this classification from a consideration of eq. (3.26) which is, introducing a superscript j for the representation,

$$P_R \phi^j_\kappa = \sum_\mu \phi^j_\mu \Gamma^j_{\mu\kappa}(R) \tag{3.35}$$

where ϕ^j_κ is the κth function in the set of basis functions forming a basis for the jth representation. In fact if we have an l-dimensional representation then it must have a set of l basis functions (one for each column of the representation matrix). The co-efficients $\Gamma^j_{\mu\kappa}(R)$ with μ varying from 1 to l_j (the number of functions in the set) form the κth column of the representation matrix $\mathbf{\Gamma}^j(R)$. Therefore a single function can be classified by the irreducible representation for which it is one of the basis functions and also by the column of the representation matrix according to which it transforms under the operator P_R.

Two problems arise at this stage, namely: (1) given an unlabelled basis function, how can we determine the irreducible representation to which it belongs and the column of the representation matrix under which it transforms? (2) given an arbitrary function, how can we construct basis functions for a particular irreducible representation?

The solution of these two problems requires the introduction of a new operator $O^j_{\mu\kappa}$ whose origin and properties we will now elucidate.

Multiplying eq. (3.35) by $\Gamma^{j'*}_{\mu'\kappa'}(R)$ and summing over R we can obtain with the aid of the orthogonality theorem (cf. eq. (2.21))

$$\begin{aligned} \sum_R \Gamma^{j'*}_{\mu'\kappa'}(R) P_R \phi^j_\kappa &= \sum_\mu \phi^j_\mu \sum_R \Gamma^{j'*}_{\mu'\kappa'}(R)\Gamma^j_{\mu\kappa}(R) \\ &= \sum_\mu \phi^j_\mu \frac{g}{l_{j'}} \delta_{jj'}\delta_{\mu\mu'}\delta_{\kappa\kappa'} \\ &= \frac{g}{l_{j'}} \phi^j_{\mu'} \delta_{jj'}\delta_{\kappa\kappa'} \end{aligned} \tag{3.36}$$

therefore

$$\frac{l_{j'}}{g}\sum_R \Gamma^{j'*}_{\mu'\kappa'}(R)P_R\phi^j_\kappa = \phi^j_{\mu'}\delta_{jj'}\delta_{\kappa\kappa'} \tag{3.37}$$

The quantity $(l_{j'}/g)\sum_R \Gamma^{j'*}_{\mu'\kappa'}(R)P_R$ is the anticipated new operator denoted by $O^{j'}_{\mu'\kappa'}$ which allows us to write eq. (3.37) as (changing the dashes for convenience)

$$O^j_{\mu\kappa}\phi^{j'}_{\kappa'} = \phi^{j'}_\mu\delta_{jj'}\delta_{\kappa\kappa'} \tag{3.38}$$

so that for $j = j'$ and $\kappa = \kappa'$ we have

$$O^j_{\mu\kappa}\phi^j_\kappa = \phi^j_\mu \tag{3.39}$$

The action of $O^j_{\mu\kappa}$ on a basis function belonging to the jth irreducible representation and the κth column of this representation produces another basis function belonging to the *same* representation but to the μth column. We can immediately see that this property of $O^j_{\mu\kappa}$ is useful because given one basis function we can generate the complete set of functions which form a basis for $\mathbf{\Gamma}^j(R)$. This process is often termed the generation of the partners of the initial basis function.

A particular case of eq. (3.39) arises when equation $\mu = \kappa$ so that

$$O^j_{\kappa\kappa}\phi^j_\kappa = \phi^j_\kappa \tag{3.40}$$

where

$$O^j_{\kappa\kappa} = \frac{l_j}{g}\sum_R \Gamma^{j*}_{\kappa\kappa}(R)P_R \tag{3.41}$$

Thus if an arbitrary function ψ exists such that $O^j_{\kappa\kappa}\psi = \psi$ then we can immediately say that it is a basis function belonging to the κth column of the jth irreducible representation showing that eq. (3.40) allows one to produce functions classified according to the column and representation for which they form basis functions.

If the jth representation happens to be one-dimensional then

$$\Gamma^{j*}_{\kappa\kappa}(R) = \chi^{j*}(R) \tag{3.42}$$

where $\chi^j(R)$ is the character of the jth representation of R, and the operator of eq. (3.41) becomes

$$O^j = \frac{1}{g}\sum_R \chi^{j*}(R)P_R \tag{3.43}$$

The usefulness of O^j will become apparent below.

3.3 Production of basis functions

As we have seen earlier in this chapter a function can be operated on by P_R thus enabling a representation of R to be determined. In fact *any* function can be used to form a representation and can be regarded as a *basis function*. For example any function of a scalar quantity must be invariant under all transformations and therefore will form the basis of the one-dimensional representation $\mathbf{\Gamma}(R)$ where $\mathbf{\Gamma}(R)$ is equal to unity for all R. (This representation is called the totally symmetric representation—i.e., the function remains the same under *all* operations of the group; as would be the case for $(x^2 + y^2 + z^2)$, i.e., the square of the length of a vector.) If we wish to determine basis functions for all the representations we must start with sufficiently arbitrary functions; otherwise it may be impossible to produce the complete set of basis functions. By sufficiently arbitrary we mean a function which has no symmetry with respect to the group so that *all* the P_R acting on this function will produce new functions. In the case of D_3 we would in this way produce six basis functions.

We now assert that any arbitrary function, on which P_R can operate, can be expressed as the sum of a set of functions, all of which are basis functions of the *irreducible representations* of the group. The basis functions of the irreducible representations will in practice be linear combinations of the functions obtained from P_R acting on the arbitrary function (the action of P_R produces basis functions but not necessarily basis functions of the irreducible representations).

Therefore an arbitrary function F can be written as

$$F = \sum_{j=1}^{c} \sum_{\kappa=1}^{l_j} \phi_\kappa^j \tag{3.44}$$

where c = number of irreducible representations of the group of P_R and l_j is the dimension of the jth representation.

If we now apply the operator $O_{\kappa\kappa}^j$ to F we obtain

$$O_{\kappa\kappa}^j F = O_{\kappa\kappa}^j \sum_{j'=1}^{c} \sum_{\kappa'=1}^{l_{j'}} \phi_{\kappa'}^{j'} = \phi_\kappa^j \tag{3.45}$$

Thus $O_{\kappa\kappa}^j$ acting on F behaves as if it is *projecting* out the part of F which belongs to the κth column of the jth irreducible representation. $O_{\kappa\kappa}^j$ is in consequence called a *projection operator*. Hence starting with an arbitrary function F and using eqs. (3.45) and (3.39) we can obtain basis functions for all the irreducible representations of the group. Naturally, before we can do this the matrix representation must be known (cf. eq. (3.41)) since $O_{\kappa\kappa}^j$ requires a knowledge of $\mathbf{\Gamma}^j(R)$. If only the character of the representation is known then another operator can be defined which is denoted by O^j (a generalization of (3.43)) where

$$O^j = \sum_{\kappa=1}^{l_j} O_{\kappa\kappa}^j = \sum_{\kappa}^{l_j} \sum_{R} \frac{l_j}{g} \Gamma_{\kappa\kappa}^{j*}(R) P_R \tag{3.46}$$

i.e.,

$$O^j = \frac{l_j}{g} \sum_R \chi^{j*}(R) P_R \tag{3.47}$$

where $\chi^{j*}(R)$ is the trace of $\mathbf{\Gamma}^{*j}(R)$.

Using eq. (3.45) we therefore obtain

$$O^j F = \sum_{\kappa}^{l_j} \phi^j_\kappa = \phi^j \tag{3.48}$$

where ϕ^j is a function which only *belongs* to the jth irreducible representation; that is all the constituents ϕ^j_κ making up the sum in eq. (3.48) are basis functions of the jth irreducible representation but ϕ^j itself is *not* a proper basis function for the jth irreducible representation.

The effect of P_R on ϕ^j is

$$\begin{aligned} P_R \phi^j = P_R \sum_\kappa \phi^j_\kappa &= \sum_\mu \sum_\kappa \phi^j_\mu \Gamma^j_{\mu\kappa}(R) \\ &= \sum_\mu \phi^j_\mu \sum_\kappa \Gamma^j_{\mu\kappa}(R) \\ &= \sum_\mu \phi^j_\mu C_\mu \end{aligned} \tag{3.49}$$

showing that $P_R \phi^j$ is a function which is a linear combination of the set of functions ϕ^j and still *belongs* to the jth irreducible representation. As we shall see later the functions ϕ^j have certain properties in common with the basis functions ϕ^j_κ.

As an example of the use of the operators $O^j_{\mu\kappa}$ let us consider the group D_3 together with the arbitrary function $F = ax + by + cx^2 + dy^2$; x and y being components along the directions $\mathbf{e}_1$ and $\mathbf{e}_2$ as shown in Fig. 3.2. Using the matrices $\mathbf{\Gamma}^3(R)$ in table 2.1 and eq. (3.21) we therefore obtain

$$\begin{aligned} P_E F &= ax + by + cx^2 + dy^2 \\ P_A F &= ax - by + cx^2 + dy^2 \\ P_B F &= \left(-\frac{a}{2} - \frac{\sqrt{3}b}{2}\right)x + \left(-\frac{\sqrt{3}a}{2} + \frac{b}{2}\right)y + \left(\frac{c}{4} + \frac{3d}{4}\right)x^2 \\ &\quad + \frac{\sqrt{3}}{2}(c-d)xy + \left(\frac{d}{4} + \frac{3c}{4}\right)y^2 \\ P_C F &= \left(-\frac{a}{2} + \frac{\sqrt{3}b}{2}\right)x + \left(\frac{\sqrt{3}a}{2} + \frac{b}{2}\right)y + \left(\frac{c}{4} + \frac{3d}{4}\right)x^2 \\ &\quad - \frac{\sqrt{3}}{2}(c-d)xy + \left(\frac{d}{4} + \frac{3c}{4}\right)y^2 \end{aligned} \tag{3.50}$$

(Equation (3.50) continued on next page.)

$$P_D F = \left(-\frac{a}{2} + \frac{\sqrt{3}b}{2}\right)x + \left(-\frac{\sqrt{3}a}{2} - \frac{b}{2}\right)y + \left(\frac{c}{4} + \frac{3d}{4}\right)x^2$$
$$+ \frac{\sqrt{3}}{2}(c - d)xy + \left(\frac{d}{4} + \frac{3c}{4}\right)y^2$$
$$P_F F = \left(-\frac{a}{2} - \frac{\sqrt{3}b}{2}\right)x + \left(\frac{\sqrt{3}a}{2} - \frac{b}{2}\right)y + \left(\frac{c}{4} + \frac{3d}{4}\right)x^2$$
$$- \frac{\sqrt{3}}{2}(c - d)xy + \left(\frac{d}{4} + \frac{3c}{4}\right)y^2$$

We now note that there are three irreducible representations of the group D_3 which have the characters shown in table 2.6. The representations labelled A_1 and A_2 are one-dimensional while E is two-dimensional. The operator O^{A_1} (a projection operator for this one-dimensional case) is given by (cf. eq. (3.47))

$$O^{A_1} = \tfrac{1}{6}(P_E + P_A + P_B + P_C + P_B + P_F) \tag{3.51}$$

Hence we can see that

$$O^{A_1}F = \frac{c + d}{2}(x^2 + y^2) \tag{3.52}$$

which gives us the basis function of the representation A_1. Similarly we can see that

$$O^{A_2}F = \tfrac{1}{6}(P_E + P_D + P_F - P_A - P_B - P_C) \tag{3.53}$$

giving

$$O^{A_2}F = 0 \tag{3.54}$$

showing that in this case F is not sufficiently arbitrary to give us a basis function of A_2. Finally we find that O^E, which is *not* a projection operator because E is two-dimensional, acting on F produces ψ given by

$$O^E F = ax + by + \frac{c - d}{2}(x^2 - y^2) = \psi \tag{3.55}$$

where ψ although it *belongs* to the irreducible representation E is actually the sum of two *basis* functions which can be found by using the projection operators O^E_{11} and O^E_{21} whose construction requires the matrices of table 2.1. Hence, using eq. (3.37) *et seq.*

$$O^E_{11} = \tfrac{1}{3}\{P_E + P_A - \tfrac{1}{2}(P_B + P_C + P_D + P_F)\} \tag{3.56}$$

giving

$$O^E_{11}F = ax + \frac{(c - d)}{2}(x^2 - y^2) = \phi^E_1 \tag{3.57}$$

and eq. (3.39) gives

$$O^E_{21}\phi^E_1 = ay - (c - d)xy = \phi^E_2 \tag{3.58}$$

The functions ϕ^E_1 and ϕ^E_2 are two possible basis functions of the representation E.

Alternatively we can use

$$O^E_{22}F = \tfrac{1}{3}\{P_E - P_A + \tfrac{1}{2}(P_B + P_C - P_D - P_F)\}F = by = \phi'^E_2 \tag{3.59}$$

and the other basis function is then obtained from

$$O^E_{12}\phi'^E_2 = bx = \phi'^E_1 \tag{3.60}$$

Thus the functions ϕ'^E_1 and ϕ'^E_2 also form a basis for the representation E.

Since the coefficients a, b, c, and d in these functions can be chosen at will, the simplest forms of the basis functions obtained by setting $(c - d) = 2$, $a = 0$ and $b = 1$ will be x and y or $(x^2 - y^2)$ and $-2xy$.

3.4 Orthogonality of basis functions

One important property of the functions ϕ^j_κ is that basis functions belonging to *different* irreducible representations are orthogonal (the scalar product of the functions is equal to zero) as are also functions belonging to different columns κ within the *same* irreducible representation.

The scalar product of ϕ^j_κ and $\phi^{j'}_{\kappa'}$ is defined (cf. eq. (A.163)) in the case of continuous functions by

$$\langle\phi^j_\kappa|\phi^{j'}_{\kappa'}\rangle \equiv \int \phi^{j*}_\kappa(\mathbf{x})\phi^{j'}_{\kappa'}(\mathbf{x})\,\mathrm{d}\mathbf{x} \tag{3.61}$$

The scalar product is preserved (invariant) under the unitary operations of P_R so we can therefore write

$$\begin{aligned}\langle\phi^j_\kappa|\phi^{j'}_{\kappa'}\rangle &= \langle P_R\phi^j_\kappa|P_R\phi^{j'}_{\kappa'}\rangle \\ &= \sum_{\mu\mu'}\langle\phi^j_\mu|\phi^{j'}_{\mu'}\rangle\Gamma^{j*}_{\mu\kappa}(R)\Gamma^{j'}_{\mu'\kappa'}(R)\end{aligned} \tag{3.62}$$

If we now sum both sides of eq. (3.62) over R and invoke the orthogonality theorem (eq. (2.21)) we finally obtain

$$\begin{aligned}g\langle\phi^j_\kappa|\phi^{j'}_{\kappa'}\rangle &= \frac{g}{l_j}\sum_{\mu\mu'}\langle\phi^i_\mu|\phi^{j'}_{\mu'}\rangle\delta_{jj'}\delta_{\kappa\kappa'}\delta_{\mu\mu'} \\ &= 0 \text{ if } j \neq j' \\ &= 0 \quad \text{if} \quad \kappa \neq \kappa' \\ &= \frac{g}{l_j}\sum_{\mu}\langle\phi^j_\mu|\phi^j_\mu\rangle \quad \text{if} \quad j = j', \kappa = \kappa'\end{aligned} \tag{3.63}$$

which proves the orthogonality of the functions ϕ^j_κ.

The functions ϕ^j for different values of j of eq. (3.48) can in a similar manner also be shown to be orthogonal, i.e.,

$$\begin{aligned} \langle\phi^j|\phi^{j'}\rangle &- \sum_{\kappa\kappa'} \langle\phi^j_\kappa|\phi^{j'}_{\kappa'}\rangle \\ &= 0 \quad \text{unless} \quad j = j' \end{aligned} \tag{3.64}$$

As an example we can see that the functions x and y which form a basis for the representation E of D_3 are orthogonal to each other since, using eq. (3.61), we find by integrating over all space

$$\begin{aligned} \langle\phi^E_1|\phi^E_2\rangle &= \lim_{a\to\infty} \int_{-a}^{a}\int_{-a}^{a} xy\,\mathrm{d}x\,\mathrm{d}y \\ &= \lim_{a\to\infty} \left[\frac{x^2}{2}\right]_{-a}^{+a}\left[\frac{y^2}{2}\right]_{-a}^{+a} \\ &= 0 \end{aligned} \tag{3.65}$$

Similarly we can show that ϕ^E_{12} are orthogonal to ϕ^{A_1} (eq. (3.52)). For example,

$$\begin{aligned} \langle\phi^E_1|\phi^{A_1}\rangle &= \lim_{a\to\infty} \int_{-a}^{+a}\int_{-a}^{+a} (x^2 + y^2)x\,\mathrm{d}x\,\mathrm{d}y \\ &= \lim_{a\to\infty}\left\{\left[\frac{x^4}{4}\right]_{-a}^{+a}[y]_{-a}^{+a} + \left[\frac{y^3}{3}\right]_{-a}^{+a}\left[\frac{x^2}{2}\right]_{-a}^{+a}\right\} \\ &= 0 \end{aligned} \tag{3.66}$$

3.5 Symmetrized basis functions

In chapter 2 we saw that a representation of a group operation is reducible provided we can find a transformation which will reduce it to block form. However, we know that we devise a representation by first specifying a basis so that the transformation which reduces a reducible representation into block form, will also cast the basis functions into sets which are associated with the component irreducible representations. This transformation of the basis functions is called a symmetrization of the basis.

The practical procedure begins with the adoption of an arbitrary set of basis functions f_i which, in general, enable us to establish a reducible representation of the group. We then determine the irreducible components by using the methods of section 2.8 and then, by applying the operators $O^j_{\kappa\kappa}$ of eq. (3.41) to one or more of the functions f_i, find the sets of functions ϕ^j_κ which form bases of the irreducible representations. This process thus transforms the original basis functions f_i into new basis functions ϕ^j_κ via the transformation $\mathbf{s}$, i.e.,

$$(\tilde{\boldsymbol{\phi}}) = (\tilde{\mathbf{f}})\mathbf{s} \tag{3.67}$$

where the row vector $(\tilde{\boldsymbol{\phi}})$ of the functions ϕ^j_κ is now ordered into sets corresponding to each irreducible representation. These sets of functions only mix amongst themselves under a symmetry operation of the group and they are said to span an irreducible sub-space of the full space spanned by the complete set of functions f_i. The matrix $\mathbf{s}$ is exactly the matrix which will bring the reducible representation into block form by means of a similarity transformation. The set of functions ϕ^j_κ are often referred to as the fully symmetrized basis and have the property that they are orthogonal between different irreducible representations and different columns within the same irreducible representation.

If, on the other hand we can only use the operators O^j, because we only possess the character table of the group, then we can only obtain a set of functions ϕ^j. This set is still a symmetrized set of functions and the matrix which transforms f_i to ϕ^j will also bring the reducible representation to block form, but the irreducible matrices on the diagonal may now not be unitary. These functions are still orthogonal between different irreducible representations, but they are not orthogonal within a representation. They are however very important because of the simplicity with which they can be obtained.

A problem arises if the reducible representation contains irreducible components which are repeated. In this case the arbitrary basis functions can still be symmetrized by use of the operators $O^j_{\kappa\kappa}$ or O^j, but now the symmetrized basis will only bring the reducible matrix into blocks associated with each *distinct* irreducible representation. Also functions belonging to the same column of the same irreducible representation will not now be orthogonal and cannot be made orthogonal on the grounds of symmetry alone.

3.6(a) Direct product groups

We now come to the final basic topic, namely direct product groups. The first three parts of this section could well have been presented in the earlier chapters; however for the sake of preserving the continuity of the present section we have decided to develop the whole theory here.

If we have two groups G_a and G_b, containing *only one* common element which must obviously be E the identity, for instance

$$G_a = E, A_2, A_3, \ldots, A_a \qquad (a \text{ elements})$$

$$G_b = E, B_2, B_3, \ldots, B_b \qquad (b \text{ elements})$$

then the set of elements obtained by multiplying all elements of G_a by all elements of G_b will form a group if and only if *all the elements of G_a commute with all the elements of G_b*. This group is called the *direct product group* of G_a and G_b and is denoted by $G_a \otimes G_b$. We can see that $G_a \otimes G_b$

is a group because the elements satisfy the group properties (cf. chapter 1), i.e.,

1. it contains the identity E,
2. the group is closed, which statement can be demonstrated by considering the product of a typical element A_jB_k with another element $A'_jB'_k$:

$$A_jB_kA'_jB'_k = A_jA'_jB_kB'_k = A''_jB''_k \tag{3.68}$$

   where we have used the facts that the sets of elements A and B form groups by themselves; and that they *commute* with each other,
3. the group is associative,
4. each element of $G_a \otimes G_b$ possesses an inverse such that

$$A_jB_kA_j^{-1}B_k^{-1} = A_j^{-1}B_k^{-1}A_jB_k = E \tag{3.69}$$

We emphasize that commutivity of the elements A and B is essential if the group properties are to be satisfied.

We have already encountered the fact that some point groups are direct product groups, so in order to illustrate this section we will pursue this topic a little further.

The inversion operator i commutes with all operations belonging to point groups which involve rotations only. If such point groups contain a principal axis then σ_h can be uniquely defined which will also commute with all the rotations. Since the identity E of course commutes with any operator we can use the groups

$$S_2: \quad E \text{ and } i$$

$$C_{1h}: \quad E \text{ and } \sigma_h$$

to form direct product groups with point groups of the type mentioned above. These form some of the most important direct product groups encountered in physical applications. Some examples are:

$$\begin{aligned} C_{4h} &= C_4 \otimes S_2, \quad D_{3d} = D_3 \otimes S_2 \\ C_{3h} &= C_3 \otimes C_{1h}, \quad D_{4h} = D_4 \otimes S_2 \\ T_h &= T \otimes S_2, \quad O_h = O \otimes S_2 \end{aligned} \tag{3.70}$$

The class structure of the direct product group can be easily inferred from the class structure of the two constituent groups. A typical element of that class of $G_a \otimes G_b$ containing the element A_jB_k can be written as

$$(A_{j'}B_{k'})^{-1}(A_jB_k)(A_{j'}B_{k'}) = (A_{j'}^{-1}A_jA_{j'})(B_{k'}^{-1}B_kB_{k'}) \tag{3.71}$$

where we have used the commutation properties. We should now note that $A_{j'}^{-1}A_jA_{j'}$ is an element of the class of A_j and $B_{k'}^{-1}B_kB_{k'}$ is an element of the class of B_k.

Thus all the elements of the class containing A_jB_k can be obtained by multiplying all the elements of the class of A_j by all the elements of the class of B_k. The number of classes in $G_a \otimes G_b$ is therefore the product of the number of classes in G_a with the number of classes in G_b.

3.6(b) Representations of direct product groups

The direct product matrices (cf. (A.2 (b))) obtained from the irreducible representation matrices of G_a and G_b form a representation of the direct product group $G_a \otimes G_b$ which can be shown to be both unitary and irreducible. Thus, if $\mathbf{\Gamma}^l(A_j)$ represents A_j in the lth irreducible representation of the group G_a and $\mathbf{\Gamma}^m(B_k)$ represents B_k in the mth irreducible representation of the group G_b then the direct product matrices $\mathbf{\Gamma}^{l\otimes m}(A_jB_k)$ given by

$$\mathbf{\Gamma}^{l\otimes m}(A_jB_k) = \mathbf{\Gamma}^l(A_j) \otimes \mathbf{\Gamma}^m(B_k) \tag{3.72}$$

form a representation of the group $G_a \otimes G_b$ as can be shown by multiplying a typical pair of elements, using the property of direct product matrices defined in eq. (A.35).

If there are c irreducible representations in the group G_a and d irreducible representations in the group G_b then there are $c \times d$ irreducible representations of $G_a \otimes G_b$. This conclusion is a consequence of eq. (2.31).

3.6(c) Characters of direct product representations

The trace of the matrix $\mathbf{\Gamma}^{l\otimes m}(A_jB_k)$ denoted by $X^{l\otimes m}$ is the character of the element A_jB_k in the representation $l \otimes m$ and is given by

$$\begin{aligned}\chi^{l\otimes m}(A_jB_k) &= \sum_{pq} \Gamma^{l\otimes m}_{pq,pq}(A_jB_k)\\ &= \sum_{pq} \Gamma^l_{pp}(A_j)\Gamma^m_{qq}(B_k)\\ &= \sum_{p} \Gamma^l_{pp}(A_j)\sum_{q} \Gamma^m_{qq}(B_k)\end{aligned} \tag{3.73}$$

and hence

$$\chi^{l\otimes m}(A_jB_k) = \chi^l(A_j)\chi^m(B_k) \tag{3.74}$$

so that the character table of the direct product group can be written out by inspection of the character tables of the constituent groups. For example the direct product group $D_{3h} = D_3 \otimes C_{1h}$ has a character table constructed from the tables contained in table 3.1 from which we can see that, since

Table 3.1 Character tables of D_3 and C_{1h}

D_3	E	$3C_2$	$2C_3$
A_1	1	1	1
A_2	1	−1	1
E	2	0	−1

C_{1h}	E	σ_h
A'	1	1
A''	1	−1

D_3 has three classes and C_{1h} has two classes, D_{3h} must have six classes and six irreducible representations. The elements of D_{3h} can be obtained by multiplying each element of D_3 by each element of C_{1h}. Thus we obtain a 6×6 character table as shown in table 3.2 which it is interesting to note has the form

$$\begin{array}{ll} 1 \times \chi(D_3), & 1 \times \chi(D_3) \\ 1 \times \chi(D_3), & -1 \times \chi(D_3) \end{array}$$

where $\chi(D_3)$ means the table of characters of the group D_3.

Table 3.2 Character table of the direct product representation

D_{3h}	E	$3C_2$	$2C_3$	σ_h	$3\sigma_v$	$2S_3$
A_1'	1	1	1	1	1	1
A_2'	1	-1	1	1	-1	1
E'	2	0	-1	2	0	-1
A_1''	1	1	1	-1	-1	-1
A_2''	1	-1	1	-1	1	-1
E''	2	0	-1	-2	0	1

The dimension of $\Gamma^{l \otimes m}(A_j B_k)$ is the product of the dimensions of $\Gamma^l(A_j)$ and $\Gamma^m(B_k)$. Note also that if one of the groups contains only one-dimensional representations then the dimensions of the representations of the direct product group are determined by the dimensions of the representations of the other group. This is particularly true for all direct product groups formed with either C_{1h} or S_2.

3.6(d) Direct product representations within a group

We can always obtain direct product representations of a group which is itself not a direct product group. If $\Gamma^j(R)$ and $\Gamma^{j'}(R)$ are irreducible representations of a group G_a which is itself not a direct product group, then the direct product of the matrices $\Gamma^j(R) \otimes \Gamma^{j'}(R)$ (denoted by $\Gamma^{j \otimes j'}(R)$) will also form a representation of the group G_a (this representation is of great value as will be seen later).

The basis functions of the representation $\Gamma^{j \otimes j'}(R)$ are the products of the basis functions of the constituent representations $\Gamma^j(R)$ and $\Gamma^{j'}(R)$. Thus if P_R operates on the basis functions ϕ^j_κ and $\phi^{j'}_{\kappa'}$ we have, according to eq. (3.35)

$$P_R \phi^j_\kappa = \sum_\mu \phi^j_\mu \Gamma^j_{\mu\kappa}(R) \tag{3.75}$$

and

$$P_R\phi^{j'}_{\kappa'} = \sum_{\mu'} \phi^{j'}_{\mu'}\Gamma^{j'}_{\mu'\kappa'}(R) \tag{3.76}$$

Hence

$$P_R\phi^{j}_{\kappa}\phi^{j'}_{\kappa'} = \sum_{\mu\mu'} \phi^{j}_{\mu}\phi^{j'}_{\mu'}\Gamma^{j}_{\mu\kappa}(R)\Gamma^{j'}_{\mu'\kappa'}(R) \tag{3.77}$$

showing that (using eq. (A.32))

$$P_R\phi^{j}_{\kappa}\phi^{j'}_{\kappa'} = \sum_{\mu\mu'} \phi^{j}_{\mu}\phi^{j'}_{\mu'}\Gamma^{j\otimes j'}_{\mu\mu'\kappa\kappa'}(R) \tag{3.78}$$

Thus $\phi^{j}_{\kappa}\phi^{j'}_{\kappa'}$ form basis functions for the direct product representation $\mathbf{\Gamma}^{j\otimes j'}(R)$.

The character $\chi^{j\otimes j'}(R)$ of the direct product representation is (cf. eq. (3.74))

$$\chi^{j\otimes j'}(R) = \chi^{j}(R)\chi^{j'}(R) \tag{3.79}$$

The direct product representation we are considering here is in fact just another representation of the group and as such *must* be either a reducible or an irreducible representation of the group for which it is formed. We can therefore write, in general, after bringing it to block form,

$$\mathbf{\Gamma}^{j\otimes j'}(R) = \sum_{k} a_{jj'k}\mathbf{\Gamma}^{k}(R) \tag{3.80}$$

where, using eq. (2.37), we can see that

$$a_{jj'k} = \frac{1}{g}\sum_{R} \chi^{k*}(R)\{\chi^{j}(R)\chi^{j'}(R)\} \tag{3.81}$$

Using the group D_3, again, and table 3.3 we can see that $\mathbf{\Gamma}^{3\otimes 3}(R)$ is reducible; and using eq. (3.81) we also see that

$$\begin{aligned} a_{331} &= \tfrac{1}{6}(4 + 0 + 2) = 1 \\ a_{332} &= \tfrac{1}{6}(4 - 0 + 2) = 1 \\ a_{333} &= \tfrac{1}{6}(8 + 0 - 2) = 1 \end{aligned} \tag{3.82}$$

Table 3.3 Character table of D_3 together with the character of the direct product representation $\mathbf{\Gamma}^{3\otimes 3}(R)$

	D_3	E	$3C_2$	$2C_3$
$\Gamma^1(R)$	A_1	1	1	1
$\Gamma^2(R)$	A_2	1	-1	1
$\Gamma^3(R)$	E	2	0	-1
$\Gamma^{3\otimes 3}(R)$		4	0	1

giving

$$\boldsymbol{\Gamma}^{3\otimes 3}(R) = \boldsymbol{\Gamma}^1(R) \oplus \boldsymbol{\Gamma}^2(R) \oplus \boldsymbol{\Gamma}^3(R) \tag{3.83}$$

By inspection we can in addition see that

$$\begin{aligned} \boldsymbol{\Gamma}^{1\otimes 1}(R) &= \boldsymbol{\Gamma}^1(R); \qquad \boldsymbol{\Gamma}^{1\otimes 2}(R) = \boldsymbol{\Gamma}^2(R); \\ \boldsymbol{\Gamma}^{1\otimes 3}(R) &= \boldsymbol{\Gamma}^3(R); \qquad \boldsymbol{\Gamma}^{2\otimes 3}(R) = \boldsymbol{\Gamma}^3(R) \end{aligned} \tag{3.84}$$

The principal application of direct product representations within a group centres on their usefulness in finding the representations to which products of basis functions belong. If for example we know the irreducible representations to which the functions x, y, and z individually belong, then we can find the representations to which *products* of x, y, and z belong.

For example in order to find the irreducible representations of the group D_2 to which the functions $x, y,$ and z belong we can, since all the representations are one-dimensional, use the operators O^j of eq. (3.43), with the aid of tables 3.4 and 3.5, to obtain

$$\begin{aligned} &O^1x = 0; \quad O^2x = x; \quad O^3x = 0; \quad O^4x = 0 \\ &O^1y = 0; \quad O^2y = 0; \quad O^3y = y; \quad O^4y = 0 \\ &O^1z = 0; \quad O^2z = 0; \quad O^3z = 0; \quad O^4z = z \end{aligned} \tag{3.85}$$

From eq. (3.40) we can now see that x is a basis function for $\boldsymbol{\Gamma}^2(R)$, y is a basis function for $\boldsymbol{\Gamma}^3(R)$, and z is a basis function for $\boldsymbol{\Gamma}^4(R)$.

Table 3.4 Character table of the group D_2 showing the irreducible representations to which the functions x, y and z belong

	E	C_2^x	C_2^y	C_2^z	Basis function
$\Gamma^1(R); A$	1	1	1	1	
$\Gamma^2(R); B_1$	1	1	-1	-1	x
$\Gamma^3(R); B_2$	1	-1	1	-1	y
$\Gamma^4(R); B_3$	1	-1	-1	1	z

Table 3.5 The action of the function operators of the group D_2 on the functions x, y and z

	x	y	z
P_E	x	y	z
$P_{C_2^x}$	x	$-y$	$-z$
$P_{C_2^y}$	$-x$	y	$-z$
$P_{C_2^z}$	$-x$	$-y$	z

Thus, for example, x^2 is a basis function of the direct product representation $\boldsymbol{\Gamma}^2(R) \otimes \boldsymbol{\Gamma}^2(R)$ (which from table 3.4 can be seen to be equal to $\boldsymbol{\Gamma}^1(R)$).

Similarly y^2 is a basis function of $\Gamma^1(R)$, z^2 is a basis function of $\Gamma^1(R)$, xy is a basis function of $\Gamma^4(R)$, (x^3y^2z) is a basis function of $\Gamma^{2\otimes2\otimes2}(R)\otimes\Gamma^{3\otimes3}(R)\otimes\Gamma^4(R)=\Gamma^3(R)$, and finally (x^2+y^2+xy) will belong to the reduced representation $\Gamma^1(R)\oplus\Gamma^4(R)$.

If we consider the product of two basis functions belonging to the *same* representation then we can introduce the concepts of symmetric and antisymmetric direct products. However, we will not pursue this topic further.

3.7 The group of the Hamiltonian

One of the central problems of quantum mechanics is to solve the eigenvalue equation

$$H\psi_n=\left[-\frac{\hbar^2}{2m}\nabla^2+V(x,y,z)\right]\psi_n=E_n\psi_n \qquad (3.86)$$

This is of course the Schrödinger equation, in which H is the Hamiltonian operator of the system under consideration and ψ_n is the eigenfunction associated with the eigenvalue E_n where E_n can take on a set of discrete values.

The Hamiltonian is composed of kinetic and potential energy terms where each term is, in general, some function of the coordinates (x, y, z). The kinetic term is of the form $-(\hbar^2/2m)\nabla^2$ where $\hbar=h/2\pi$ and h is Planck's constant, m is the mass of a particle and

$$\nabla^2\equiv\frac{\partial^2}{\partial x^2}+\frac{\partial^2}{\partial y^2}+\frac{\partial^2}{\partial z^2}$$

The latter operator clearly remains invariant under a unitary transformation of coordinates. The potential energy term $V(x, y, z)$ possesses a symmetry peculiar to the physical system under consideration. For example electrons in certain solids will experience a $V(x, y, z)$ with cubic symmetry.

In view of what we have just said we can regard H as possessing a symmetry so that if we now consider the transformation of H to H' due to a unitary change of coordinate system then (3.86) goes to

$$H'\psi'_n=E_n\psi'_n \qquad (3.87)$$

where

$$\psi'_n=P_R\psi_n \qquad (3.88)$$

P_R being the function operator introduced in section 3.1. Therefore

$$H'(P_R\psi_n)=E_n(P_R\psi_n) \qquad (3.89)$$

showing that

$$H'=P_RHP_R^{-1} \qquad (3.90)$$

If P_R is a symmetry operator under which the Hamiltonian remains invariant then $H' = H$, a fact which immediately shows that

$$P_R H = H P_R \tag{3.91}$$

The set of operators P_R which commute with H form a group which is called the group of the Hamiltonian. For the rest of the discussion P_R will always be taken as an element of the group of the Hamiltonian.

If P_R is an operator which commutes with H then it follows that

$$P_R H\psi_n = H(P_R\psi_n) = E_n(P_R\psi_n) \tag{3.92}$$

Thus the function $P_R\psi_n$ is a solution of the Schrödinger equation. $P_R\psi_n$ is therefore either simply ψ_n or *another* eigenfunction ψ_n' of H associated with the *same* eigenvalue E_n. In the latter case ψ_n' is said to be *degenerate* with ψ_n. Clearly if an eigenvalue is degenerate, the set of eigenfunctions associated with it must form a basis for a representation of P_R. Formally then P_R acting on one of a set of l degenerate eigenfunctions must produce the same eigenfunction or a linear combination of the members of the set, i.e.,

$$P_R\psi_{nj} = \sum_{i=1}^{l} \psi_{ni}a_{ij} \tag{3.93}$$

Thus by eq. (3.26) we can see that the matrix with elements a_{ij} forms a l-dimensional representation of the group of P_R.

For any Hamiltonian there are a number of different eigenvalues E_n and eigenfunctions ψ_{nj} so that we can write

$$\begin{aligned} H\psi_{1j} &= E_1\psi_{1j} \\ H\psi_{2j} &= E_2\psi_{2j} \\ &\vdots \\ H\psi_{nj} &= E_n\psi_{nj} \end{aligned} \tag{3.94}$$

Equations (3.94) can alternatively be written in matrix form by collecting the functions ψ_{nj} into the form of a row vector in which we *order* the eigenfunctions into degenerate sets belonging to eigenvalues E_1, E_2, E_3, and so on. Naturally eigenstates are often non-degenerate, but we are assuming degeneracy to demonstrate a particular point. The eigenvalues can be collected into a diagonal matrix, the diagonal elements being the eigenvalues and all the other elements being zero. The matrix form of (3.94) is therefore

$$H(\tilde{\psi}) = (\tilde{\psi})\mathbf{E} \tag{3.95}$$

E in eq. (3.95) is by analogy to eq. (3.27) the representation of the Hamiltonian operator using the eigenfunctions as a basis. In this basis the representation of H is diagonal. The representation of H is not diagonal in a basis where the functions are not eigenfunctions. If degeneracy occurs some of the eigenvalues will be equal to each other but even in this case the matrix **E** will still be diagonal.

Now $P_R\psi_{1j}$ is a linear combination of the l_1 functions ψ_{1i} where l_1 is the degeneracy of E_1. This linear combination of functions clearly does not include any function belonging to a different eigenvalue which means that the representation of P_R, using *all* the eigenfunctions, collected into the ordered set $(\tilde{\psi})$, as a basis, must be in block form, for example

$$\boldsymbol{\Gamma}(R) = \left[\begin{array}{cc|c|cc} \multicolumn{2}{c|}{\multirow{2}{*}{\mathbf{A}}} & 0 & 0 & 0 \\ & & 0 & 0 & 0 \\ \hline 0 & 0 & \mathbf{B} & 0 & 0 \\ \hline 0 & 0 & 0 & \multicolumn{2}{c}{\multirow{2}{*}{\mathbf{C}}} \\ 0 & 0 & 0 & & \end{array}\right] \tag{3.96}$$

in which **A** is an $l_1 \times l_1$ square matrix with elements a_{ij} where l_1 is the degeneracy of the eigenvalue E_1 and **B** is an $l_2 \times l_2$ square matrix with elements b_{ij} where l_2 is the degeneracy of E_2, etc.

For example let us assume that there are only six eigenfunctions of H belonging to three eigenvalues, so that eq. (3.95) becomes

$$H(\tilde{\psi}) = (\tilde{\psi})\begin{bmatrix} E_1 & 0 & 0 & 0 & 0 & 0 \\ 0 & E_1 & 0 & 0 & 0 & 0 \\ 0 & 0 & E_1 & 0 & 0 & 0 \\ 0 & 0 & 0 & E_2 & 0 & 0 \\ 0 & 0 & 0 & 0 & E_2 & 0 \\ 0 & 0 & 0 & 0 & 0 & E_3 \end{bmatrix} \tag{3.97}$$

where $\psi_{11}, \psi_{12}, \psi_{13}$ belong to E_1 (three-fold degenerate); ψ_{21}, ψ_{22} belong to E_2 (two-fold degenerate), and ψ_{31} belongs to E_3 (singly degenerate). Therefore

$$P_R(\tilde{\psi}) = (\tilde{\psi})\begin{bmatrix} a_{11} & a_{12} & a_{13} & 0 & 0 & 0 \\ a_{21} & a_{22} & a_{23} & 0 & 0 & 0 \\ a_{31} & a_{32} & a_{33} & 0 & 0 & 0 \\ 0 & 0 & 0 & b_{11} & b_{12} & 0 \\ 0 & 0 & 0 & b_{21} & b_{22} & 0 \\ 0 & 0 & 0 & 0 & 0 & c_{11} \end{bmatrix} \tag{3.98}$$

This equation will give $P_R\psi_{1j}$ as a linear combination of ψ_{11}, ψ_{12}, and ψ_{13}, $P_R\psi_{2j}$ as a linear combination of ψ_{21} and ψ_{22} and so on.

$\mathbf{\Gamma}(R)$ is in block form and is therefore a reducible representation containing a certain number of irreducible representations. The dimensions of the blocks in $\mathbf{\Gamma}(R)$ are the same as the dimensions of the irreducible representations of the group of P_R and therefore the degeneracy of the eigenvalues of the Hamiltonian is determined *entirely* by the nature of P_R, the group of the Hamiltonian.

The l functions of an l-times degenerate eigenvalue being linearly independent can only mix amongst themselves under the operations P_R; hence they span an invariant subspace of the space of all the eigenfunctions and form a basis for an l-dimensional *irreducible* representation of the group of P_R. Finally we emphasize that functions which form bases for an irreducible representation have the same eigenvalue whilst functions belonging to *different* irreducible representations must have different eigenvalues; however, the converse of the latter statement is not necessarily true. If functions from *different* irreducible representations belong to the *same* eigenvalue this is called accidental degeneracy.

3.8 Quantum numbers

Any eigenfunction must be a basis function for one of the irreducible representations of the group of the Hamiltonian. It can therefore be labelled using eq. (3.40) according to the irreducible representation to which it belongs and to the column of the matrix by which it is transformed under P_R. These two labels completely specify the eigenfunction and are therefore said to be good quantum numbers with which to describe the state of the system.

For example the spherical harmonics $Y_l^m(\theta, \phi)$ where θ and ϕ are the polar angles are eigenfunctions of L^2 where L is the angular momentum operator and the group of this operator is the rotation group R_3 (cf. section 1.9). This latter group is the group of all rotations of a sphere (the specification of a single rotation in this group requires three parameters; namely two polar angles defining the axis and the angle of rotation about that axis). These harmonics are therefore basis functions for a representation of the rotation group and the action of an operator P_R from this group of R_3 yields

$$P_R Y_l^m(\theta, \phi) = \sum_{m'} Y_l^{m'}(\theta, \phi) D^l_{m'm}(R) \tag{3.99}$$

where $\mathbf{D}^l(R)$ is, of course, the matrix representing P_R in the lth irreducible representation of the rotation group and m denotes its column. This representation will be given a further discussion in chapter 9. Thus l and m are good quantum numbers with which to label the angular momentum eigenstates of a centrally symmetric system and are, of course, the familiar orbital and magnetic quantum numbers used in atomic physics. However, we must emphasize that l and m are particular to the rotation group and are therefore not relevant to systems such as molecules; in these systems the quantum numbers are the labels of irreducible representations of other groups.

3.9 Matrix representation of the Hamiltonian

In the matrix formulation of quantum theory the Hamiltonian operator H is represented by a matrix in the same way as the operators P_R are in eq. (3.27). If we actually choose the eigenfunctions of the Hamiltonian as basis functions for the representation, then the matrix representing H will be in diagonal form with the eigenvalues along the diagonal (cf. eq. (3.97)).

However this is not usually the procedure we can adopt because in the eigenvalue equation

$$H\psi_n = E_n\psi_n \tag{3.100}$$

only the Hamiltonian is known. Instead we select as basis functions some complete set of orthonormal functions ϕ_i which are not the real solutions of the problem and obtain a representation of H from the equation

$$H\phi_i = \sum_{j'} \phi_{j'} H_{j'i} \tag{3.101}$$

where $H_{j'i}$ is the matrix element on the j'th row and the ith column of the matrix representing H. Naturally if we use an infinite set of basis functions $[H_{ij}]$ will be an infinite matrix. In practice, however, we often only require a finite set of basis functions (cf. chapter 8). We will show later how group theory will enable us to determine how many elements H_{ij} are *identically zero* solely on the grounds of symmetry.

Since ϕ_i have been chosen to be an orthonormal set we, of course, have

$$\langle \phi_i | \phi_j \rangle = \delta_{ij} \tag{3.102}$$

so that

$$\langle \phi_j | H | \phi_i \rangle = \sum_{j'} \langle \phi_j | \phi_{j'} \rangle H_{j'i} = \sum_{j'} \delta_{jj'} H_{j'i} \tag{3.103}$$

therefore

$$\langle \phi_j | H | \phi_i \rangle = H_{ji} \tag{3.104}$$

We can always express the true eigenfunctions ψ_i in terms of the selected basis functions ϕ_i. For example the ith eigenfunction can be written as

$$\psi_i = \sum_j \phi_j U_{ji} \tag{3.105}$$

so that we can write the row vector constructed from the true functions ψ_i as

$$(\tilde{\boldsymbol{\psi}}) = (\tilde{\boldsymbol{\phi}})\mathbf{U} \tag{3.106}$$

Now the matrix representation of H, using the true eigenfunctions ψ_i as basis functions, is, as we have already shown, the *diagonal* matrix **E**, i.e.,

$$H(\tilde{\boldsymbol{\psi}}) = (\tilde{\boldsymbol{\psi}})\mathbf{E} \tag{3.107}$$

On the other hand the matrix representation of H using the arbitrary basis functions ϕ_i is **H** which is *not* diagonal (having elements given by eq. (3.104)), i.e.,

$$H(\tilde{\boldsymbol{\phi}}) = (\tilde{\boldsymbol{\phi}})\mathbf{H} \tag{3.108}$$

Thus using eqs. (3.106), (3.107), and (3.108) we obtain

$$H(\tilde{\boldsymbol{\phi}})\mathbf{U} = (\tilde{\boldsymbol{\phi}})\mathbf{U}\mathbf{E} \tag{3.109}$$

so that

$$(\tilde{\boldsymbol{\phi}})\mathbf{H}\mathbf{U} = (\tilde{\boldsymbol{\phi}})\mathbf{U}\mathbf{E} \tag{3.110}$$

Hence,

$$\mathbf{E} = \mathbf{U}^{-1}\mathbf{H}\mathbf{U} \tag{3.111}$$

Eigenvalue problems therefore amount to diagonalizing the representation matrix **H**. Ideally, then, we begin with a set of arbitrary basis functions ϕ_i and construct the representation **H**. We then diagonalize **H** by constructing the matrix **U** (cf. appendix A.18). The knowledge of **U**, as implied by the relationship (3.106), then enables us to determine the eigenfunctions. The process of diagonalizing **H** can, however, involve a great deal of labour and may, for some problems, be virtually impossible. Any steps that can be taken towards a *systematic* simplification of **H** must therefore be of great value. We will now see how we can use group theory to effect such a simplification.

Group theory enables us to choose the functions ϕ_i in such a way that many of the matrix elements H_{ij} are zero.

This choice is made by constructing the operators O^j of eq. (3.47) for each irreducible representation of the group of the Hamiltonian. Naturally, we could in principle use the projection operators $O^j_{\kappa\kappa}$ of eq. (3.41) but in order to do this we would have to construct *all* the matrix representations for all the symmetry operations of the group of the Hamiltonian. The advantage of using O^j lies in the fact that it makes use of the readily available character tables (cf. appendix C).

The operation of O^j on the set ϕ_i will generate a new set of *symmetrized* basis functions $\{\xi_i\} = \{\phi^j\}$ for all j, cf. eq. (3.48) belonging to the irreducible representations $\boldsymbol{\Gamma}^j(R)$. The functions ξ_i can be related to the initial choice of basis functions ϕ_i by some matrix **S**, say, where

$$(\tilde{\boldsymbol{\xi}}) = (\tilde{\boldsymbol{\phi}})\mathbf{S} \tag{3.112}$$

where we deliberately order the row vector $(\tilde{\boldsymbol{\xi}})$ into sets corresponding to each irreducible representation, i.e.,

$$(\tilde{\boldsymbol{\xi}}) = [\underbrace{\xi_1, \xi_2, \xi_3}_{\boldsymbol{\Gamma}^i(R)}, \underbrace{\xi_4, \xi_5}_{\boldsymbol{\Gamma}^j(R)}, \ldots] \tag{3.113}$$

The representation of H using the new basis functions ξ_i is therefore $\mathbf{H}'$ where

$$H(\tilde{\boldsymbol{\xi}}) = (\tilde{\boldsymbol{\xi}})\mathbf{H}' \tag{3.114}$$

so that from eqs. (3.112) and (3.108) we find that

$$\mathbf{H}' = \mathbf{S}^{-1}\mathbf{H}\mathbf{S} \tag{3.115}$$

This representation is in block form as will now be demonstrated.

The symmetrized functions are distributed over the irreducible representations $\boldsymbol{\Gamma}^j(R)$ of the group of the Hamiltonian so that it seems intuitively reasonable that a function ξ_i, say, belonging to the jth irreducible representation is in fact a linear combination of only those true eigenfunctions which belong to the jth irreducible representation. Thus the function $H\xi_i$ for a particular value of j, must be some other function *still* belonging to the jth irreducible representation. That is

$$H\xi_s = \xi_t \tag{3.116}$$

where ξ_s and ξ_t both belong to the jth irreducible representation. Hence the matrix element H'_{rs} between functions belonging to different irreducible representations j and j' is as a consequence of eq. (3.63)

$$\begin{aligned} H'_{rs} &= \langle \xi_{rj'} | H | \xi_{sj} \rangle \\ &= \langle \xi_{rj'} | \xi_{tj} \rangle \\ &= 0 \text{ if } j \neq j' \end{aligned} \tag{3.117}$$

or otherwise some constant. The matrix $\mathbf{H}'$ must, therefore, be in block form.

If we could construct all the projection operators $O^j_{\kappa\kappa}$ then we could produce $\{\Phi_i\} = \{\phi^j_\kappa\}$ for all the possible values of j and κ, that is the set of orthonormal *basis functions* for the irreducible representations of the group of the Hamiltonian. These functions given by

$$(\tilde{\boldsymbol{\Phi}}) = (\tilde{\boldsymbol{\phi}})\mathbf{Q} \tag{3.118}$$

lead to a representation $\mathbf{H}''$ of H where

$$\mathbf{H}'' = \mathbf{Q}^{-1}\mathbf{H}\mathbf{Q} \tag{3.119}$$

$\mathbf{H}''$ is of course now diagonal. Hence $\mathbf{H}''$ is equal to $\mathbf{E}$ and ϕ^j_κ are the true eigenfunctions. In practice, as we have mentioned before, character tables are readily available while representations for all the group operations are not. We are therefore more likely to reach the stage of eq. (3.115) than eq. (3.119). However, on reaching the stage of eq. (3.115) we can often invoke arguments designed to support the neglect of many off-diagonal elements (cf. chapter 8). Also we could, having simplified $\mathbf{H}$ to $\mathbf{H}'$ using group theory, attempt to diagonalize $\mathbf{H}'$ directly using the method described in section A.18.

3.10 The splitting of eigenvalues

As we shall see in chapter 6, the degeneracy of an eigenvalue of a system can be 'lifted' by some additional influence which lowers the symmetry; this means that the energy levels of, say, an atom in an environment with cubic symmetry, will not have the same degeneracy as the free atom. In fact the eigenfunctions of a previously irreducible representation associated with a particular energy level, become, in the system of lower symmetry the basis functions of a reducible representation. The latter now has several irreducible components each of which is associated with a distinct eigenvalue, and we are thus led to the conclusion that a lowering of the symmetry of the system causes a level to split. Obviously, by splitting, we cannot obtain more levels, from a particular level, than there are irreducible components in the reducible representation.

A nice example of such a splitting is provided by the well-known Zeeman effect in which the energy levels of an atom are separated by the action of a magnetic field. This is because the direction of the field is an axis of symmetry, so that the Hamiltonian of the atom is now invariant under the point group C_∞ whereas it was, in the absence of a field, invariant under the point group R_3; now C_∞ is a cyclic group, which means that all its representations are one-dimensional, and hence *all* the degeneracy must be lifted in the presence of a magnetic field.

3.11 Selection rules

If a system initially in equilibrium has its Hamiltonian changed, by an amount V, due to the action of some external perturbing agency then it is raised from the ground state to some excited state. It can then make transitions between two quantum states with eigenfunctions ϕ_1 and ϕ_2 with a probability proportional to the square of the matrix element $\langle\phi_1|V|\phi_2\rangle$ where V is the perturbation operation. If the system is an atom, say, illuminated with electromagnetic radiation, then so-called electric dipole transitions can occur (cf. chapter 8) via a V given by

$$V = e\langle r|E^0\rangle = e(xE_x^0 + yE_y^0 + zE_z^0) \tag{3.120}$$

where $|E^0\rangle$ is the vector amplitude of the incident radiation and e is the charge on the electron. The matrix element $\langle\phi_1|V|\phi_2\rangle$ in this case becomes

$$E_x^0\langle\phi_1|x|\phi_2\rangle + E_y^0\langle\phi_1|y|\phi_2\rangle + E_z^0\langle\phi_1|z|\phi_2\rangle \qquad (3.121)$$

If a transition is allowed then at least one of these scalar products must be non-zero which means that because of the fact that basis functions are orthogonal at least one of the functions $x\phi_2$, $y\phi_2$, or $z\phi_2$ must belong to the same irreducible representation as ϕ_1, or a reducible representation which contains the irreducible representation of ϕ_1 as a component.

We can now interpret this state of affairs using the properties of direct product representations of a group (cf. section 3.6 (d)). Let us therefore suppose that ϕ_1^* belongs to the representation $\mathbf{\Gamma}^{j*}(R)$, x belongs to $\mathbf{\Gamma}^k(R)$, and that ϕ_2 belongs to $\mathbf{\Gamma}^l(R)$. Then the function $(x\phi_2)$ must belong to the direct product representation $\mathbf{\Gamma}^{k\otimes l}(R)$. In order that a matrix element be non-zero $\mathbf{\Gamma}^{k\otimes l}(R)$ must be either $\mathbf{\Gamma}^j(R)$ or (if reducible) contain $\mathbf{\Gamma}^j(R)$. Also, if $\mathbf{\Gamma}^{k\otimes l}(R)$ contains $\mathbf{\Gamma}^j(R)$, then $\mathbf{\Gamma}^{j*\otimes k\otimes l}(R)$ must contain $\mathbf{\Gamma}^{j*\otimes j}(R)$. Now $\mathbf{\Gamma}^{j*\otimes j}(R) = \mathbf{\Gamma}^{j*}(R) \otimes \mathbf{\Gamma}^j(R)$, as shown below, *always* contains $\mathbf{\Gamma}^1(R)$ the totally symmetric representation, i.e., (cf. equation 3.81)

$$a_{jj1} = \frac{1}{g}\sum_R \chi^{1*}(R)\chi^{j*}(R)\chi^j(R) = 1 \qquad (3.122)$$

where the last equality is a result of the eq. (2.28).

It follows, therefore, that $\langle\phi_1|x|\phi_2\rangle$ will equal zero unless $\mathbf{\Gamma}^{j*\otimes k\otimes l}(R)$ contains $\mathbf{\Gamma}^1(R)$. This, in fact is quite a general argument which can be universally applied to many physical problems. We therefore assert that matrix elements $V_{ij} = \langle\phi_i|V|\phi_j\rangle$ will vanish unless the direct product representation contains $\mathbf{\Gamma}^1(R)$, the totally symmetric representation.

Returning now to dipole transitions we will determine the selection rules associated with electric dipole transitions between states which are odd or even basis functions for the group S_2. Indeed if the group of the Hamiltonian is a direct product group with S_2, then all eigenfunctions of this group must be odd or even.

The character table of the group S_2 is:

S_2	E	i	Basis functions
$\Gamma^g(R)$	1	1	even, i.e., $P_i f(x) = f(x)$
$\Gamma^u(R)$	1	-1	odd, i.e., $P_i f(x) = -f(x)$

Also, x, y, and z all belong to $\mathbf{\Gamma}^u(R)$ since they change sign under inversion thus $\mathbf{\Gamma}^{j*\otimes u\otimes l}(R)$ must contain $\mathbf{\Gamma}^g(R)$ if a transition between ϕ_1 and ϕ_2 is allowed. (N.B. $\mathbf{\Gamma}^1(R)$ is in this case simply $\mathbf{\Gamma}^g(R)$.)

Now from the character table we can see that

$$\Gamma^{u\otimes u}(R) = \Gamma^{g\otimes g}(R) = \Gamma^{g}(R) \tag{3.123}$$

$$\Gamma^{u\otimes g}(R) = \Gamma^{g\otimes u}(R) = \Gamma^{u}(R) \tag{3.124}$$

so that if ϕ_1 belongs to $\Gamma^u(R)$ then $j = u$. $\Gamma^{j^*\otimes u\otimes l}(R)$ therefore becomes $\Gamma^{u^*\otimes u\otimes l}(R) = \Gamma^{g\otimes l}(R)$. Thus if a transition is allowed l must be equal to g. Similarly if ϕ_1 belongs to $\Gamma^g(R)$ then l must be equal to u for an allowed transition. We therefore conclude that the only permissible electric dipole transitions are from odd to even states or vice versa.

The use of direct products to determine selection rules is given a further discussion in chapter 8.

3.12 Problems

3.1 Consider two basis vectors $\mathbf{e}_1$ and $\mathbf{e}_2$ drawn from the centre of the equilateral triangle of Fig. 3.2 in such a way that they point towards the apices a and b. Using this basis construct a two-dimensional representation of the group D_3. Show that this is not a unitary representation and explain why.

3.2 Set up a three-dimensional basis whose vectors are parallel to the three mutually perpendicular sides of a square prism having D_{4h} symmetry. Using this basis construct a three-dimensional representation of the group D_{4h} (remember that D_{4h} is a direct product group $D_4 \otimes S_2$ and that you can therefore produce a representation of D_{4h} from the representations of D_4 and S_2).

3.3 (a) Cut out a square piece of paper. Now take the diagonal corners and fold two of them into the centre on top of the square and two into the centre underneath the square. Finally lift the triangles so formed so that they are perpendicular to the plane of the square. Determine the symmetry group to which the object belongs.
(b) Assuming an orthogonal co-ordinate system with x and y axes parallel to the sides of the object construct a two-dimensional representation of its group.
(c) Now use the function $f(x)$ to construct another representation of the group of the object using the method of section 3.1.
(d) Construct the projection operators $O^j_{\mu\kappa}$ (cf. eq. (3.37)) of the group of the object and by using the function

$$F = ax + by + cx^2 + dy^2 + exy$$

determine basis functions for the irreducible representation of the group.
(e) Show that these basis functions obey the orthogonality relationship of eq. (3.63).

3.4 Construct the set of direct product matrices $\Gamma^3(R) \otimes \Gamma^4(R)$ using table 2.1. By considering some members of this set of matrices show that they obey the same multiplication table as that of table 1.3.

3.5 Construct the projection operators O^j, defined in eq. (3.47), for the group D_{4h}. Using the function e^{iz} show that the only symmetrized functions produced are $\cos(z)$ and $i \sin(z)$.

3.6 Set up a table showing the behaviour of the Cartesian coordinates x, y, and z under the operations of the group C_4 (choosing z as the principal axis). Now using the character table of C_4 and the operators O^j show that
(a) The function z belongs to A.

(b) The functions $x \pm iy$ belong to the one-dimensional representations labelled E.
(c) By adding the characters of the two representations of E under each element of the group we can produce a set of characters for a two-dimensional representation. Using these characters show that x and y belong to this representation.
(d) Show that $x^2 + y^2$ belongs to A and that $x^2 - y^2$ belongs to B.
(e) Find the representations to which z^2, xy, xz, and yz belong.

4 Crystal symmetry

4.0 Crystals

We have only to see some beautifully formed crystal, such as calcite or gypsum, to realize that symmetry is a facet of Nature. Indeed the word 'crystal' is taken to mean a solid possessing some kind of external symmetrical form (or habit). However, even though there are many possible habits, even for the same substance, depending upon the method of growth, the external forms must be essentially manifestations of some sort of internal order, i.e., some internal symmetrical arrangement of atoms or molecules. We should, therefore, from considerations of symmetry be able to determine the types of arrangements of atoms and molecules which are possible within the solid.

In order to achieve this goal it is necessary to apply a knowledge of group theory to the study of the crystalline state; where the word *crystalline* will not now refer merely to the external habit of the solid but rather to the *ordered* state of the atoms and molecules within the solid.

A crystal then (even if its external form belies the fact) is composed of a regular array of atoms or groups of atoms in such a way that, from a microscopic viewpoint, a section through the solid (crystal) would reveal a pattern containing a periodically repeated motif; the sort of pattern we have become accustomed to associate with wallpapers. *Periodicity*, therefore, is clearly an important property of the pattern. In fact, periodicity implies that spatial translations through distances equal to the intervals between identical sites of the atoms must leave the pattern unchanged.

Let us now be more specific and consider, for the time being, a periodic array of structureless points which are situated in space in such a manner that each point has an *identical environment*. Such an array will be called an *empty lattice*. (A consideration of lattices containing other than structureless points at each lattice site will be undertaken a little later.)

If we consider an arbitrary lattice point as the origin and choose three non-coplanar vectors joining this point to three other lattice points then it is obvious that we can reach every other point in the lattice by a linear combination of these three vectors. In particular if we can reach every point in the lattice by means of a vector translation $\mathbf{R} = m_1\mathbf{a}_1 + m_2\mathbf{a}_2 + m_3\mathbf{a}_3$ where m_1, m_2, and m_3 are positive or negative *integers* then $\mathbf{a}_1$, $\mathbf{a}_2$, and $\mathbf{a}_3$ are known as basic or primitive translation vectors. The triad of vectors $\mathbf{a}_i$ actually defines an important fundamental volume of the space filled by the lattice which is associated with *only one* lattice site. This volume is known as a *primitive* cell of the lattice and its significance is that the whole of the space filled by the lattice can be reproduced by systematically stacking the primitive cells together. However, the primitive cell may not always be a convenient entity with which to fill the space covered by the lattice, especially if it does not possess all the symmetry of the lattice.

In the latter case we can choose other vectors that define symmetric cells, but we may then have to face the fact that these are associated with *more than one* lattice site, and therefore non-primitive. In the choice of another cell, we also make use of orthogonal vectors whenever this is possible. Thus in Fig. 4.1 we see that the primitive cell associated with a simple cubic lattice can be chosen as a cube. However, from Fig. 4.2, we can see that the primitive cell of a face-centred cubic lattice defined by the triad of lattice vectors $\mathbf{a}_i$, is rhombohedral in shape, and although such a cell only contains one lattice site it possesses an awkward shape; the rectangular cell which contains four lattice sites is more useful because it reflects the inherent symmetry of the lattice. In any event the fundamental cell associated with a lattice, whether it be primitive or otherwise, will be called the *unit cell*. In spite of the latter statement certain problems require the use of an alternative *primitive cell* which, while still associated with only one lattice site, possesses the *full* symmetry of the lattice. A cell of this

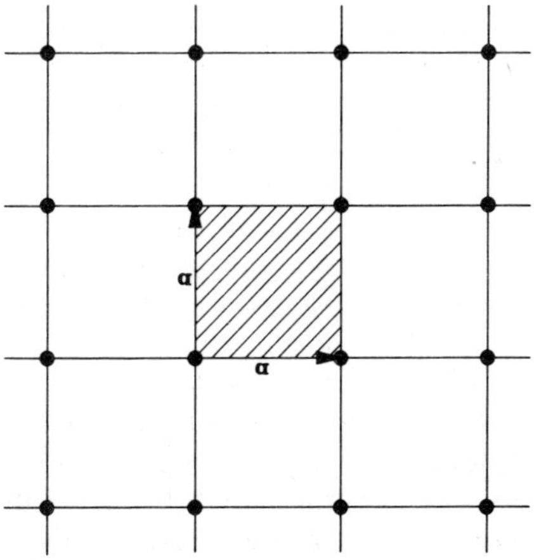

Fig. 4.1 *Section through a (simple) cubic cell: $a_1 = a_2 = a_3 = a$. Shaded area shows a primitive cell. ● denotes a lattice site.*

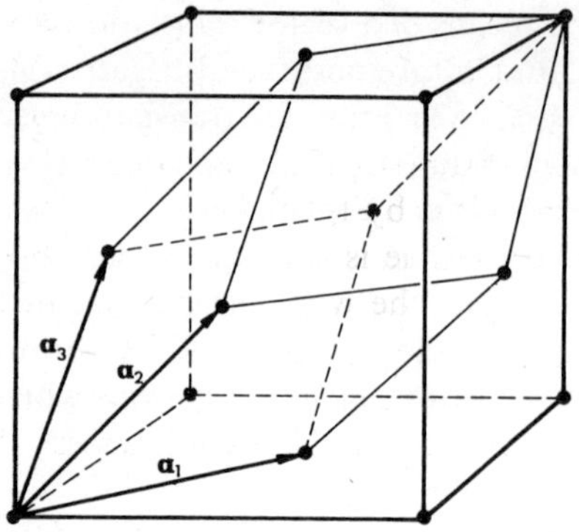

Fig. 4.2 *Rhombohedral primitive cell of a (face-centred) cubic lattice.*

type can be constructed by perpendicularly bisecting, with planes, all lines joining a given lattice site to its neighbouring lattice sites; the appropriate cell being the smallest volume enclosed by the intersecting planes. The latter type of primitive cell is known as the Wigner–Seitz cell and is illustrated in Fig. 4.3 for some lattices with cubic symmetry, i.e., those that are symmetric under the operations of the group O_h (cf. section 1.9(12)).

4.1 Translational symmetry

The vectors $\mathbf{a}_i$, which we have called primitive translations, are defined in such a way that a translation through a *lattice vector*

$$\mathbf{R} = m_1\mathbf{a}_1 + m_2\mathbf{a}_2 + m_3\mathbf{a}_3 \tag{4.1}$$

where m_i are integers, will bring the lattice into coincidence with itself. If we now define an operation T_i which translates the lattice through a vector $\mathbf{a}_i$ then it is clear that the translation (4.1) amounts to operating m_1 times with T_1, m_2 times with T_2, and m_3 times with T_3. A translation such as (4.1) can therefore be effected by the operation

$$\mathscr{T} = T_1^{m_1}T_2^{m_2}T_3^{m_3} \tag{4.2}$$

Experimental observations are made on *real* crystals, which are, of course, finite; but it is convenient in purely theoretical work to imagine that we are dealing with a rather special kind of entity known as a *cyclic* crystal. One reason for adopting such a model is the avoidance of introducing boundary conditions which are complicated to apply. If we took into account the *real surfaces* of a crystal then a calculation of the bulk properties of the crystal (e.g., the electrical resistivity) would become exceedingly complicated. However, a bulk property (i.e., a property of the whole crystal) does not, experimentally, depend very greatly on the

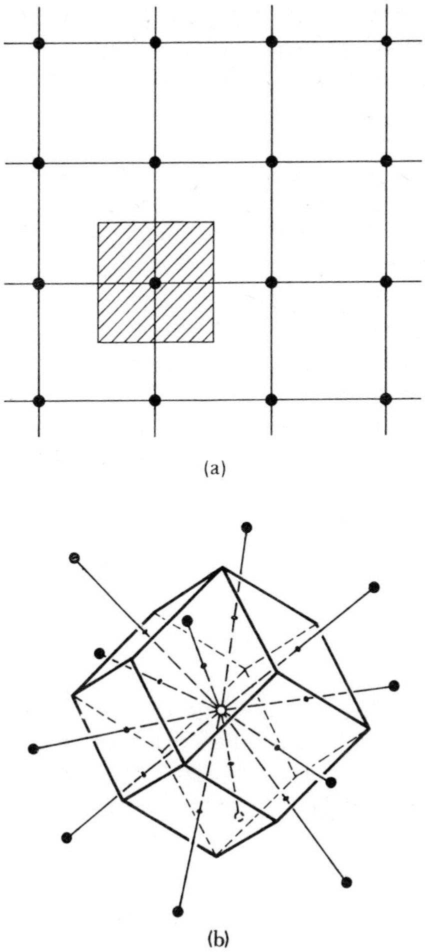

Fig. 4.3 *(a) Shaded area shows a section through Wigner–Seitz cell of a (simple) cubic lattice. (b) Wigner–Seitz cell of a (face-centred) cubic lattice.* ● *denotes a lattice site.*

surface properties (unless we use a crystal in which one of the dimensions has been made exceedingly small such as in a thin film) so that, apart from the sheer mathematical convenience of avoiding awkward boundary conditions, the model we are going to discuss makes good physical sense too!

From the group theoretical point of view a more important reason for the adoption of the concept of a cyclic crystal is that this is necessary for the operations T_i^m to form a finite group. For the group to be closed there must be some value of $m = N_i$, such that $T_i^{N_i} = E$, the identity operator.

Thus cyclic boundary conditions amount to recognizing that the operator $T_i^{N_i}$, which is the displacement through a vector $N_i\mathbf{a}_i$, brings us back to the *same* point. The operation is therefore equivalent to a displacement of zero.

A cyclic crystal then should properly be regarded as a part of an infinite crystal, i.e., some slice of an infinite crystal for which the cyclic boundary conditions discussed above are fulfilled, where the size of the slice will be defined by $N_1\mathbf{a}_1$, $N_2\mathbf{a}_2$, and $N_3\mathbf{a}_3$. We should note that it is impossible in a three-dimensional space to satisfy such conditions with the surfaces of a three-dimensional crystal; for a one-dimensional crystal we can always bend the crystal round, in a second dimension, into a ring; also a two-dimensional crystal can be turned into a three-dimensional torus. Nevertheless, the insistence on cyclic boundary conditions enables us to forget about broken symmetry at crystal surfaces and allows us to use the properties of a *cyclic group* to develop representations for the operators T_i. The boundary conditions for $N_1N_2N_3$ unit cells are therefore formally,

$$\begin{aligned} T_1^{N_1} &= E \\ T_2^{N_2} &= E \qquad (4.3) \\ T_3^{N_3} &= E \end{aligned}$$

Obviously, as we discussed in section 1.6 the elements T_i^m (where m is any integer from 1 to N_i) are members of a cyclic group; also although we have previously discussed the properties of the latter, it does not seem inappropriate to itemize them again for our immediate use. These properties are:

(a) A cyclic group is Abelian.
(b) A consequence of (a) is that each element is in a class of its own (cf. section 1.7).
(c) A consequence of (b) is that an Abelian group of order N has N classes and N irreducible representations (cf. eq. (2.31)).
(d) Since the sum, over the irreducible representations, of the squares of the dimensions of the irreducible representations is equal to the order of the group (cf. equation (2.23)) the conditions (a), (b), and (c) imply that all irreducible representations of the cyclic group are one-dimensional.

We now adopt the symbol $\mathbf{\Gamma}^{\nu_j}(T_j)$ to denote the matrix of T_j in the ν_jth irreducible representation of the group of translational operators T_j so that we can write the boundary conditions (4.3) considering, for the moment, only one dimension, as

$$\{\mathbf{\Gamma}^{\nu_1}(T_1)\}^{N_1} = 1 \qquad (4.4)$$

$\mathbf{\Gamma}^{\nu_1}(T_1)$ is therefore one of the N_1th roots of unity which means that

$$\mathbf{\Gamma}^{\nu_1}(T_1) = e^{2\pi i \nu_1 / N_1} \tag{4.5}$$

If we count the irreducible representations from $\nu_1 = 0$ then there are N_1 irreducible representations; one for each value of ν_1 from 0 to $N_1 - 1$. Also, since the representations $\mathbf{\Gamma}^{\nu_1}(T_1)$ are actually one-dimensional, the character of the representation is also $e^{2\pi i \nu_1 / N_1}$.

The determination of the representation of the operation (4.2) requires us to find $\mathbf{\Gamma}^{\nu_1}(T_1^{m_1})$. This latter quantity, as can be seen from equation (4.5), is clearly

$$\mathbf{\Gamma}^{\nu_1}(T_1^{m_1}) = \{\mathbf{\Gamma}^{\nu_1}(T_1)\}^{m_1} = e^{2\pi i \nu_1 m_1 / N_1} \tag{4.6}$$

The sets of translations $T_1^{m_1}$, $T_2^{m_2}$, and $T_3^{m_3}$ which describe a three-dimensional crystal each form a group whose elements commute with each other so that the group of operators $\mathscr{T}$ of eq. (4.2) is the direct product group of $T_1^{m_1}$, $T_2^{m_2}$, and $T_3^{m_3}$ (cf. section 3.6(a)). The irreducible representation of $\mathscr{T}$ is therefore the direct product matrix (eq. (3.72))

$$\mathbf{\Gamma}^{\nu_1}(T_1^{m_1}) \otimes \mathbf{\Gamma}^{\nu_2}(T_2^{m_2}) \otimes \mathbf{\Gamma}^{\nu_3}(T_3^{m_3}) = e^{2\pi i[(\nu_1 m_1/N_1) + (\nu_2 m_2/N_2) + (\nu_3 m_3/N_3)]} = e^{i\mathbf{k}.\mathbf{R}} \tag{4.7}$$

where in eq. (4.7) we have set

$$2\pi\left(\frac{\nu_1 m_1}{N_1} + \frac{\nu_2 m_2}{N_2} + \frac{\nu_3 m_3}{N_3}\right) = \mathbf{k}\,.\,\mathbf{R} \tag{4.8}$$

Now $\mathbf{R}$ is the arbitrary lattice vector of eq. (4.1), i.e., $\mathbf{R} = m_1\mathbf{a}_1 + m_2\mathbf{a}_2 + m_3\mathbf{a}_3$, and what we have done is to introduce a new kind of vector

$$\mathbf{k} = \frac{\nu_1}{N_1}\mathbf{b}_1 + \frac{\nu_2}{N_2}\mathbf{b}_2 + \frac{\nu_3}{N_3}\mathbf{b}_3 \tag{4.9}$$

where clearly for $\mathbf{k}$ to satisfy eq. (4.8) we require the condition

$$\mathbf{a}_i\,.\,\mathbf{b}_j = 2\pi\delta_{ij} \tag{4.10}$$

$\mathbf{b}_j$ are obviously basis vectors for the vector $\mathbf{k}$ and are reciprocal to the basis vectors $\mathbf{a}_i$ (cf. section A.10(c)). The new vectors $\mathbf{k}$ exist in a space which is usually referred to as reciprocal space. Also as we shall see in chapter 6 the representation $e^{i\mathbf{k}.\mathbf{R}}$ will have basis functions which are of the form of waves, $e^{i\mathbf{k}.\mathbf{r}}$, where $\mathbf{r}$ is an arbitrary vector in direct space. Thus, although group theoretically $\mathbf{k}$ labels the irreducible representations of the group, it also has the *physical meaning* of wave-number. It can, in fact, be identified as the wave-number of an excitation propagating in a periodic lattice. As a consequence of this, as we shall see in chapter 6, $\mathbf{k}$ *labels* the state of an electron propagating through a periodic lattice.

4.2 Proper rotations of a crystal lattice

Naturally a crystal may, apart from translational symmetry, possess axes of rotational symmetry (we are now referring to the concepts introduced in chapter 1, where we discussed the possibility of rotating a body through an angle $2\pi/n$ say (cf. section 1.8) about an axis of symmetry; the effect of such a rotation being to bring the body into coincidence with itself).

However, whereas an arbitrary isolated body could have arbitrary n-fold axes, i.e., n could take any value, a crystal lattice is a different matter because the additional requirement of translational symmetry will clearly restrict the values that n can assume.

Thus, if translational symmetry has also to be satisfied in the presence of a rotation then a lattice vector must, after some rotation θ, end up as another lattice vector. This requirement, as we have said, places severe restrictions on the possible values that θ can take.

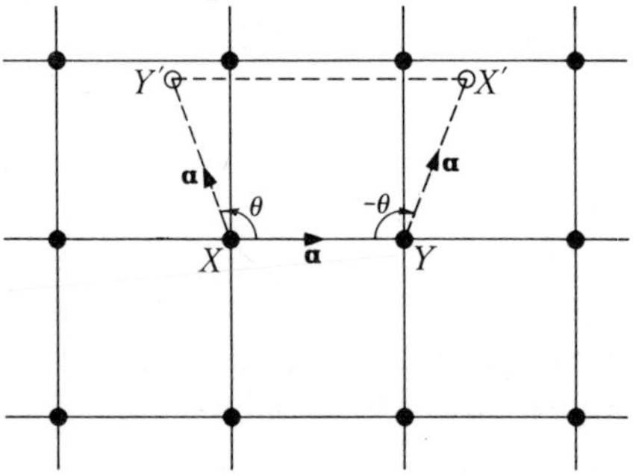

Fig. 4.4 *Rotation of a lattice vector a through angles θ and $-\theta$ about points X and Y respectively. ● denotes a lattice site. ○ denotes the new position of X or Y after the rotation.*

In order to discover these values let us refer to Fig. 4.4 which is a section through part of an infinite lattice. Let us now imagine that we rotate this lattice about an axis, perpendicular to the plane of the paper, through the lattice site X. If the angle of rotation is some arbitrary angle θ, then we can see that the lattice site at Y would go to Y' (for convenience we have not drawn the new positions of all the lattice sites). Similarly, the same rotation, in the opposite sense, about an axis through Y would take X to X'. Such an arbitrary rotation does not automatically take the lattice into coincidence with itself. Indeed the diagram is deliberately drawn in a manner which illustrates this point. Nevertheless, we can now ask what the length $Y'X'$ should be if translational invariance is to be preserved. In order to satisfy this invariance $Y'X'$ must, since it is parallel to the line joining X and Y, be some integral multiple, m, of the distance a. Thus from Fig. 4.4 we require

$$ma = a + 2a \sin(\theta - \pi/2) \tag{4.11}$$

A re-arrangement of (4.11) then gives us

$$\cos\theta = (1 - m)/2 \tag{4.12}$$

which, since $\cos\theta$ must lie between -1 and 1, implies that

$$-1 \leqslant (1 - m)/2 \leqslant 1 \tag{4.13}$$

Now only the integers $m = -1, 0, 1, 2,$ and 3 can satisfy (4.13) which means that θ can *only* take on the values 360°, 60°, 90°, 120°, or 180°, respectively. Therefore, since the above argument applies to any lattice, the only possible n-fold rotational symmetry axes *any* crystal can possess are 1, 2, 3, 4, and 6-fold in character.

A cube, for example, possesses (amongst others) four-fold symmetry axes of rotation which are normal to its faces, i.e., four consecutive turns of $\pi/2$ will bring a face back to its original position so that in the case of the lattice shown in Fig. 4.4 it is clear that the symmetry axes of rotation are actually four-fold axes.

4.3 Space groups

As we have already stated, a crystal possesses a translationally invariant lattice of points; a fact which imposes certain restrictions on the types of rotations we can perform on a lattice in order to bring it into coincidence with itself. Apart from proper rotations, crystals can also be invariant under reflections and improper rotations; indeed the complete set of operators of these types constitute point groups (as discussed in chapter 1). Therefore the full group of symmetry operators of a crystal must contain operations which *combine* elements of a point group with translation operations. Since we are including translations in *space* we talk now not in terms of the point group, whose operators leave at least one point in the system unmoved, but in terms of the *space group* of a lattice whose operators can be written symbolically as $\{\boldsymbol{\alpha}|\mathbf{t}\}$ where $\boldsymbol{\alpha}$ represents rotation and reflection operations and $\mathbf{t}$ represents a translation which does not necessarily have to be equal to a lattice vector. Indeed as we shall see later space group operators exist which involve *non-primitive* translations. The symbol $\{\boldsymbol{\alpha}|\mathbf{t}\}$ thus defines an operation consisting of $\boldsymbol{\alpha}$ followed by a translation $\mathbf{t}$. If we wish to denote a pure translation, i.e., a space group operation in the absence of a rotation or reflection, then we write $\{E|\mathbf{t}\}$. The identity element of the space group is therefore $\{E|0\}$.

Having now defined the space group in terms of the symbols let us examine some of its properties before going on to consider such matters as the types of crystal lattices which can exist; and the relationship of the point groups to actual crystals.

The manner in which $\{\alpha|\mathbf{t}\}$ operates on a position vector $\mathbf{r}$ say, is to transform $\mathbf{r}$ to $(\alpha\mathbf{r} + \mathbf{t})$. This transformation is illustrated in Fig. 4.5 which

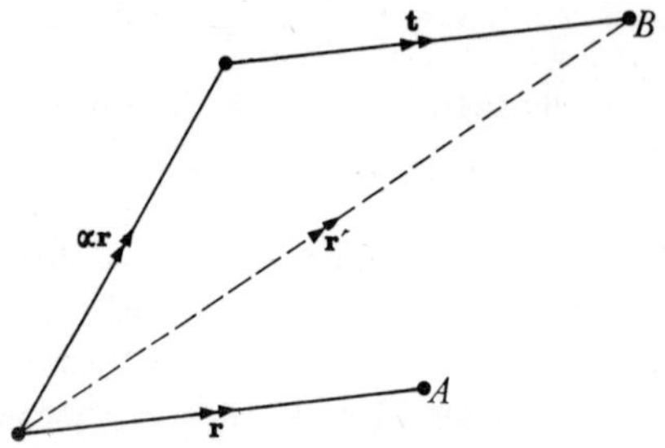

Fig. 4.5 *The operation of* $\{\alpha|\mathbf{t}\}$ *on the vector* $\mathbf{r}$.

shows that $\mathbf{r} \to \mathbf{r}'$ via the operations α and $\mathbf{t}$. Thus if we operate with $\{\alpha|\mathbf{t}\}$ on $\mathbf{r}$ we produce

$$\mathbf{r}' = \alpha\mathbf{r} + \mathbf{t} \tag{4.14}$$

Similarly the operation of $\{\alpha'|\mathbf{t}'\}$ on $\mathbf{r}'$ produces

$$\mathbf{r}'' = \alpha'\alpha\mathbf{r} + \alpha'\mathbf{t} + \mathbf{t}' \tag{4.15}$$

so we conclude that

$$\{\alpha'|\mathbf{t}'\}\{\alpha|\mathbf{t}\} = \{\alpha'\alpha|\alpha'\mathbf{t} + \mathbf{t}'\} \tag{4.16}$$

The inverse of the operation $\{\alpha|\mathbf{t}\}$ is

$$\{\alpha|\mathbf{t}\}^{-1} = \{\alpha^{-1}| - \alpha^{-1}\mathbf{t}\} \tag{4.17}$$

which can be verified by operating on (4.17) with $\{\alpha|\mathbf{t}\}$ and making use of eq. (4.16), i.e.,

$$\begin{aligned} \{\alpha|\mathbf{t}\}\{\alpha|\mathbf{t}\}^{-1} &= \{\alpha\alpha^{-1}|-\alpha\alpha^{-1}\mathbf{t} + \mathbf{t}\} = \{E|0\} \\ \{\alpha|\mathbf{t}\}^{-1}\{\alpha|\mathbf{t}\} &= \{\alpha^{-1}\alpha|\alpha^{-1}\mathbf{t} - \alpha^{-1}\mathbf{t}\} = \{E|0\} \end{aligned} \tag{4.18}$$

thus showing (4.17) to be the correct form for the inverse.

4.4 Crystal classes and systems

The most general form of a space group operator is

$$\{\alpha|\mathbf{t}\} = \{E|\mathbf{R}\}\{\alpha|V(\alpha)\} \tag{4.19}$$

where $\mathbf{R}$ is the lattice vector of eq. (4.1) and $V(\alpha)$ is a *non-primitive* translation vector, e.g., $\frac{1}{4}\mathbf{a}_1 + \frac{1}{4}\mathbf{a}_2 + \frac{1}{4}\mathbf{a}_3$ associated with α. $\{E|\mathbf{R}\}$ is the group of pure translations and is in fact a subgroup of the full space group. The set of operators $\{\alpha|V(\alpha)\}$ can be written in general as

$$\{\alpha|V(\alpha)\} = \{\alpha'|0\} + \{\alpha''|V(\alpha'')\} \tag{4.20}$$

where $\{\alpha'|0\}$ is a point group which is a subgroup of the space group. The set $\{\alpha''|V(\alpha'')\}$ is the set of operators which involve a point group operation α'' coupled with a non-primitive translation $V(\alpha'')$. If the set $\{\alpha''|V(\alpha'')\}$ is an empty set then the entire point group $\{\alpha|0\}$ is a subgroup of the space group.

As an example consider the diamond lattice shown in Fig. 4.10 consisting of two interpenetrating face-centred cubic lattices which are separated by a non-primitive vector $\frac{1}{4}(\mathbf{a}_1 + \mathbf{a}_2 + \mathbf{a}_3)$. Here the entire point group is O_h, i.e., $\{\alpha|0\} = \{\alpha'|0\} + \{\alpha''|0\}$ form the group O_h; however $\{\alpha'|0\}$ forms the group T_d containing the operations E, $8C_3$, $3C_4^2$, $6S_4$, $6\sigma_d$ (cf. Appendix C), which is a subgroup of O_h, whereas the set $\{\alpha''|0\}$ do not form a group and indeed cannot exist in the space group without an associated $V(\alpha'')$. In this case although the entire point group is O_h it is only T_d, a subgroup of O_h, which is a subgroup of the space group. Naturally $\{\alpha'|0\}$ must be consistent with the restrictions, on the rotations of a lattice, discussed in section 4.2.

In Chapter 1, during the discussion of the point groups, we did not have in mind any particular restrictions; yet, as we have been at pains to emphasize, in this chapter, a restriction must be placed upon α when dealing with a crystal lattice. Thus, of all the possible point groups one could devise, it is a consequence of the necessity of symmetry axes being either 1, 2, 3, 4, or 6-fold in nature that only a rather limited number of point groups can be used in a discussion of crystal symmetry.

If we are restricted to 1, 2, 3, 4, and 6-fold rotations then it can be shown that the number of permitted point groups that can be associated with crystal lattices is 32; and a space group whose entire point group is one of these 32 point groups is said to belong to one of the 32 crystal classes (not to be confused with the concept of class used in group theory).

If restrictions have to be placed on the operations in order to maintain a translationally invariant lattice then it is equally apparent that only certain types of lattice will be compatible with each of the 32 crystal classes.

By the word 'lattice' we still mean an array of structureless points which can be generated by primitive translations (i.e., an empty lattice). The limitation to 32 possible point groups implies that, in three dimensions, only 14 distinct lattices can exist. These are known as the Bravais lattices (after Bravais who discovered them in 1848) and are shown in Fig. 4.6. In this figure, we have actually drawn the conventional unit cells of the Bravais lattices, i.e., the Bravais lattice is generated by systematically stacking these cells together. The shapes of these cells were, originally.

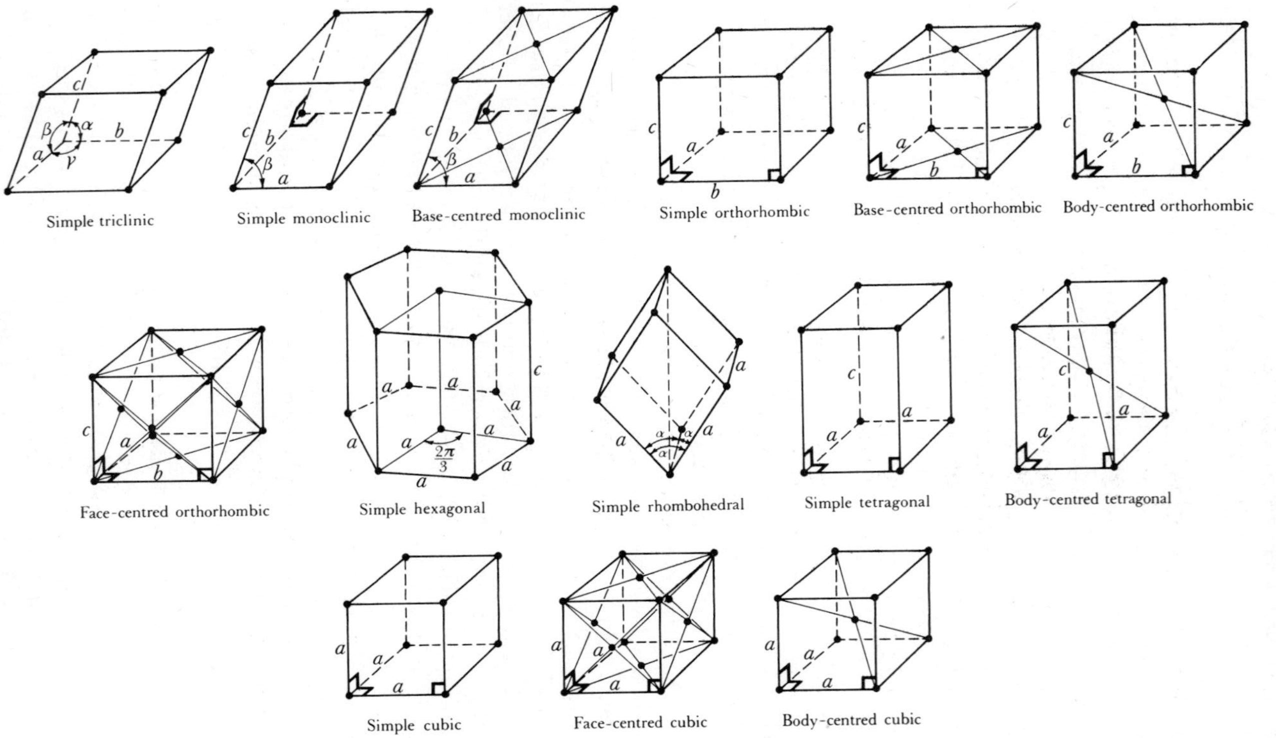

Fig. 4.6 *The Bravais lattices a, b, and c denote the lengths of the sides of the unit cell. α, β, γ are angles which are not right angles.*

selected to make them look, as far as possible, like the external form (habit) of actual crystals. We could of course, since the choice is arbitrary, use other shapes of cells.

In a Bravais lattice each site must have an *identical environment*; because the whole lattice can be generated by primitive translations from an arbitrary point in the lattice. Also the point group of each Bravais lattice must contain the inversion operator, i.e., the lattice has a centre of symmetry. However, if we have a solid such as diamond (see Fig. 4.10) we *cannot* reach all atoms, in the lattice by primitive translations through **R** of eq. (4.1). We must therefore in this case associate *two* atoms with *each* Bravais lattice site, i.e., the primitive cell, although containing *one* lattice site, will now contain two atoms.

If we continue to think of the diamond lattice as a lattice of single structureless points then it is *not* a Bravais lattice. Rather it consists of two interpenetrating face-centred cubic Bravais lattices. On the other hand, as we stated above, we could think of the diamond lattice as a single face-centred cubic Bravais lattice with *two* atoms associated with each lattice site (i.e., a lattice with a *basis* of two atoms). Another well-known example of what we have just discussed is NaCl (sodium chloride) which could be considered to be made up of two interpenetrating face-centred cubic lattices; one composed of sodium ions and the other composed of chlorine ions (cf. Fig. 4.7).

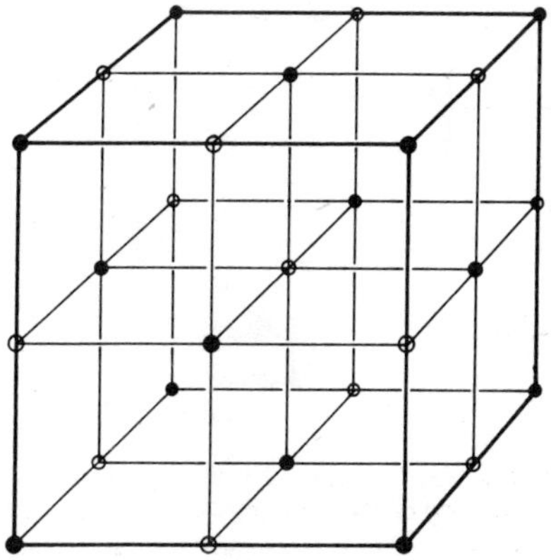

Fig. 4.7 *The sodium chloride lattice. ● denotes a chlorine ion. ○ denotes a sodium ion.*

If we have structureless points at the lattice sites then since such points have full rotational symmetry every operation of the 32 point groups will be a symmetry operation of the lattice site. The symmetry of a crystal of structureless points is, therefore, entirely determined by the symmetry of the Bravais lattice. If we place molecules, say, or other collections of atoms at each lattice site then not all the symmetry operations allowed for an empty Bravais lattice will now be permissible, i.e., the symmetry of a *crystal* can be lower than that of the empty Bravais lattice (cf. Fig. 4.8(a)).

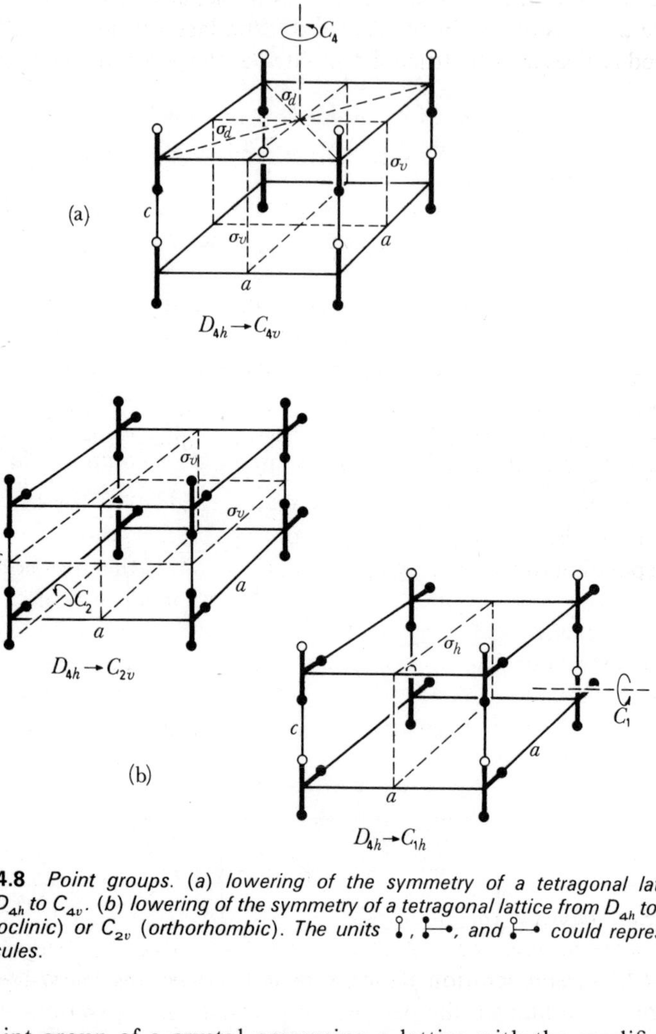

Fig. 4.8 *Point groups. (a) lowering of the symmetry of a tetragonal lattice from D_{4h} to C_{4v}. (b) lowering of the symmetry of a tetragonal lattice from D_{4h} to C_{1h} (monoclinic) or C_{2v} (orthorhombic). The units , , and could represent molecules.*

The point group of a crystal possessing a lattice with the modification discussed above could now well be only a *subgroup* of the point group of the empty lattice. Thus, if we begin with an empty lattice belonging to a particular point group and then modify it by introducing groups of atoms of lower symmetry than the original lattice sites then the resulting system will have a point group which will be a subgroup of the original one.

If we consider the point groups of the 14 empty Bravais lattices then we find that only seven point groups of the 32 possible ones are needed in order to define their symmetry. These seven groups are called the *crystal systems* and are listed in table 4.1.

We can see from Fig. 4.6 that we can have more than one type of lattice associated with a crystal system. For example, face-centred lattices have lattice points situated in the centre of each face of the unit cell; body-centred lattices have a lattice site situated at the geometrical centre of the unit cell; a base-centred lattice is of the form shown for a monoclinic lattice. In all cases, where more than one lattice type is assigned to a crystal system, each of the lattices in that system transforms in the same way under the action of the point group of the empty lattice. There is thus, for example, an essential identity between the simple and base-centred monoclinic lattices. If only seven point groups are required to describe the symmetry properties of a lattice, how is it that there are 32 crystal classes? We have already answered this question in our discussion of lattice contents. For instance, in an empty cubic lattice, O_h may be reduced to T_d by suitably introducing groups of atoms which *lower* the symmetry. Indeed the symmetry may be so lowered that we pass from one crystal system to another (cf. Fig. 4.8(b)). Table 4.1 shows how the 32 *crystal classes* are distributed over the *crystal systems.*

We should observe from the last column of table 4.1 that the triclinic system has the lowest symmetry. It only possesses symmetry elements corresponding to the identity operation $C_1 = E$ and the inversion operator $S_2 = \sigma_h C_2 = i$. The next crystal system is monoclinic and possesses a somewhat higher symmetry; its Bravais lattices, namely the second and third diagrams of Fig. 4.6, possess symmetry elements which involve a two-fold rotational symmetry axis and a reflection plane perpendicular to this axis. The point group of a monoclinic lattice is therefore C_{2h} which has subgroups C_{1h} and C_2.

The space group of a crystal, as we have learned, is a set of symmetry elements which takes into account the actual symmetry of the lattice *and* its contents. The enumeration of the number of possible space groups is well documented, so we do not feel that it is justifiable to go into any significant detail in this book. We can, however, in a simple-minded way, make some progress provided we are willing to accept the facts listed in table 4.1. The enumeration of the total number of permissible space groups requires us to link up the permissible translational operations with the allowed point group operations. A simple way of doing this is to multiply each operator $\{E|\mathbf{R}\}$ of the translation group by each operator $\{\boldsymbol{\alpha}|0\}$ of the point group. This set of elements forms a space group.

If the point group operations are E, C_2, and σ_h, say, and the translational operations are E, T_1, $T_1^2, \ldots, T_2$, $T_2^2, \ldots, T_3, T_3^2, \ldots,$ where T_1, T_2, and T_3 are operations which produce translations along the primitive

Table 4.1 32 crystal classes distributed amongst the seven crystal systems

Crystal system	Types of lattice	Point group of lattice	Associated classes (subgroups of point group of lattice)	Example
Triclinic	Simple	$S_2 = i$	C_1	Copper sulphate $CuSO_4 \cdot 5H_2O$
Monoclinic	Simple, base-centred	C_{2h}	C_{1h}, C_2	Borax $Na_2B_4O_7 \cdot 10H_2O$
Orthorhombic	Simple, base-centred, body-centred, face-centred	D_{2h}	C_{2v}, D_2	Acetic Acid CH_3CO_2H
Tetragonal	Simple, body-centred	D_{4h}	C_4, C_{4v}, C_{4h}, S_4, D_4, D_{2d}	Rutile TiO_2
Rhombohedral	Simple	D_{3d}	C_3, S_6, C_{3v}, D_3	Calcite $CaCO_3$
Hexagonal	Simple	D_{6h}	C_6, C_{3h}, C_{6h}, C_{6v}, D_6, D_{3h}	Beryl $Be_3Al_2Si_6O_{18}$
Cubic	Simple, face-centred, body-centred	O_h	T, T_h, T_d, O	Galena PbS

lattice directions, then the elements of the space group are $\{E|0\}$, $\{E|T_1^2\}$, ..., $\{C_2|0\}$, $\{C_2|T_1^2\}$, ..., $\{\sigma_h|0\}$, $\{\sigma_h|T_1^2\}$, ... Some space groups can therefore be constructed by using the translational operations for a particular lattice and combining them with the set of point group operations associated with it. At first sight it would appear that we could take the translational operations of the cubic system and combine them with the point group C_4 of the tetragonal system; but the space group we would obtain would not be distinguishable from that formed from the tetragonal system.

Actually we can place any structure at a lattice site of any crystal system. Indeed if we placed structures with no symmetry at the sites of a cubic lattice we would destroy all its symmetry properties and the resultant structure would not be of higher symmetry than triclinic. Similarly if we place, at lattice sites, objects of very high symmetry we cannot produce a space group corresponding to a symmetry greater than that possessed by the empty lattice.

Now table 4.1 shows that the triclinic system transforms under the operations contained in the point group C_1 or S_2 and possesses one type of lattice; making up a total of two space groups. The monoclinic system has two types of lattice and is associated with the point groups (classes) C_{2h}, C_{1h}, C_2, producing a total of six space groups. As we go from the triclinic system to the cubic system we can readily deduce that each system gives rise to 2, 6, 12, 14, 5, 7, and 15 space groups respectively. Thus proceeding in this simple manner, we have now reached a total of 61 possible space groups; a number which falls far short of the known 230 space groups.

The full total of 230 cannot be accounted for without introducing the additional concepts of glide plane and screw axis. In the above discussion we have only considered space group operations which can always be expressed in terms of point group operations which do not themselves produce a translation. If a separation between the operations $\boldsymbol{\alpha}$ and $\mathbf{t}$ can be effected then we are dealing with the so-called point space groups. If a translation is *associated* with a point group operation then, as in equation (4.19), the space group symbol can be formally written as

$$\{\boldsymbol{\alpha}|V(\boldsymbol{\alpha}) + \mathbf{R}\} = \{E|\mathbf{R}\}\{\boldsymbol{\alpha}|V(\boldsymbol{\alpha})\} \tag{4.21}$$

where $\mathbf{R}$ is a lattice translation and $V(\boldsymbol{\alpha})$ is a non-primitive translation, associated with $\boldsymbol{\alpha}$, i.e., it is *not* a full lattice translation. $V(\boldsymbol{\alpha})$ can be shown to correspond to glide and screw operations. These are illustrated in Fig. 4.9. A glide plane operation consists of a reflection across a plane followed by a non-primitive translation *parallel* to that plane. A screw operation consists of a rotation about an axis accompanied by a translation *parallel* to that axis. A glide plane and a screw axis are both illustrated in Fig. 4.10 for the case of a diamond lattice. In this case the lattice can be brought into coincidence with itself by a reflection in the glide plane followed by a

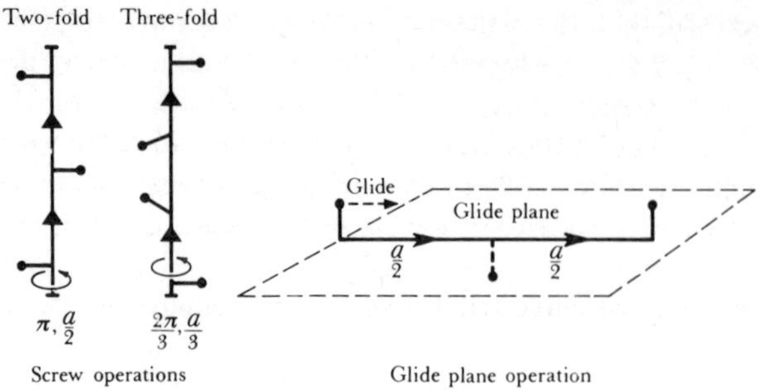

Fig. 4.9 *Illustration of glide and screw operations. In the screw operations we list the angle of rotation and the translation produced, e.g., a rotation of π produces a translation through a/2, i.e., half a lattice vector. In this glide plane operation (known as an axial glide) we translate through a distance a/2 between each reflection.*

translation of $(a/4, a/4)$. This is known as a diamond glide. An axial glide plane operation involves a translation through half a primitive vector; while a screw operation involves a translation through $1/n$ of a lattice vector, where n is an integer. These values are implied by the facts that (a) a glide operation involves a reflection hence *two* successive glide plane operations produce a pure translation which must of course be at least a

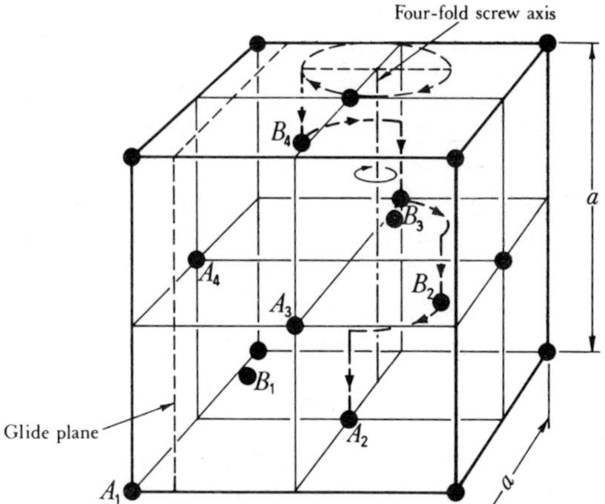

Fig. 4.10 *Diamond lattice illustrating glide plane and screw axis symmetry elements.* ● *denotes the position of a carbon atom. The lattice is face-centred cubic and has a basis of the two atoms A and B. If the origin (0, 0, 0) is chosen to be any A_i atom, a B_i atom will be located at (a/4, a/4, a/4). Each B atom lies at the centre of a regular tetrahedron with A atoms at its vertices.*

lattice vector. For this reason the non-primitive translation in a glide operation must always be some *rational* fraction of a lattice vector. (b) A screw axis repeats a motif after a rotation of $2\pi/n$ and a translation which is a fraction of lattice vector; hence n applications of a screw operator must again produce a translation through at least one lattice vector. The allowance for glide and screw operations leads to a total of 230 space groups.

In this chapter we have tried to highlight some of the basic concepts which are needed to discuss the symmetry of crystals. The mathematical detail has been deliberately kept down to a minimum in order not to obscure the central ideas. Some things have naturally had to be assumed but it is hoped that the contents of this chapter will form a guide to further study.

4.5 Problems

4.1 Show, from $\mathbf{a}_i . \mathbf{b}_j = 2\pi\delta_{ij}$, that

$$\mathbf{b}_1 = \frac{2\pi(\mathbf{a}_2 \times \mathbf{a}_3)}{\mathbf{a}_1 \cdot (\mathbf{a}_2 \times \mathbf{a}_3)}$$

$$\mathbf{b}_2 = \frac{2\pi(\mathbf{a}_3 \times \mathbf{a}_1)}{\mathbf{a}_1 . (\mathbf{a}_2 \times \mathbf{a}_3)}$$

$$\mathbf{b}_3 = \frac{2\pi(\mathbf{a}_1 \times \mathbf{a}_2)}{\mathbf{a}_1 . (\mathbf{a}_2 \times \mathbf{a}_3)}$$

If the origin of a face-centred cubic lattice is at the corner of a cube of side a show that the basis vectors are, using a right-handed coordinate system,

$$\mathbf{a}_1 = \frac{a}{2}(\mathbf{k} - \mathbf{i}), \qquad \mathbf{a}_2 = \frac{a}{2}(\mathbf{j} + \mathbf{k}), \qquad \mathbf{a}_3 = \frac{a}{2}(\mathbf{j} - \mathbf{i})$$

where **i**, **j**, **k** are unit vectors. Hence show that the reciprocal lattice is body-centred with basis vectors

$$\mathbf{b}_1 = \frac{2\pi}{a}(-\mathbf{i} - \mathbf{j} + \mathbf{k})$$

$$\mathbf{b}_2 = \frac{2\pi}{a}(\mathbf{i} + \mathbf{j} + \mathbf{k})$$

$$\mathbf{b}_3 = \frac{2\pi}{a}(-\mathbf{i} + \mathbf{j} - \mathbf{k})$$

4.2 Show that a space lattice vector $\mathbf{R} = m_1\mathbf{a}_1 + m_2\mathbf{a}_2 + m_3\mathbf{a}_3$ can be written as $\mathbf{A}|m\rangle$ where $\mathbf{A}$ is a matrix and

$$|m\rangle = \begin{pmatrix} m_1 \\ m_2 \\ m_3 \end{pmatrix}$$

is a vector formed from the components of the lattice vector. Show also that a reciprocal lattice vector $\mathbf{k} = s_1\mathbf{b}_1 + s_2\mathbf{b}_2 + s_3\mathbf{b}_3$ can be written as $\langle s|\mathbf{B}$ where $\mathbf{B}$ is again a matrix and $\langle s| = (s_1 s_2 s_3)$ where s_1, s_2, s_3 are the components of the vector $\mathbf{k}$. Prove that, for face-centred and body-centred cubic lattices with lattice constant a are respectively,

$$\mathbf{A}_1 = \begin{bmatrix} 0 & a/2 & a/2 \\ a/2 & 0 & a/2 \\ a/2 & a/2 & 0 \end{bmatrix}, \quad \mathbf{A}_2 = \begin{bmatrix} -a/2 & a/2 & a/2 \\ a/2 & -a/2 & a/2 \\ a/2 & a/2 & -a/2 \end{bmatrix}$$

Hence show that $\mathbf{a}_i \,.\, \mathbf{b}_j = 2\pi\delta_{ij}$ implies that $\mathbf{BA} = 2\pi\mathbf{1}$, where $\mathbf{1}$ is the unit matrix, and establish the matrices $\mathbf{B}_1$ and $\mathbf{B}_2$. (N.B. The matrices $\mathbf{A}_1$ and $\mathbf{A}_2$ are not unique since they depend upon the choice of coordinate system.)

4.3 Prove that a one-dimensional empty lattice transforms under a point group that contains the symmetry elements E and i. If the lattice is composed of four atoms, such that cyclic boundary conditions apply, derive the eight elements of the space group and show that they, indeed, form a group.

4.4 Find the space group of the one-dimensional lattice shown in the figure. If the symmetry operations are restricted to E, C_2, i, σ_v, and σ_h show that extra units of the form can be placed at the lattice sites, thus generating a total of five space groups.

Draw the possible pictures of the lattice as they would look if the horizontal plane is a glide plane. (N.B. a glide plane operation is in this case σ_h followed by a translation of half a primitive lattice vector.)

4.5 An invariant subgroup of a group consists of one or more classes of the group. Derive the group of operations corresponding to a glide along a line using E as the identity, σ a reflection operation, and τ a translation through half a lattice spacing. Show that the group consists of the invariant subgroup $(E, T, T^2, T^3, \ldots)$ and the product of this group with $\sigma\tau$ where T is a lattice translation.

4.6 Give the reasoning leading to the conclusion that the only five possible lattices in two-dimensions have monoclinic, orthorhombic, hexagonal, and tetragonal symmetry. Show using the motif of problem 4.4 that 13 space-groups can be constructed (neglecting screw or glide operations).

4.7 Show that in two dimensions the basic cell of a hexagonal lattice is a 120° rhombus. Now show that such a lattice would be adequate for systems having three or six-fold rotational symmetry and in addition show that in two dimensions only three types of lattice are required if we are only taking into account rotational symmetry. Prove that, in the 120° rhombus, the only possible position for a six-fold rotation axis is at a corner. If the sides of the rhombus are denoted by vectors $\mathbf{a}_1$ and $\mathbf{a}_2$, demonstrate that the three possible non-equivalent three-fold axes are $(0, 0)$, $(\frac{1}{3}\mathbf{a}_1, \frac{2}{3}\mathbf{a}_2)$, and $(\frac{2}{3}\mathbf{a}_1, \frac{1}{3}\mathbf{a}_2)$. Hence show that a rhombohedral lattice can be constructed by starting from a plane hexagonal lattice, placing lattice sites at the corners of triangles and then introducing successive layers whose lattice sites lie above the centres of the triangles.

5 Tensors and symmetry

5.0 The nature of tensors

The concept of a *scalar* quantity, i.e., an undirected magnitude is fairly easy for us to understand. However, many physical quantities possess both magnitude and direction; leading us from the concept of a scalar to that of a *vector*. Examples of the latter which spring readily to mind, are quantities such as velocity or electric field; or simply a force.

Although it is clear that the characteristics of many phenomena are readily expressible in terms of scalar and vector quantities it is found that these concepts are limited in scope. For example the electric current in some solids is not related to the applied electric field by means of a single number called the electrical conductivity; again the stress in a solid may not be expressible solely in terms of a magnitude and a direction, indeed, in general, the stress at a point in the solid may have a different magnitude across every plane constructed through that point.

Stating the situation in more detail, we first of all note that substances which do not, for example, possess an electrical conductivity represented by a scalar quantity are said to be electrically *anisotropic*. We then observe that the application of an electric field $\mathbf{E} = (E_1, E_2, E_3)$ to an anisotropic material produces a current density $\mathbf{j} = (j_1, j_2, j_3)$ given, in general, by

$$\begin{aligned} j_1 &= \sigma_{11}E_1 + \sigma_{12}E_2 + \sigma_{13}E_3 \\ j_2 &= \sigma_{21}E_1 + \sigma_{22}E_2 + \sigma_{23}E_3 \\ j_3 &= \sigma_{31}E_1 + \sigma_{32}E_2 + \sigma_{33}E_3 \end{aligned} \tag{5.1}$$

which can be written using the summation convention as $j_i = \sigma_{ij}E_j$ (cf. section A.21). Such a substance clearly has an electrical conductivity which has nine components each of which is labelled by two suffices. The electrical conductivity is now not simply a scalar but a new entity σ_{ij} with suffices i and j in the range 1 to 3.

Now there can be no doubt that quantities such as the electrical conductivity σ_{ij} and stress are physical entities; but the problem is how to specify them mathematically. The answer is that they are tensors, as indeed are j_i and E_j; and we must now consider the criteria which delineate physical entities as tensors and physical laws as tensor equations.

We shall restrict our attention to rectilinear, rectangular coordinate systems (non-curved orthogonal axes) which means that we shall be considering Cartesian tensors only. We do this in order to avoid having to make a distinction between covariant and contravariant tensors; indeed we shall refrain from any further mention of these classifications although they are briefly referred to in Appendix A, section A.21. Although the above restriction has been adopted for convenience the Cartesian reference frame is by far the most commonly required frame so that our subsequent remarks are not really weakened by this choice.

In mathematical terms a tensor must remain invariant under a group of linear transformations. By stating the definition in this way we have introduced some fresh ideas which we can begin to investigate by showing that the set of linear transformations does indeed form a group.

Suppose we effect a transformation, in an n-dimensional space, from a coordinate system x_i (where we immediately adopt the suffix notation, i.e., x_i stands for the independent variables $(x_1, x_2, \ldots, x_n)$ of a rectangular, rectilinear coordinate system) to a new coordinate system x'_i. Such a transformation can be written using a matrix $\mathbf{A} = [a_{ij}]$ as

$$x'_i = a_{ij}x_j \tag{5.2}$$

Now, if a_{ij} is independent of the variables, (5.2) is a *linear* transformation. Successive transformations $x_i \to x'_i$, followed by $x'_i \to x''_i$ gives us

$$x''_k = a'_{ki}x'_i \tag{5.3}$$

which becomes

$$x''_k = a'_{ki}a_{ij}x_j = a''_{kj}x_j \tag{5.4}$$

thus establishing the relationship

$$a''_{kj} = a'_{ki}a_{ij} \tag{5.5}$$

Therefore the result of a succession of linear transformations is itself a linear transformation. We now recall that one of the properties of a group is that products of group elements are themselves members of the group. (N.B. this is a continuous group since a_{ij} contains parameters such as angles of rotation, which are continuous functions.) This being the case we can appreciate that the result given by eq. (5.5) is a manifestation of the group properties of linear transformations, however the complete list of group characteristics must include the existence of inverse elements and an identity element. Assuming therefore that the inverse matrix $\mathbf{A}^{-1}$ exists we can write

$$x_i = a_{ij}^{-1}x'_j \tag{5.6}$$

so that on substituting x'_j from eq. (5.2) we obtain

$$x_i = a_{ij}^{-1} a_{jk} x_k \tag{5.7}$$

Since we are now back in the *original* coordinate system it is clear that

$$a_{ij}^{-1} a_{jk} = 1, \qquad i = k \tag{5.8}$$

The left-hand side of eq. (5.8) is usually defined as the Kronecker delta δ_{ik} which is unity provided that $i = k$, and zero otherwise. δ_{ik} is thus the unit element of the set of transformations. We therefore conclude that the set of linear transformations forms a group.

The idea of invariance can be readily grasped with the aid of the time-honoured example of a vector (a tensor with one suffix) considered respectively in two coordinate systems displaced from one another by a rotation through an angle θ. This linear transformation is shown in Fig. 5.1 where

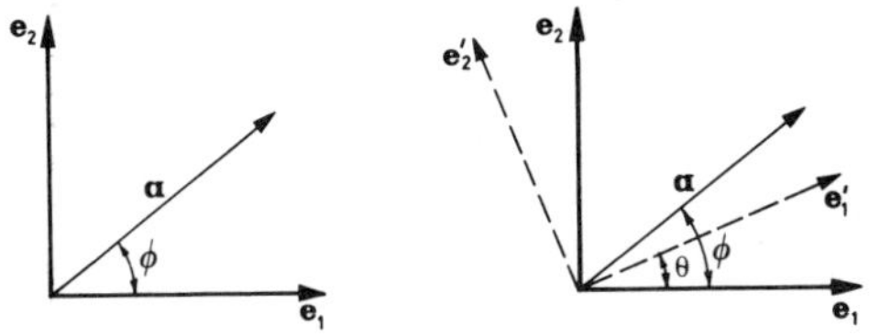

Fig. 5.1 *Linear transformation which leaves the vector* **a** *unchanged.*

the *vector* **a** remains *invariant*, even though its components in the new coordinate system **e**′ are *different* from those in the old coordinate system **e**, (cf. appendix A.16). We can appreciate this by observing that the vector **a** is

$$\mathbf{a} = a_1\mathbf{e}_1 + a_2\mathbf{e}_2 = a'_1\mathbf{e}'_1 + a'_2\mathbf{e}'_2 \tag{5.9}$$

showing that **a** retains an absolute identity, while its *description* varies with the choice of coordinate system. Returning now to the fundamental definition of a general tensor we re-iterate that, in a similar manner, it is an entity which remains invariant under the operations of the group of linear transformations. If the linear transformations did not form a group then the inverse of **A** may not exist, in which case, having transformed the tensor components using **A**, we would never be able to return to the original coordinate system. From a physical point of view such a transformation would be irreversible and hence totally inadmissible. The absence of the other group properties would also cause similar difficulties.

In a practical application of tensor analysis we must *always* ensure that the transformations are admissible. For example let us consider the magnetic field vector **B** which can be regarded as a vector which, however, does not change sign on reflection in the same way as an ordinary vector would. We can see this by considering the elementary example of a horizontal loop of wire, carrying a current, situated above a horizontal plane. If we now reflect the loop in this plane the image appears to be identical, i.e., the current is still travelling in the same direction; showing that **B** has *not reversed.* **B** can therefore only be considered as a one-suffix tensor under a restricted (admissible) set of transformations which does *not* include reflections. A vector such as **B** is called a pseudovector (pseudotensor) or an axial vector. Vectors which behave conventionally are called polar vectors. Thus we see that we can only transform from one reference frame to another in a manner commensurate with the character of the problem we are investigating.

Having stated, in mathematical terms, the properties of a tensor we can now put the arguments in more physical terms. An immediate consequence of the definition of a tensor is that once a tensor equation (such as eq. (5.1)) describing a physical system has been established it will *retain* the same form for all physically analogous systems. In other words the form of the equation will remain invariant under an admissible linear transformation of coordinates (change of basis—cf. eq. (A.131)). Naturally as we go from one frame of reference to another, the *components* of a tensor will change but the tensors themselves remain unchanged. Indeed we must regard a tensor as an *observable physical entity* which is unaffected by the coordinate system changes that we have previously discussed. We must also regard all physical laws as tensor equations such as eq. (5.1).

The simplest form of tensor is a scalar which is said to be a tensor of rank zero (the rank is a technical word referring merely to the number of suffices the tensor possesses). The invariant nature of a scalar can be readily understood because, being simply a number, it possesses the same value no matter what the coordinate system happens to be (within the limitations we have already laid down). For example suppose we measure the temperature of a body then its value clearly ought not to depend upon the frame of reference.

We now proceed from a scalar to a vector which is probably one of the most frequently occurring tensors in science; and is a tensor of rank one because its components possess only one suffix. In the same way as a scalar, a first rank tensor also remains invariant under a group of linear transformations. However, its components change as we go from one frame of reference to another.

Apart from scalars and vectors we can have tensors of higher rank. Second rank tensors, as we can see from eq. (5.1), can relate first rank tensors (vectors) to each other so, generalizing, we should always expect higher rank tensors to relate tensors of smaller rank to each other, e.g., stress and strain (second rank tensors) are related to each other by a fourth rank tensor.

Before we pass on to specific applications of tensor analysis let us first consider the merits of using a tensor formalism at all. A basic advantage is that, because of the invariant nature of tensors, equations involving tensors will also be invariant so that large numbers of physically analogous systems can be characterized by a single equation. We can imagine that a given physical system sustains a current density $j_i = \sigma_{ij}E_j$ where σ_{ij} is the electrical conductivity tensor and E_j is the electric field and that it is always possible to devise alternative, physically analogous, systems which sustain a current density $j'_i = \sigma'_{ij}E'_j$ where the dash denotes the alternative systems. This construction of alternative physically analogous systems is really what we mean by a linear transformation of coordinates. Thus it is possible to expose the structure of a whole class of problems with the utmost economy of symbols. Also all that is necessary to solve complicated problems is to select and analyse the *simplest* system physically analogous to the *actual* system. Then because of the fact that we are dealing with tensors we only require a suitable transformation matrix in order to deduce, from the simple system, the behaviour of the actual system. Beautiful applications of these concepts have been made by engineers to electrical network problems.

5.1 Examples of second and third rank tensors

In elementary physics we tend to become indoctrinated with the idea that the physical properties of solids, for example, are, in the main, isotropic. Thus there is a tendency to regard the transport of heat in solids, for instance, as being characterized by a single scalar quantity, namely the thermal conductivity. Naturally if the solid were indeed isotropic it would not matter in which direction the flow of heat were measured; we would expect to obtain the same value for the conductivity for all possible directions. However, solids in the crystalline state are in general anisotropic and we recognize that measurement of, say, the thermal conductivity in different directions may well produce *different* answers. A scalar physical property such as density is, of course, always isotropic and some crystals, in the cubic class, give the appearance of being isotropic if we only measure their electrical and thermal conductivities but are, however, anisotropic for many other physical properties. We will now by way of example examine the problem of heat conduction across an arbitrary crystal in the form of a flat plate (as shown in Fig. 5.2). If a temperature gradient grad $(T) \equiv \partial T/\partial x_i$ exists, where T is temperature, then a heat current $\mathbf{Q} \equiv Q_i$ (defined as the rate of flow of heat per unit cross-section) will flow across the plate. The heat flow and the temperature gradient are first rank tensors and if we employ the repeated suffix convention the general relationship between Q_i and $\partial T/\partial x_i$ is

$$Q_i = -\kappa_{ij}\frac{\partial T}{\partial x_j} \tag{5.10}$$

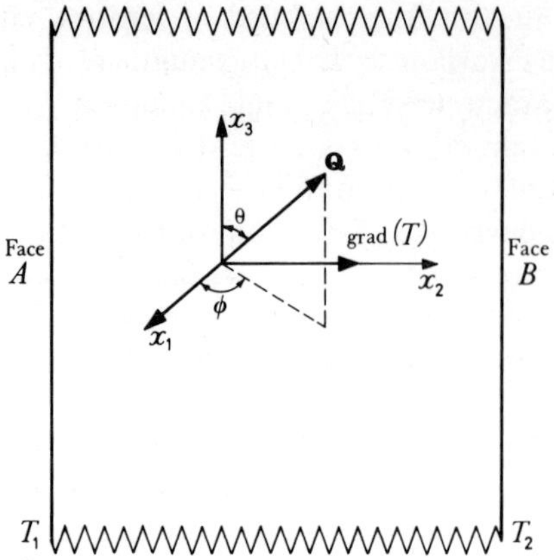

Fig. 5.2 *Cross-section of a plate with a temperature gradient grad (T) between the faces A and B. The Cartesian co-ordinate system (x_1, x_2, x_3) is such that x_2 and x_3 lie in the plane of the section.*

where we now have a conductivity *tensor* κ_{ij}. (N.B. this equation has the same *form* as eq. (5.1).) If the solid happens to be isotropic then $\kappa_{ij} = \kappa\delta_{ij}$ where κ is a scalar quantity and **Q** will be parallel to grad (T); however, as shown in Fig. 5.2 if the solid is anisotropic then **Q** will *not* be parallel to grad (T). We can perhaps appreciate the foregoing statement a little better by writing out eq. (5.10) in matrix form, i.e.,

$$\begin{bmatrix} Q_1 \\ Q_2 \\ Q_3 \end{bmatrix} = -\begin{bmatrix} \kappa_{11} & \kappa_{12} & \kappa_{13} \\ \kappa_{21} & \kappa_{22} & \kappa_{23} \\ \kappa_{31} & \kappa_{32} & \kappa_{33} \end{bmatrix} \begin{bmatrix} \dfrac{\partial T}{\partial x_1} \\ \dfrac{\partial T}{\partial x_2} \\ \dfrac{\partial T}{\partial x_3} \end{bmatrix} \tag{5.11}$$

The heat flux across a plate possessing only a temperature gradient $\partial T/\partial x_1$ is therefore given by

$$\begin{aligned} Q_1 &= -\kappa_{11}\frac{\partial T}{\partial x_1} \\ Q_2 &= -\kappa_{21}\frac{\partial T}{\partial x_1} \\ Q_3 &= -\kappa_{31}\frac{\partial T}{\partial x_1} \end{aligned} \tag{5.12}$$

Equation (5.12) now shows that, apart from a flux along x_1, we also have *transverse* fluxes along the directions x_2 and x_3. In an isotropic solid κ_{21} and κ_{31} would, of course, be identically zero.

Now eq. (5.10) is a tensor equation which should remain invariant with respect to an admissible linear transformation of the coordinates. In order to show this let us perform an orthogonal linear transformation on eq. (5.10) which transforms Q_i to

$$Q_i' = a_{ij}Q_j \tag{5.13}$$

and the temperature gradient to

$$\frac{\partial T}{\partial x_i'} = a_{ij}\frac{\partial T}{\partial x_j} \tag{5.14}$$

The heat flow equation thus transforms to

$$Q_i' = -a_{ij}\kappa_{jk}\frac{\partial T}{\partial x_k} = -a_{ij}\kappa_{jk}a_{mk}\frac{\partial T}{\partial x_m'} \tag{5.15}$$

(N.B. we have performed an orthogonal transformation for which $a_{km}^{-1} = a_{mk}$—cf. section A.4(k).) The heat flow equation in the new coordinate system is therefore

$$Q_i' = \kappa_{im}'\frac{\partial T}{\partial x_m'} \tag{5.16}$$

where the transformed conductivity tensor is

$$\kappa_{im}' = a_{ij}a_{mk}\kappa_{jk} \tag{5.17}$$

Equation (5.10) is therefore preserved (remains invariant) and κ_{ij} transforms as a second rank tensor (cf. Appendix A.21).

There are actually many physical entities in physics which are second rank tensors. The dielectric tensor, the permittivity tensor, and the electrical conductivity tensor are among those which immediately come to mind. Indeed second rank tensors seem to be such common currency that there is a danger, because they have two suffices, of regarding them simply as elements of matrices without probing into their tensor character. However when we come to tensors of rank greater than two we cannot think of them in terms of a matrix in an ordinary way.

A tensor of rank three possesses three suffices and arises quite naturally in the mathematical formulation of the piezoelectric effect. The latter effect is essentially the appearance of a polarization charge P_i per unit area (electric moment per unit volume) on the surface of a solid, such as quartz, after the application of a stress s_{ij}. Now since the stress s_{ij} is a second rank tensor it can only be related to P_i by means of a third rank tensor which we will call d_{ijk}; the piezoelectric tensor. The foregoing statement can perhaps be more readily understood by realizing that P_1 is related to *all* nine components of s_{ij} so that

$$P_1 = d_{1ij}s_{ij} \tag{5.18}$$

Equation (5.18) therefore implies that

$$P_i = d_{ijk}s_{jk} \tag{5.19}$$

where d_{ijk} is a third rank tensor transforming as

$$d'_{ijk} = a_{il}a_{jm}a_{kn}d_{lmn} \tag{5.20}$$

A general picture of crystal properties and the variables associated with them is shown in Fig. 5.3.

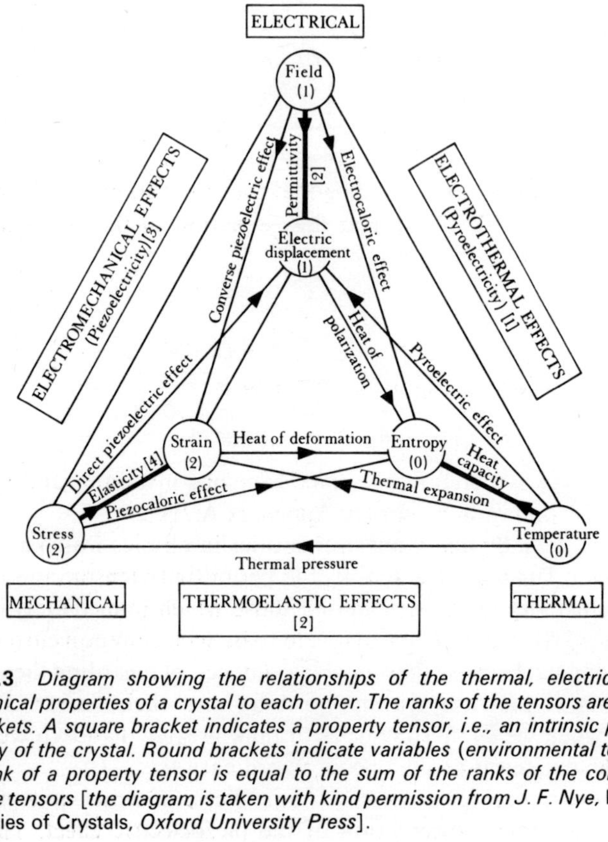

Fig. 5.3 *Diagram showing the relationships of the thermal, electrical, and mechanical properties of a crystal to each other. The ranks of the tensors are shown in brackets. A square bracket indicates a property tensor, i.e., an intrinsic physical property of the crystal. Round brackets indicate variables (environmental tensors). The rank of a property tensor is equal to the sum of the ranks of the connected variable tensors [the diagram is taken with kind permission from J. F. Nye,* Physical Properties of Crystals, *Oxford University Press].*

5.2 The role of crystal symmetry

We are now going to consider the role of crystal symmetry by examining the behaviour of tensors, representing *physical properties*, under the operations of the point groups. Tensors such as stress tensors will be excluded because they are not inherent properties of matter. We begin by implicitly assuming that the symmetry elements of the group of operators under which a particular physical property of a crystal remains unchanged includes at least all the elements of the crystal point group. This assumption is known as Neumann's principle. In fact we are going to show how a knowledge of the crystal symmetry will enable us to determine *exactly* which elements of a tensor are identically zero; and which are equal to each other. In order to do this we must obviously refer the tensor to a set of axes which, for convenience, we take to be orthogonal. However, the results we are going to obtain using group theory will be independent of the orientation of these axes. Intuitively we feel at the outset, that the number of non-zero elements of a given tensor ought to *decrease* as we pass through the seven crystal systems from triclinic to cubic.

When we refer to the group of symmetry operations of a crystal we strictly mean space group operations; but since we shall be concerned with some measured *macroscopic property* we need only to consider point group operations—changes due to translations associated with glide and screw operations are *microscopic* phenomena and would not be detectable in a macroscopic observation of for example the electrical conductivity. The microscopic (space group) symmetry elements could, of course, be detected by say X-ray diffraction.

The point group operations can naturally be carried out in the laboratory. For instance the measurement of the thermal conductivity of a solid could be taken along different axes of the crystal. In fact either the crystal *or* the apparatus could be oriented in a variety of aspects; thus simulating the rotational operations of the point group. If necessary the apparatus could be entirely re-built to simulate reflections or inversion.

If the components of the tensor defining the property of a crystal are measured for one position of the crystal and then we perform a symmetry operation by moving the crystal or the apparatus, we must obtain *exactly* the same values because the point group operations leave the components of the tensor unchanged.

As an example we will now consider the effect of a point group operation on the thermal current Q_i which transforms it to

$$Q_i' = a_{ij}Q_j \tag{5.21}$$

The components of the conductivity tensor κ_{im}' in the new coordinate system are therefore given by (5.17); however a_{ij} in this case, represents a point group operation so we can drop the dash in equation (5.17) and write

$$\kappa_{im} = a_{ij}a_{mk}\kappa_{jk} \tag{5.22}$$

Obviously this statement can be generalized to encompass tensors of higher ranks.

The quantities a_{ij} are of course the elements of matrices as opposed to tensors, and represent the operations of the point group; in fact since **A** relates column vectors it is the inverse of a representation matrix. If we concentrate entirely on transformations which only involve rotations then we can determine the appropriate a_{ij} from the general expression (cf. eq. (A.183)).

$$\mathbf{A}=\begin{bmatrix} \cos\psi + l^2(1-\cos\psi) & lm(1-\cos\psi)+n\sin\psi & ln(1-\cos\psi)-m\sin\psi \\ lm(1-\cos\psi)-n\sin\psi & \cos\psi + m^2(1-\cos\psi) & mn(1-\cos\psi)+l\sin\psi \\ ln(1-\cos\psi)+m\sin\psi & mn(1-\cos\psi)-l\sin\psi & \cos\psi+n^2(1-\cos\psi) \end{bmatrix} \tag{5.23}$$

which represents a general rotation of the coordinate system through an angle ψ about an axis, inclined to the axes (x, y, z), with direction cosines l, m, and n respectively.

We will now consider some of the symmetry elements of the group O_h, i.e., the symmetry group of a cubic crystal.

A four-fold C_4 rotation about the axis is represented by $\mathbf{A}$ evaluated for a rotation of $\psi = \pi/2$ about an axis for which $l = 1$, $m = 0$, $n = 0$. This matrix is therefore

$$\mathbf{A} = \begin{bmatrix} 1 & 0 & 0 \\ 0 & 0 & 1 \\ 0 & -1 & 0 \end{bmatrix} \tag{5.24}$$

The direction cosines of the (111) axis are $l = 1/\sqrt{3}$, $m = 1/\sqrt{3}$, $n = 1/\sqrt{3}$ so that a three-fold rotation about a (111) axis (i.e., $\psi = 2\pi/3$) is represented by

$$\mathbf{A} = \begin{bmatrix} 0 & 1 & 0 \\ 0 & 0 & 1 \\ 1 & 0 & 0 \end{bmatrix} \tag{5.25}$$

Thus from eq. (5.22) using the matrix, representing C_4, of eq. (5.24) we can write, e.g.

$$\begin{aligned} \kappa_{33} &= a_{3j}a_{3k}\kappa_{jk} \\ &= a_{31}a_{31}\kappa_{11} + a_{31}a_{32}\kappa_{12} + a_{31}a_{33}\kappa_{13} + a_{32}a_{31}\kappa_{21} \\ &\quad + a_{32}a_{32}\kappa_{22} + a_{32}a_{33}\kappa_{23} + a_{33}a_{31}\kappa_{31} + a_{33}a_{32}\kappa_{32} \\ &\quad + a_{33}a_{33}\kappa_{33} \\ &= a_{32}a_{32}\kappa_{22} \end{aligned} \tag{5.26}$$

which shows immediately that

$$\kappa_{33} = \kappa_{22} \tag{5.27}$$

The symmetry element C_2 namely a two-fold rotation ($\psi = \pi$) about the (110) axis ($l = 1/\sqrt{2}, m = 1/\sqrt{2}, n = 0$) is represented by

$$\mathbf{A} = \begin{bmatrix} 0 & 1 & 0 \\ 1 & 0 & 0 \\ 0 & 0 & -1 \end{bmatrix} \tag{5.28}$$

so that using eqs. (5.28) and (5.22) we obtain

$$\kappa_{11} = \kappa_{22} \tag{5.29}$$

The use of C_2 also leads, in a similar manner, to

$$\kappa_{12} = \kappa_{21} \tag{5.30}$$

while the use of C_4 leads to $\kappa_{31} = -\kappa_{21}$, $\kappa_{21} = \kappa_{31}$, $\kappa_{12} = \kappa_{13}$, all of which show that $\kappa_{31} = 0, \kappa_{13} = 0, \kappa_{21} = 0, \kappa_{12} = 0$. Therefore the thermal conductivity tensor of a cubic crystal is given by

$$\boldsymbol{\kappa} = \begin{bmatrix} \kappa_{11} & 0 & 0 \\ 0 & \kappa_{11} & 0 \\ 0 & 0 & \kappa_{11} \end{bmatrix} \tag{5.31}$$

showing that it is isotropic being characterized by only one coefficient. We can soon show in a similar manner that *any* property described by a second rank tensor is isotropic in materials possessing cubic symmetry. However, we must emphasize that although in this instance a cubic crystal presents itself as an isotropic substance it is in fact *not* isotropic for properties defined in terms of tensors of higher rank. Indeed a measurement of its elastic properties, which are described by a fourth rank tensor, would reveal that it is markedly anisotropic. We could now continue our study of second rank tensors and go through the crystal classes and determine the number of independent coefficients which exist. The results we would obtain are summarized by Fig. 5.4. We can set some tensor elements equal to each other because all second rank tensors are symmetrical, as a consequence of Onsager's principle. This principle is based on time-reversal symmetry a concept which we do not intend to pursue in this book. Notice also that, as the symmetry is lowered the number of non-zero tensor elements increases so that a practical consequence of lowering the symmetry would be to increase the number of measurements necessary to completely characterize the material under study.

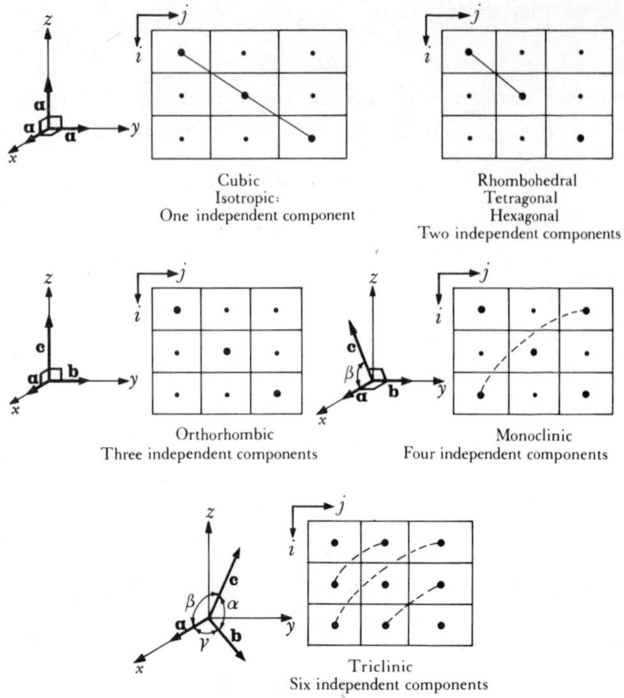

Fig. 5.4 *Representation of second rank tensors with suffices ij. The boxes in the array contain • if the component is zero and ● if the component is non-zero. The tensors are referred to the orthogonal axes (xyz) and the relationship of these axes to the crystallographic axes is as shown; the rhombohedral, tetragonal, and hexagonal cases can be worked out from Fig. 4.6. ●—● denotes components which are equal because of the point group symmetry. ●⋯● denotes components which are equal because of Onsager's principle.*

5.3 Conclusion

We have attempted in the previous sections, to understand the influence of crystal symmetry on the physical properties of a crystal. We have emphasized that a tensor is a physical entity which is invariant under a change of reference frame, i.e., a physical entity whose components may change but which is, itself, transmitted from one frame to another, without change. Also we have stressed the fact that physical laws are tensor equations and therefore also invariant.

The section on the role of crystal symmetry shows that the labour involved in a calculation or an experiment can be reduced by first examining the tensor describing a particular property to see which elements are zero and so on.

5.4 Problems

5.1 Establish that the vector product $\mathbf{C} = \mathbf{A} \times \mathbf{B}$ can be written as

$$C_i = \varepsilon_{ijk} A_j B_k$$

where ε_{ijk} is 1 if ijk are 312, 231, or 123, -1 if ijk are 132, 321, or 213, and 0 if two or more indices are identical. Show also that the determinant of the transformation matrix is given by

$$|a|\varepsilon_{pqr} = \varepsilon_{ijk} a_{pi} a_{qj} a_{rk}$$

Hence prove that the transformation law for C_i is

$$C'_i = |a| a_{ij} C_j$$

to show that the vector product is a pseudovector (axial vector) because it changes sign under an improper rotation.

5.2 Some crystals, such as tourmaline, exhibit the phenomenon of pyroelectricity, which is the appearance of an electric polarization P_i, when a change of ΔT in the mean temperature occurs. The tensor equation describing this phenomenon is

$$P_i = p_i \Delta T$$

showing that the crystal property in this case is represented by a vector $\mathbf{p} = (p_1, p_2, p_3)$. Show that the pyroelectric effect cannot exist in any crystal possessing a centre of symmetry; and cannot exist for any of the cubic crystal classes. Determine the crystal system for which all the components of $\mathbf{p}$ exist.

5.3 Show with the aid of eq. (5.23) that (a) a three-fold rotation about the z axis is described by the matrix

$$\begin{bmatrix} -\frac{1}{2} & \frac{\sqrt{3}}{2} & 0 \\ -\frac{\sqrt{3}}{2} & -\frac{1}{2} & 0 \\ 0 & 0 & 1 \end{bmatrix}$$

(b) a three-fold rotation about a (111) axis is represented by

$$\begin{bmatrix} 0 & 1 & 0 \\ 0 & 0 & 1 \\ 1 & 0 & 0 \end{bmatrix}$$

Also, show that inversion is described by

$$\begin{bmatrix} -1 & 0 & 0 \\ 0 & -1 & 0 \\ 0 & 0 & -1 \end{bmatrix}$$

and that a three-fold rotation–inversion about z is described by

$$\begin{bmatrix} \frac{1}{2} & -\frac{\sqrt{3}}{2} & 0 \\ \frac{\sqrt{3}}{2} & \frac{1}{2} & 0 \\ 0 & 0 & -1 \end{bmatrix}$$

5.4 One of the rhombohedral classes is C_{3v}. Using the appropriate matrices **A** show that all second rank tensor crystal properties of crystals in a rhombohedral class are diagonal with only two independent components.

5.5 Develop an argument to show that any phenomenon described by a tensor of odd rank cannot be observed in a crystal possessing a centre of symmetry.

If stress is applied to a piezoelectric crystal charge separation takes place and a polarization occurs. If the stress tensor is symmetrical produce a *physical argument* to show that a crystal with a centre of symmetry cannot be piezoelectric.

6 Energy bands in solids

6.0 Electrons in a perfect solid

In this chapter we are going to consider some important aspects of the behaviour of charge carriers in solids but we will at the outset restrict our attention to perfect solids consisting of a regular lattice of ions immersed in a gas of electrons. The behaviour of these electrons can best be understood by adopting the so-called one-electron approximation in which it is assumed that the ions are, effectively, at rest and that the electrons move independently of each other. In fact all the complex electron–electron, electron–lattice site, and lattice site–lattice site interactions are in practice transformed into an effective electron potential energy $V(\mathbf{r})$. The eigenfunctions $\psi_{\mathbf{k}}(\mathbf{r})$ and the energy eigenvalues $E(\mathbf{k})$ of an electron in a perfect solid are then obtained by solving the time-independent Schrödinger equation:

$$\left[-\frac{\hbar^2}{2m}\nabla^2 + V(\mathbf{r})\right]\psi_{\mathbf{k}}(\mathbf{r}) = E(\mathbf{k})\psi_{\mathbf{k}}(\mathbf{r}) \tag{6.1}$$

where m is the electron mass, $\hbar = h/2\pi$, and h is Planck's constant (cf. eq. (3.86)). The quantity $\mathbf{k}$ labels the state of the electron and $\hbar\mathbf{k}$ is interpreted as the momentum of the electron. However, in making this assertion, we are saying that although the energy $E(\mathbf{k})$, of an electron in a solid, may be a complicated function of $\mathbf{k}$ the electron will nevertheless respond to an external electric field, say, as a classical particle with momentum $\hbar\mathbf{k}$.

The electron potential energy $V(\mathbf{r})$ will, naturally, be periodic if the lattice of the solid is periodic and in general $V(\mathbf{r})$ will possess the symmetry of the crystal (solid) lattice. $V(\mathbf{r})$ is, of course, phenomenological and has to be constructed to suit the solid under study; a task that is often very difficult, as is the task of finding general solutions of eq. (6.1). The labour involved in the solution of eq. (6.1) can be somewhat reduced with the aid of group theory. Indeed it is often essential to consider the problem from a group theoretical point of view if an *efficient* means of solution is to be found.

It is the aim of this chapter to highlight the relevant solid state concepts concerned with energy bands and to show, by making use of a simple example, how group theory has the power of providing exact but qualitative conclusions, before the really detailed task of solving eq. (6.1) begins.

We will begin by considering the behaviour of electrons in a one-dimensional lattice-free solid and we will set $V(\mathbf{r})$ equal to zero. The model will then be extended to three dimensions and a simple lattice will be introduced. However, the electron potential energy will be maintained at zero throughout the early part of our discussions. Subsequently we shall see, using group theory, what happens to the electron energy eigenvalues ($E(\mathbf{k})$) when we 'switch on' a finite crystal field, $V(\mathbf{r})$, of cubic symmetry.

6.1 Electrons in a ring

A most convenient one-dimensional system consists of a ring of circumference L. The choice of a ring is dictated by the fact that we wish to exploit the cyclic boundary conditions which we have already discussed in chapter 4 (section 4.1).

The eigenfunction $\psi_k(x)$ of an electron in this ring is the solution of the one-dimensional form of eq. (6.1) with $V(x) = 0$, i.e.,

$$\frac{\mathrm{d}^2\psi_k(x)}{\mathrm{d}x^2} + \frac{2m}{\hbar^2}E(k)\psi_k(x) = 0 \tag{6.2}$$

and has the form, taking the $e^{i\alpha x}$ solution,

$$\psi_k(x) = A\,e^{i\alpha x} \tag{6.3}$$

where

$$\alpha^2 = \frac{2m}{\hbar^2}E(k) \tag{6.4}$$

and A is an arbitrary constant which can be obtained by normalization.

After the particle has traversed a distance L it returns to its starting point in which case the electron eigenfunctions at points x and $x + L$ must be identical to each other (this is a cyclic boundary condition).

Therefore, we have $\psi_k(x + L) = \psi_k(x)$ and hence

$$\mathrm{e}^{i\alpha L} = 1 \tag{6.5}$$

which yields

$$\alpha = \left(\frac{2\pi}{L}\right)\nu, \qquad \nu = 0, \pm 1, \pm 2, \ldots. \tag{6.6}$$

The eigenvalues of the electron are therefore determined by eq. (6.6) and can be written as

$$E(k) = \frac{\hbar^2}{2m}\left(\frac{2\pi\nu}{L}\right)^2 = \frac{\hbar^2 k^2}{2m} \tag{6.7}$$

Thus the energy of the electron is quantized and each state is specified by a *quantum number* k defined as $2\pi\nu/L$. Also if we remember that $E = p^2/2m$, where p is momentum, we can see that $p = \hbar k$.

If the electron moved in a one-dimensional atomic-chain of length L with lattice sites at intervals a, then L would be equal to Na where N is some integer. If the lattice were such that $V(x) = 0$ then the electron would then be moving in what we call an *empty lattice* and the energy would still be given by eq. (6.7).

The extension of the foregoing one-dimensional model to three dimensions is straightforward provided we accept that cyclic boundary conditions can be used even in three dimensions. Nevertheless we can perhaps (cf. discussion in chapter 4), intuitively, see that it ought not to matter whether cyclic boundary conditions can physically be achieved or not provided the volume to surface ratio of the solid is large. In other words results based on cyclic boundary conditions should be valid provided the internal properties of the lattice do not depend strongly on the surface properties. Let us therefore regard a cyclic crystal here, as a block with sides of length $N_1 a_1$, $N_2 a_2$, and $N_3 a_3$ respectively, contained within an infinite solid where N_1, N_2, and N_3 are integers and a_i are the moduli of the primitive lattice vectors.

The eigenvalues of an electron in such a three-dimensional empty lattice can therefore be readily obtained by generalizing eq. (6.7) to

$$E(\mathbf{k}) = \frac{\hbar^2}{2m}\left[\left(\frac{2\pi\nu_1}{N_1 a_1}\right)^2 + \left(\frac{2\pi\nu_2}{N_2 a_2}\right)^2 + \left(\frac{2\pi\nu_3}{N_3 a_3}\right)^2\right] \tag{6.8}$$

where ν_i are integers. The state of the electron is now specified by the quantum numbers (cf. eq. (4.9))

$$(k_1, k_2, k_3) = 2\pi\left[\frac{\nu_1}{N_1 a_1}, \frac{\nu_2}{N_2 a_2}, \frac{\nu_3}{N_3 a_3}\right] \tag{6.9}$$

6.2 k-Space

The fact that we label the state of an electron moving in a crystal lattice by quantum numbers (k_1, k_2, k_3) is true whether we are dealing with an empty lattice or not. This being the case we should now emphasize that the conclusions drawn in *this section* are quite general and are in no way restricted to an empty lattice model.

The *vector* $\mathbf{k}$ with components (k_1, k_2, k_3) is, according to eq. (6.9), related to the reciprocal of the lattice spacing. It is therefore convenient to regard $\mathbf{k}$ as a vector in a reciprocal space (cf. eq. (4.10)), often referred to as $\mathbf{k}$-space. If the crystal possesses a finite number of primitive cells $\mathbf{k}$ takes on discrete values; but since in practical cases a typical crystal sample contains a huge number (i.e., $N_1N_2N_3$) of primitive cells we may for the most part safely assume that $\mathbf{k}$ is continuous.

As we have shown in chapter 4 every perfect lattice is invariant under the operation of a space group. The irreducible representations of the translational subgroup of this space group are given by (cf. eq. (4.7))

$$\Gamma(\mathscr{T}) = \mathrm{e}^{i\mathbf{k}.\mathbf{R}} \tag{6.10}$$

where $\mathbf{k}$ is given by eq. (4.9) which we can re-write as

$$\mathbf{k} = q_1\mathbf{b}_1 + q_2\mathbf{b}_2 + q_3\mathbf{b}_3 \tag{6.11}$$

where

$$q_i = v_i/N_i, \qquad v_i = 0, 1, \ldots, N_i - 1; \qquad i = 1, 2, 3 \tag{6.12}$$

We should now observe that eq. (6.11) and eq. (6.9) are formally identical; however, v_i in eq. (6.9) can assume *any* integral values but v_i in eq. (6.12) is *restricted* because $\mathbf{k}$ labels the irreducible representations; the number of which is finite. We shall return to this point later. We now emphasize the fact that the vectors $\mathbf{b}_i$ span a space in which distances have dimensions of inverse length. Also, just as the vectors $\mathbf{a}_i$ define a primitive cell in the space of the real lattice the vectors $\mathbf{b}_i$ define a primitive cell in $\mathbf{k}$-space.

Since the vector $\mathbf{k}$ labels the irreducible representations of the group of pure translations we can see that if we have $N_1N_2N_3$ primitive cells in the crystal lattice then this is also the number of irreducible representations. Also as we have already mentioned, in most discussions of solid-state phenomena we can assume that $N_1N_2N_3$ is very large indeed making q_i virtually continuous and lying in the range

$$0 \leqslant q_i < 1 \tag{6.13}$$

where we count a state on the surfaces $q_1 = 0, q_2 = 0, q_3 = 0$ but not on the surfaces $q_1 = 1$, $q_2 = 1$, and $q_3 = 1$, as will be shown below. The irreducible representations of the pure translation group therefore correspond to points within the volume of $\mathbf{k}$-space (defined by the vectors $\mathbf{b}_i$) and on the surfaces $q_1 = 0$, $q_2 = 0$, and $q_3 = 0$.

It should be clear by now that the $\mathbf{k}$ which labels the irreducible representations of the translation group is simply a vector in $\mathbf{k}$-space lying within the primitive cell or on one of its surfaces. Hence if we consider a vector $\mathbf{k}'$ equal to $\mathbf{k} + \mathbf{K}$ then $\mathbf{k}'$ labels *another* irreducible representation

$$\Gamma^{\mathbf{k}'}(\mathscr{T}) = e^{i(\mathbf{k}+\mathbf{K}).\mathbf{R}} \tag{6.14}$$

This representation is unique unless

$$\mathbf{K}\,.\,\mathbf{R} = 2n\pi, \qquad n = 0, 1, 2, \ldots \tag{6.15}$$

in which case $\Gamma^{\mathbf{k}'}(\mathscr{T}) = \Gamma^{\mathbf{k}}(\mathscr{T})$. However, this equation can only be true provided

$$\mathbf{K} = s_1\mathbf{b}_1 + s_2\mathbf{b}_2 + s_3\mathbf{b}_3 \tag{6.16}$$

where s_i are integers, as can be seen from eq. (4.1) and (4.10). Hence $\mathbf{K}$ must be interpreted as a reciprocal lattice vector. Indeed $\mathbf{b}_i$ form a basis for a reciprocal lattice in the same way as $\mathbf{a}_i$ form a basis for a lattice in real space. Any state $\mathbf{k}$ for which any $q_i \geqslant 1$ lies outside the primitive cell, or on one of the surfaces $q_i = 1$, and will contain one or more reciprocal lattice vectors. The states on the surfaces $q_i = 1$, for example, are equivalent to the states on the surfaces $q_i = 0$.

6.3 The Brillouin zone; The reduced zone scheme

The primitive cell of the reciprocal lattice is defined by the vectors $\mathbf{b}_i$ but, as discussed in chapter 4, this cell may not possess the full point-group symmetry of the lattice so it would be better to choose a symmetric cell in the same way that the Wigner–Seitz cell of the real space lattice was chosen. This symmetric cell, drawn in Fig. 6.1 for a face-centred cubic real lattice, is known as the Brillouin zone. It is interesting to note that the reciprocal lattice of a face-centred cubic real lattice is body-centred and

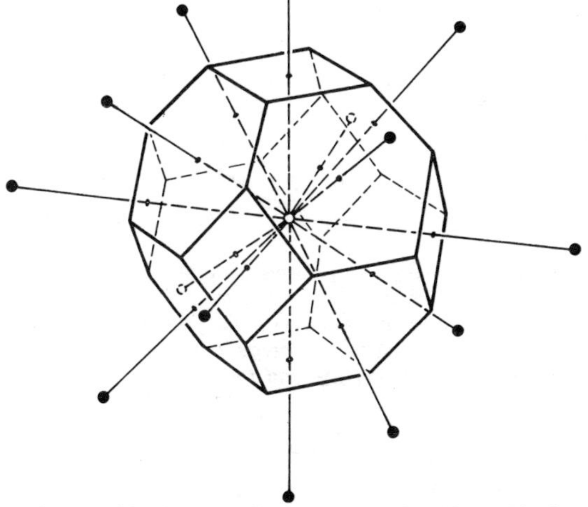

Fig. 6.1 *Brillouin zone of a face-centred cubic real lattice.*

vice-versa. Thus the Brillouin zone of one is the Wigner–Seitz cell of the other. Conventionally, solid-state texts refer to the symmetric cell in **k**-space as the *first* Brillouin zone. The Brillouin zone thus defines a volume in **k**-space such that every point inside it or on its surface corresponds to an irreducible representation of the group of pure translations. However some points on the surface differ only by **K** and will therefore be equivalent.

The free electron energy–momentum relationship (i.e. the plot of $E(\mathbf{k})$ versus **k**) is shown in Fig. 6.2 for a one-dimensional lattice with $a_1 = a$ (cf. eq. (6.8)). However, since the full significance of the Brillouin zone lies in the fact that states **k** and $\mathbf{k} + \mathbf{K}$ label the same irreducible representations we can work *entirely* with vectors **k** which are restricted to the Brillouin zone. Hence Fig. 6.2 can be transformed to the equivalent figure, by bodily translating pieces of the $E(\mathbf{k})$ curve to the Brillouin zone. This is called a *reduced zone* scheme because, as we mentioned above, the volume we call the Brillouin zone is actually the first zone; other zones, equal in volume to the first zone, can be constructed around the first zone in such a way that we can fill the whole of **k**-space. If we fill the whole of **k**-space by stacking *first* Brillouin zones together we adopt what is known as a *repeated zone* scheme.

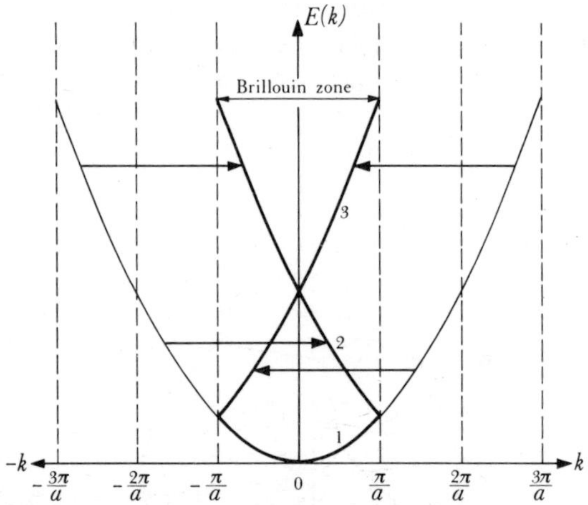

Fig. 6.2 *Free electron energy $E(k)$ versus momentum k (in units of $\hbar$) for a one-dimensional lattice with constant equal to a. The equation of the curve is $E(k) = \hbar^2 k^2/2m$ where m is the electron mass. The production of the reduced zone scheme from the free electron energy curve is also shown. The region $-\pi/a \leqslant k \leqslant \pi/a$ is the Brillouin zone and the numbers 1, 2, 3, ... label energy bands.*

The penalty of using the reduced zone procedure is that many values of E now exist for each value of **k**. The part of the curve in Fig. 6.2 which corresponds to vectors lying in the range $-\pi/a$ to π/a is called *band* 1, the pieces of the curve between $-2\pi/a$ and $-\pi/a$, and π/a and $2\pi/a$ make up band 2 and so on. Thus for a three-dimensional solid, we can label an electron state by the quantum numbers n, k_x, k_y, k_z where n is called the *band index*.

6.4 Bloch functions

The eigenfunction of an electron in a periodic solid will now be denoted by $\psi_{n,\mathbf{k}}(\mathbf{r})$. This function must be a basis function for the **k**th irreducible representation of the translation group $\mathscr{T}$ so that according to eq. (3.27) we have

$$\mathscr{T}\psi_{n,\mathbf{k}}(\mathbf{r}) = \psi_{n,\mathbf{k}}(\mathbf{r})\boldsymbol{\Gamma}^{\mathbf{k}}(\mathscr{T}) \tag{6.17}$$

The irreducible representation $\boldsymbol{\Gamma}^{\mathbf{k}}(\mathscr{T})$ is $e^{i\mathbf{k}.\mathbf{R}}$ so that since the operation $\mathscr{T}$ translates **r** through a lattice vector **R** we can see that

$$\mathscr{T}\psi_{n,\mathbf{k}}(\mathbf{r}) = \psi_{n,\mathbf{k}}(\mathbf{r} + \mathbf{R}) = e^{i\mathbf{k}.\mathbf{R}}\psi_{n,\mathbf{k}}(\mathbf{r}) \tag{6.18}$$

We can now re-arrange (6.18) by multiplying each side by $e^{-i\mathbf{k}.(\mathbf{r}+\mathbf{R})}$ to give

$$e^{-i\mathbf{k}.\mathbf{r}}\psi_{n,\mathbf{k}}(\mathbf{r}) = e^{-i\mathbf{k}.(\mathbf{r}+\mathbf{R})}\psi_{n,\mathbf{k}}(\mathbf{r} + \mathbf{R}) \tag{6.19}$$

Equation (6.19) now shows that translational symmetry demands that $e^{-i\mathbf{k}.\mathbf{r}}\psi_{n,\mathbf{k}}(\mathbf{r})$ be preserved under a lattice translation **R**. The quantity $e^{-i\mathbf{k}.\mathbf{r}}\psi_{n,\mathbf{k}}(\mathbf{r})$ which we will define as $u_{n,\mathbf{k}}(\mathbf{r})$ must therefore be periodic, i.e.,

$$u_{n,\mathbf{k}}(\mathbf{r}) = u_{n,\mathbf{k}}(\mathbf{r} + \mathbf{R}) = e^{-i\mathbf{k}.\mathbf{r}}\psi_{n,\mathbf{k}}(\mathbf{r}) \tag{6.20}$$

The eigenfunction of an electron, in a periodic lattice, is therefore

$$\psi_{n,\mathbf{k}}(\mathbf{r}) = u_{n,\mathbf{k}}(\mathbf{r})\, e^{i\mathbf{k}.\mathbf{r}} \tag{6.21}$$

The functions defined in eq. (6.21) are the celebrated Bloch functions introduced into solid state physics by Bloch in 1928. Bloch functions are, of course, the eigenfunctions of the Hamiltonian of charge carriers in a periodic lattice. However, we have obtained (6.21) without considering the Hamiltonian; rather we have demanded that the eigenfunctions must be basis functions for the irreducible representations of the translation group. This then is a very good example of the power of group theory.

6.5 Empty simple cubic lattice

The substitution of the Bloch function $\psi_{n,\mathbf{k}}(\mathbf{r})$ into the Schrödinger eq. (6.1) results in an equation for $u(\mathbf{r})$ (we omit the suffices n and **k** for ease of notation) namely

$$\nabla^2 u - k^2 u + 2i\mathbf{k}\,.\,\nabla u = -\frac{2m}{\hbar^2}[E(\mathbf{k}) - V(\mathbf{r})]u \tag{6.22}$$

We cannot, in general, find an analytical solution for $u(\mathbf{r})$ if $V(\mathbf{r})$ is not zero everywhere. On the other hand if $V(\mathbf{r})$ happens to be zero then $\psi(\mathbf{r})$ will be a plane wave state even when modulated by $u(\mathbf{r})$. This is because there can be no scattering from the ions at the lattice sites if $V(\mathbf{r})$ is zero. Let us therefore for the time being resort to this *empty lattice* device in which we set $V(\mathbf{r})$

equal to zero but retain the lattice symmetry. The eigenfunctions are therefore plane wave Bloch functions and $u(\mathbf{r})$ must be of the form $e^{i\mathbf{k}'.\mathbf{r}}$. Equation (6.20) now requires that

$$e^{i\mathbf{k}'.\mathbf{r}} = e^{i\mathbf{k}'.(\mathbf{r}+\mathbf{R})} \tag{6.23}$$

which, as can be seen by examining eq. (6.15), can only be satisfied if $\mathbf{k}'$ is a reciprocal lattice vector. Therefore the eigenfunction of an electron in an empty lattice is

$$\psi(\mathbf{r}) = e^{i(\mathbf{k}+\mathbf{K}).\mathbf{r}} \tag{6.24}$$

where $\mathbf{K}$ is a reciprocal lattice vector and $\mathbf{k}$ is restricted to the Brillouin zone.

The use of the eigenfunction of eq. (6.24) in eq. (6.22) or eq. (6.1) leads to

$$E(\mathbf{k}) = \frac{\hbar^2}{2m}(\mathbf{k} + \mathbf{K})^2 = \frac{\hbar^2}{2m}(\mathbf{k} + \mathbf{K}).(\mathbf{k} + \mathbf{K}) \tag{6.25}$$

where, as we shall see, the presence of $\mathbf{K}$ leads to the appearance of energy bands with zero band gap.

At this stage we will for convenience transform eqs. (6.24) and (6.25) to dimensionless form. We can begin this task by writing the components of $\mathbf{K}$ and $\mathbf{k}$ as

$$\begin{aligned} (p, q, r) &= \frac{a}{2\pi}(K_x, K_y, K_z) \\ (\alpha, \beta, \gamma) &= \frac{a}{2\pi}(k_x, k_y, k_z) \end{aligned} \tag{6.26}$$

where a is the lattice constant of the simple cubic lattice, p, q, and r are *integers*; and α, β, and γ are *numbers* lying in the range $-0{\cdot}5$ to $0{\cdot}5$ because, for $\mathbf{k}$ to lie within the Brillouin zone (a cube of side $2\pi/a$), the maximum value k_x, k_y, or k_z can assume is π/a; taking the origin at the zone centre. We can also define a dimensionless energy parameter

$$\xi = \frac{2ma^2}{\hbar^2}E(\mathbf{k}) \tag{6.27}$$

Equation (6.25) in dimensionless form is therefore

$$\xi = [(\alpha + p)^2 + (\beta + q)^2 + (\gamma + r)^2] \tag{6.28}$$

and the eigenfunction of eq. (6.24) now becomes

$$\psi(\mathbf{r}) = e^{(2\pi i/a)[(\alpha+p)x+(\beta+q)y+(\gamma+r)z]} \tag{6.29}$$

Equation (6.28) shows that ξ, being a continuous function of α, β, and γ, can, provided p, q, and r are specified, be plotted along various directions of the Brillouin zone. The latter, since we are considering a simple cubic lattice is just the cube illustrated in Fig. 6.3. The labelling of the special symmetry points and lines follows the convention adopted, over the years, by most workers in solid state physics. The features we are going to examine in detail are the zone centre Γ (not to be confused with the symbol for matrix representation), the axis, Δ, and the centre of the zone face crossing this axis, namely X.

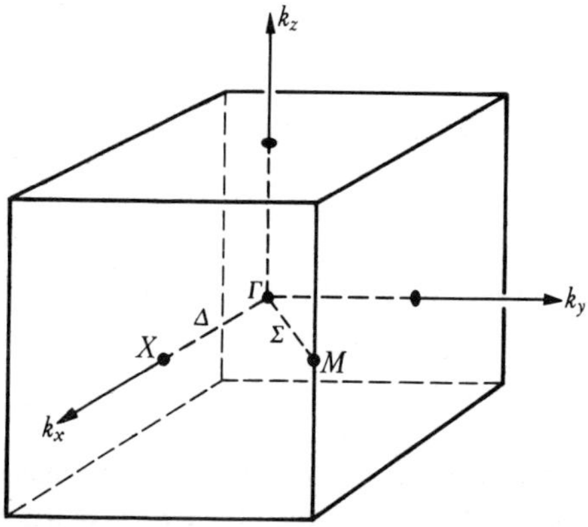

Fig. 6.3 *Brillouin zone of simple cubic lattice showing the symmetry points Γ, X, and M; and the symmetry lines Δ and Σ.*

In order to determine the energy ξ as a function of (α, β, γ) we must clearly specify (p, q, r). If $(p, q, r) = (0, 0, 0)$ we obtain the first energy band. Also since ξ depends on (α, β, γ) we can only plot ξ in a particular direction. Arbitrarily choosing this direction to be $(\alpha, 0, 0)$ we obtain for the first energy band

$$\xi = \alpha^2 \tag{6.30}$$

The direction $(|\alpha|, 0, 0)$ is equivalent to $(-|\alpha|, 0, 0)$, $(0, \pm|\beta|, 0)$, and $(0, 0, \pm|\gamma|)$ and hence ξ will have the same functional form for each of these directions.

A second set of curves is defined by $(p, q, r) = (\mp 1, 0, 0)$, $(0, \mp 1, 0)$, or $(0, 0, \mp 1)$ because this adds on one lattice vector; the first of these choices for (p, q, r) leads to

$$\xi = (1 - \alpha)^2, \quad (p, q, r) = (-1, 0, 0) \tag{6.31}$$

or

$$\xi = (1 + \alpha)^2, \quad (p, q, r) = (1, 0, 0) \tag{6.32}$$

The second and third choices for (p, q, r) lead to

$$\xi = 1 + \alpha^2 \tag{6.33}$$

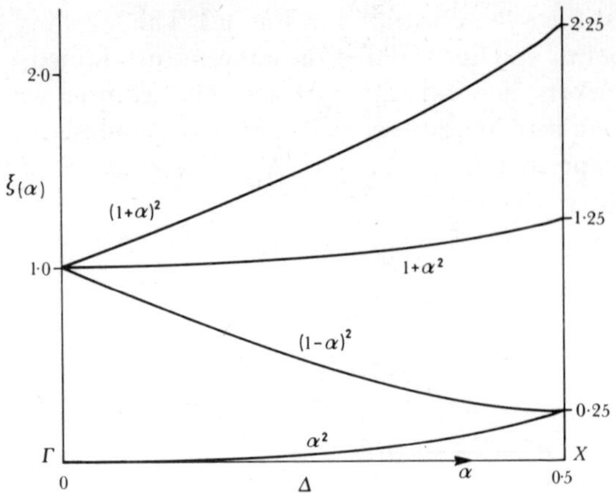

Fig. 6.4 *Band structure of electrons in an empty simple cubic lattice plotted along the symmetry line* Δ.

These curves are plotted in Fig. 6.4 and are collectively known as part of the *band structure* of the crystal (obviously further choices of (p, q, r) lead to many more bands). A summary of these results is given in Table 6.1.

Table 6.1 Parameters and variables of the energy bands in the Γ–X direction. The last two columns give the number of eigenfunctions associated with the same energy level

Free electron energy bands in Γ–X direction						
$a\mathbf{K}_i/2\pi$	$a\mathbf{k}/2\pi$	$\xi(\alpha)$	$\psi(x, y, z)$	$\xi(0)$	Degeneracy at Γ	Degeneracy at $0 < \alpha < 0{\cdot}5$
(0 0 0)	(α 0 0)	α^2	$e^{2\pi i\alpha x/a}$	0	1	1
(−1 0 0)	(α 0 0)	$(1-\alpha)^2$	$e^{(2\pi i/a)(\alpha-1)x}$	1		1
(1 0 0)	(α 0 0)	$(1+\alpha)^2$	$e^{(2\pi i/a)(\alpha+1)x}$	1	6	1
(0 ±1 0)	(α 0 0)	$1+\alpha^2$	$e^{(2\pi i/a)(\alpha x \pm y)}$	1		4
(0 ±1 0)	(α 0 0)	$1+\alpha^2$	$e^{(2\pi i/a)(\alpha x \pm z)}$	1		

From this table we can also see that the energies at Γ, i.e., $\xi(0)$, are, for the bands we have considered, respectively zero and unity. The lower value is singly degenerate (i.e., has only one eigenfunction associated with it); and the upper value is six-fold degenerate. As we go from Γ along the line Δ towards X we shall find that the singly degenerate level remains singly degenerate while the six-fold degenerate divides into three levels of which two are singly degenerate the other being four-fold degenerate.

This 'lifting' of degeneracy is clearly associated with the *lowering* of the symmetry as we go from Γ to X. In the next section we will use our knowledge of group theory to determine the ultimate degeneracy of eigenstates when the symmetry is lowered by the introduction of a *finite* cubic field, $V(\mathbf{r})$, into the lattice.

6.6 Effect of point group symmetry on the eigenvalues at Γ and X

At this stage we have only considered one way of discussing *free* electrons moving in an empty lattice. However if we wish to consider electrons moving in an environment where they experience the electric field which the lattice sites produce, we could solve the problem *exactly* by introducing a potential energy $V(\mathbf{r})$, with cubic symmetry say, into the Hamiltonian. On the other hand we can avoid detailed computations and make considerable progress towards an understanding of the real band structure by classifying the eigenvalues of the free electron model in accordance with the full symmetry of the lattice. Since, as we have seen earlier, each point $\mathbf{k}$ inside the Brillouin zone labels only an irreducible representation of the translation group we must now, in order to account for the whole symmetry, impose the point group symmetry of the lattice on the basis functions of these representations, i.e., these basis functions must also be basis functions of the irreducible representations of the point group. This procedure of classifying the free electron eigenstates is a means of determining what degeneracy remains when the actual symmetry of the environment is accounted for. It is therefore equivalent to examining the qualitative effects of introducing a perturbation $V(\mathbf{r})$ with appropriate symmetry into the Hamiltonian (cf. section 3.10).

If a lattice in real space is invariant under certain point group operations then its reciprocal lattice is also invariant under the same operations. Thus, if, for example, the real space lattice has cubic symmetry, the reciprocal lattice will also have cubic symmetry. Also because we have carefully elected to use the symmetric primitive cell (the Brillouin zone) to characterize the reciprocal lattice it will possess the full symmetry of the point group. Therefore in the work we are now going to describe we regard the Brillouin zone as a symmetric object and the state $\mathbf{k}$ labelling an irreducible representation of the translation group as some *arbitrary* point inside or on the surface of this object. Certain points inside the zone will obviously have more symmetry than others. Γ, for example, being the centre of the zone does not move under *any* of the operations of the full point group (i.e., the point group of the lattice) and is clearly the point of highest symmetry. An arbitrary point will move to a different position for every operation of the point group. We are thus led to the concept of the group of $\mathbf{k}$ which obviously must be a subgroup of the full point group, and is that subgroup of operators which leaves the point **k** invariant. If we imagine performing all the group operations on a point labelled by **k** then each operation may take **k** to a new position hence generating an array of vectors **k** emanating from the origin. This array is called the 'star' of **k**. The fewer the points in the star the higher the symmetry of the point **k**, e.g., the star of X has only three points. The lowest symmetry point would have the same number of points as there are operations in the full point group of the lattice. In fact the order of the group of **k** is equal to the order of the full point group divided by the number of points in its star.

The eigenfunctions $\psi(\mathbf{r})$ must form basis functions for the group of **k** and hence the degeneracy of eigenvalues will be determined by the irreducible representations of this group. In other words there will be one distinct energy value for each irreducible representation, and the degeneracy of a particular energy value will be equal to the number of basis functions belonging to the associated irreducible representation (cf. section 3.7).

Let us consider the six-fold degenerate level $\xi(0)$ at the point Γ (cf. table 6.1). The group of **k** at this point is the point group O_h and therefore the six functions in column four of table 6.1 after setting $\alpha = 0$ must form a basis for a representation of the group O_h which, since a six-dimensional irreducible representation of O_h does not exist, must be reducible. This reducible representation of O_h can be constructed by considering the effect of the symmetry operations of the group on the vector $\mathbf{r} = (x, y, z)$. The results of these operations are shown in table 6.2. Now the six basis functions are $e^{\pm 2\pi i(x/a)}$, $e^{\pm 2\pi i(y/a)}$, and $e^{\pm 2\pi i(z/a)}$ which, following the procedure of chapter 3 can be denoted by a row vector

$$(\tilde{\mathbf{f}}) = [e^{Bx}, e^{-Bx}, e^{By}, e^{-By}, e^{Bz}, e^{-Bz}] \tag{6.34}$$

where for convenience we have written $B = 2\pi i/a$. Thus using eq. (3.27) the representation matrix $\mathbf{\Gamma}(R)$ of an operation P_R on $(\tilde{\mathbf{f}})$ can be found from

$$P_R(\tilde{\mathbf{f}}) = (\tilde{\mathbf{f}})\mathbf{\Gamma}(R) \tag{6.35}$$

Obviously $\mathbf{\Gamma}(E)$ is a six-dimensional unit matrix. We can find $\mathbf{\Gamma}(C_4^2)$ from the fact that the operation C_4^2 about the z-axis, as recorded in table 6.2, changes x to $-x$, y to $-y$, and leaves z unchanged. That is,

$$P_{C_4^2}(\tilde{\mathbf{f}}) = [e^{Bx}, e^{-Bx}, e^{By}, e^{-By}, e^{Bz}, e^{-Bz}] \left[\begin{array}{cc|cc|cc} 0 & 1 & 0 & 0 & 0 & 0 \\ 1 & 0 & 0 & 0 & 0 & 0 \\ \hline 0 & 0 & 0 & 1 & 0 & 0 \\ 0 & 0 & 1 & 0 & 0 & 0 \\ \hline 0 & 0 & 0 & 0 & 1 & 0 \\ 0 & 0 & 0 & 0 & 0 & 1 \end{array}\right] \tag{6.36}$$

Table 6.2 The operations of the cubic group in terms of their action on a *vector* with components $(x\ y\ z)$ (J is the conventional solid state notation for i)

Class	Result of the Operation			Rotation	Details of operation
E	x	y	z	None	None
C_4^2	$-x$	$-y$	z	π about the three	Rotation of π about z
	x	$-y$	$-z$	cubic (100) axes	Rotation of π about x
	$-x$	y	$-z$		Rotation of π about y
C_4	$-y$	x	z	$\pm\pi/2$ about the	Rotation of $+\pi/2$ about z
	y	$-x$	z	three cubic (100)	Rotation of $-\pi/2$ about z
	x	z	$-y$	axes (N.B. rotation	Rotation of $+\pi/2$ about x
	x	$-z$	y	of vector is	Rotation of $-\pi/2$ about x
	$-z$	y	x	opposite to	Rotation of $+\pi/2$ about y
	z	y	$-x$	rotation of coords.)	Rotation of $-\pi/2$ about y
C_2	y	x	$-z$	π about the six	x and y reflected across
	z	$-y$	x	two-fold (110) axes	(110) plane: z reversed,
	$-x$	z	y		etc.
	$-y$	$-x$	$-z$		
	$-z$	$-y$	$-x$		
	$-x$	$-z$	$-y$		
C_3	z	x	y	$\pm(2\pi/3)$ about the	$x = y = z$ rotate through
	$-y$	z	$-x$	4 three-fold (111)	$+120°$
	$-z$	x	$-y$	axes	$x = y = z$ rotate through
	y	$-z$	$-x$		$-120°$, etc.
	y	z	x		
	z	$-x$	$-y$		
	$-y$	$-z$	x		
	$-z$	$-x$	y		
$i = J$	$-x$	$-y$	$-z$	None	Inversion
$\sigma = JC_4^2$	x	y	$-z$	π about (100) axes	Rotation followed by
	$-x$	y	z		inversion.
	x	$-y$	z		Reflection in (100) plane.
$S_4 = JC_4$	y	$-x$	$-z$	$+\pi/2$ about (100)	Rotation followed by
	$-x$	z	$-y$	axes	inversion
		etc.			
$\sigma = JC_2$	$-y$	$-x$	z	π about (110) axes	Rotation followed by
	$-z$	y	$-x$		inversion $\equiv$ reflection
		etc.			
$S_6 = JC_3$	$-z$	$-x$	$-y$	$\pm(2\pi/3)$ about	Rotation followed by
		etc.		(111) axes	inversion.

Similarly, $\boldsymbol{\Gamma}(C_4)$, for a rotation about the z-axis, satisfies

$$P_{C_4}^{(z)}(\tilde{\mathbf{f}}) = (\tilde{\mathbf{f}})\left[\begin{array}{cc|cc|cc} 0 & 0 & 1 & 0 & 0 & 0 \\ 0 & 0 & 0 & 1 & 0 & 0 \\ \hline 0 & 1 & 0 & 0 & 0 & 0 \\ 1 & 0 & 0 & 0 & 0 & 0 \\ \hline 0 & 0 & 0 & 0 & 1 & 0 \\ 0 & 0 & 0 & 0 & 0 & 1 \end{array}\right] \tag{6.37}$$

In order to find the component irreducible representations we need only consider the character of the reducible representation as can be appreciated from eq. (2.37) namely

$$a_j = \frac{1}{g}\sum_R \chi^{j*}(R)\chi^{\text{red}}(R) \tag{6.38}$$

where a_j is the number of times the jth irreducible representation appears in the reducible representation and $\chi^{\text{red}}(R)$, the character of the reducible representation, is simply the trace of the matrix $\boldsymbol{\Gamma}(R)$, e.g., the representations of eqs. (6.36) and (6.37) have characters equal to 2. Actually, since we are only concerned with diagonal elements we can, very quickly, from table 6.2, determine the value of the traces of $\boldsymbol{\Gamma}(R)$ for *all* the classes of O_h. (N.B. all elements in a class have the same character.) We simply look for the number of the components $\pm x$, $\pm y$, or $\pm z$ which remain invariant under a typical group operation of each class. This number must be the trace of the matrix. Reference to table 6.2 shows that the character of $\boldsymbol{\Gamma}(R)$ is the set 6, 2, 2, 0, 0, 0, 4, 0, 2, 0 arranged in order of the classes set out in the first column.

Using eq. (6.38) and the character table (table 6.3) of the group O_h we find that the only non-zero a_j's are

$$\begin{aligned} a_1 &= \tfrac{1}{48}[6 + 6 + 12 + 0 + 0 + 0 + 12 + 0 + 12 + 0] = 1 \\ a_{12} &= \tfrac{1}{48}[12 + 12 + 0 + 0 + 0 + 0 + 24 + 0 + 0 + 0] = 1 \\ a_{15} &= \tfrac{1}{48}[18 - 6 + 12 + 0 + 0 + 0 + 12 + 0 + 12 + 0] = 1 \end{aligned} \tag{6.39}$$

where instead of using $j = 1, 2$, etc., we have adopted the conventional solid state notation for labelling the irreducible representations of O_h. The fact that the other coefficients are zero can be readily checked; for example

$$a'_{25} = \tfrac{1}{48}[18 - 6 - 12 + 0 + 0 + 0 - 12 + 0 + 12 + 0] = 0 \tag{6.40}$$

Thus we can see that

$$\boldsymbol{\Gamma}^{\text{red}}(R) = \boldsymbol{\Gamma}_1 \oplus \boldsymbol{\Gamma}_{12} \oplus \boldsymbol{\Gamma}_{15} \tag{6.41}$$

Table 6.3 Character table of the group of **k** at Γ i.e. the group O_h. The notation of the irreducible representations is conventional but will make some sense when the compatibility relations are developed

O_h	E	$3C_4^2$	$6C_4$	$6C_2$	$8C_3$	J	$3JC_4^2$	$6JC_4$	$6JC_2$	$8JC_3$
Γ_1	1	1	1	1	1	1	1	1	1	1
Γ_2	1	1	−1	−1	1	1	1	−1	−1	1
Γ_{12}	2	2	0	0	−1	2	2	0	0	−1
Γ'_{15}	3	−1	1	−1	0	3	−1	1	−1	0
Γ'_{25}	3	−1	−1	1	0	3	−1	−1	1	0
Γ'_1	1	1	1	1	1	−1	−1	−1	−1	−1
Γ'_2	1	1	−1	−1	1	−1	−1	1	1	−1
Γ'_{12}	2	2	0	0	−1	−2	−2	0	0	1
Γ_{15}	3	−1	1	−1	0	−3	1	−1	1	0
Γ_{25}	3	−1	−1	1	0	−3	1	1	−1	0

where $\oplus$ denotes the direct sum. [N.B. if we wish to refer to the label of an irreducible representation we will simply use Γ_1, Γ_{12}, etc.].

It can be seen from eq. (6.41) that the irreducible representations we require are $\mathbf{\Gamma}_1$, $\mathbf{\Gamma}_{12}$, and $\mathbf{\Gamma}_{15}$ which have dimensions of 1, 2, and 3 respectively.

The implication of these group theoretical deductions, as we have previously stated, is that the number of irreducible components of the reducible representation is equal to the number of distinct eigenvalues that the state we are considering possesses. Also the degeneracy of each eigenstate is equal to the dimension of the associated irreducible representation. Thus we can now state that the six-fold degenerate electron eigenvalue at Γ which exists for an empty lattice 'splits' into three distinct levels when we introduce a $V(\mathbf{r})$ with cubic symmetry. One level is singly degenerate and is labelled Γ_1, one level, labelled Γ_{12} is doubly degenerate, and one level, labelled Γ_{15}, is triply degenerate.

Although the above conclusions are exact they are not quantitative. By this we mean that, although we have shown rigorously that the level 'splits' in the fashion described, we cannot attach any numerical value to the size of the splitting or even say which of the Γ_1, Γ_{12}, or Γ_{15} levels has the highest value. This information can only be obtained from a detailed numerical solution of eq. (6.1).

The eigenvalues at X can be treated in a similar manner to those at Γ. However, we will only consider the lowest eigenvalue at X, i.e., $\xi(0{\cdot}5) = 0{\cdot}25$ which, in the empty lattice, is doubly degenerate with wave functions $e^{\pm i(\pi/a)x}$. We can now with the aid of table 6.4, which is the character table of the group of **k** at X, use the same procedure as before to show that a *real* simple cubic crystal field will split this level into X_1 and X_4 states. Thus the curves $\xi = \alpha^2$ and $\xi = (1 - \alpha)^2$ which in the free-electron empty lattice model meet at X must separate when a finite $V(\mathbf{r})$ is 'switched on'.

Table 6.4 Character table of the group of **k** at X i.e. $D_{4h} = D_4 \otimes S_2$ (N.B. the x axis has been chosen as the C_4 principal axis and is denoted by $||$; the y and z axes are denoted by $\perp$. Also the class notation is that of the group O_h)

D_{4h}	E	$2C_{4\perp}^2$	$C_{4\parallel}^2$	$2C_{4\parallel}$	$2C_2$	J	$2JC_{4\perp}^2$	$JC_{4\parallel}^2$	$2JC_{4\parallel}$	$2JC_2$
X_1	1	1	1	1	1	1	1	1	1	1
X_2	1	1	1	−1	−1	1	1	1	−1	−1
X_3	1	−1	1	−1	1	1	−1	1	−1	1
X_4	1	−1	1	1	−1	1	−1	1	1	−1
X_5	2	0	−2	0	0	2	0	−2	0	0
X'_1	1	1	1	1	1	−1	−1	−1	−1	−1
X'_2	1	1	1	−1	−1	−1	−1	−1	1	1
X'_3	1	−1	1	−1	1	−1	1	−1	1	−1
X'_4	1	−1	1	1	−1	−1	1	−1	−1	1
X'_5	2	0	−2	0	0	−2	0	2	0	0

The splitting of the energy levels at Γ and X is of course simply the appearance of the familiar energy gaps and forbidden bands.

6.7 Compatibility relations

Having established the behaviour of the electron eigenvalues at Γ and X, under the influence of a crystal field of cubic symmetry, we will now consider how to connect X with Γ by bands whose symmetry is determined by the group of **k** along the line Δ (cf. Fig. 6.4) whose character table is given in table 6.5. The objective therefore is to discover whether the bands which leave Γ_1, Γ_{12}, Γ_{15}, X_1, and X'_4 have symmetry Δ_1, Δ_2, Δ'_2, Δ'_1, or Δ_5.

Table 6.5 Character table of the group of **k** along the line Δ i.e. C_{4v}. The class notation is that of the group O_h.

C_{4v}	E	$C_{4\parallel}^2$	$2C_4$	$2JC_{4\perp}^2$	$2JC_2$
Δ_1	1	1	1	1	1
Δ_2	1	1	−1	1	−1
Δ'_2	1	1	−1	−1	1
Δ'_1	1	1	1	−1	−1
Δ_5	2	−2	0	0	0

As we proceed from Γ to X along the line Δ the symmetry is lowered and the group of **k** changes from O_h to C_{4v}, a sub-group of O_h. This means that the irreducible representation of O_h associated with the triply degenerate level Γ_{15} becomes reducible as we progress from Γ towards X. We must therefore seek the irreducible components of this reducible representation with the aid of the character table of C_{4v}.

We begin by noting that the *irreducible* representation matrices of those symmetry elements in O_h, identical to those in C_{4v}, are the *reducible* representation matrices for the group C_{4v}. The characters of these representations, associated with the three states are selected from table 6.3 and are shown in table 6.6.

Table 6.6 Characters of reducible representations of C_{4v} associated with the line Δ near to the Γ point

State	E	C_4^2	$2C_4$	$2JC_4^2$	$2JC_2$
Γ_1	1	1	1	1	1
Γ_{12}	2	2	0	2	0
Γ_{15}	3	-1	1	1	1

Thus, using eq. (6.38) and the character table 6.5, we find that

$$\begin{aligned} \Gamma_1 &\to \Delta_1 \\ \Gamma_{12} &\to \Delta_1 + \Delta_2 \\ \Gamma_{15} &\to \Delta_1 + \Delta_5 \end{aligned} \tag{6.42}$$

Similarly we can show that, as we move away from X along Δ,

$$\begin{aligned} X_1 &\to \Delta_1 \\ X_4' &\to \Delta_1 \end{aligned} \tag{6.43}$$

We do not use equalities in eqs. (6.42) and (6.43) because we are only indicating that we pass from one symmetry state to another as we leave a Γ or an X point. The relationships in eqs. (6.42) and (6.43) are actually known as *compatibility* relationships because we have established which energy curves are compatible with Γ or X. The notation for the irreducible representations of Γ now appears to make a little more sense, in so far as we can accept that Γ_{15} is a suitable label because Γ_{15} splits into Δ_1 and Δ_5. Figure 6.5 summarizes, in a schematic way, our principal results and

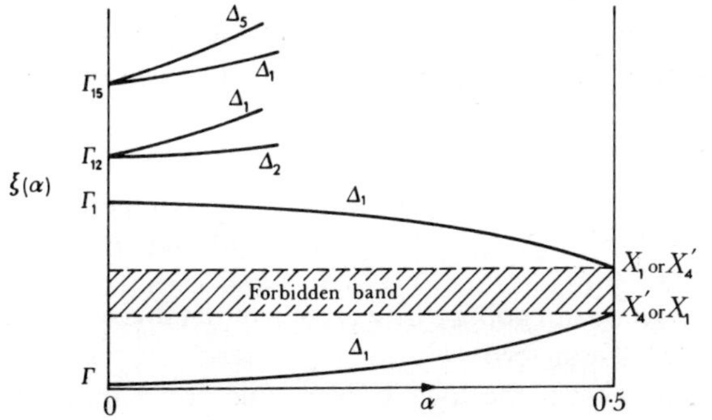

Fig. 6.5 *Schematic effect of lowering the symmetry on the energy curves of Fig. 6.4 due to the introduction of a non-zero simple cubic lattice field.*

therefore shows, as we have previously discussed, the effect of a crystal field on the free-electron eigenstates of a simple cubic solid. We must emphasize, however, that Fig. 6.5 is very incomplete because we have left many *higher* bands out of the discussion. Also since group theory does not give us a quantitative knowledge of the energy levels we can not place them in any particular order. Thus the levels shown in Fig. 6.5 are only a possible interpretation of the band structure. In spite of this reservation group theory does enable us to produce a scheme in which the only degeneracies remaining are those demanded by the symmetry.

6.8 Symmetrized electron eigenfunctions

As we discussed in chapter 3 we can develop eigenfunctions belonging to the irreducible representations of the group of **k**. This can be done using eqs. (3.47) and (3.48) which state that the eigenfunction belonging to the jth irreducible representation is

$$\psi^j = \frac{l_j}{g} \sum_R \chi^j(R) P_R F \tag{6.44}$$

where F is an arbitrary function.

At the zone centre, for $(p, q, r) = (0, 0, 0)$, we have a non-degenerate eigenfunction corresponding to the first row of table 6.1. The function $\psi = 1$ can now be used as the arbitrary function F and what we have to do is to determine how many non-zero ψ^j exist. We know at the outset that the zone centre can only be associated with the totally symmetric state Γ_1 but we can formally establish this using eq. (6.44), i.e., we obtain, using the character table of O_h,

$$\Gamma_1: \quad \psi^1 = \tfrac{1}{48}[1 + 3 + 6 + 6 + 8 + 1 + 3 + 6 + 6 + 8] = 1 \tag{6.45}$$

where the numbers arise because the class of C_4^2 has three operations associated with it, that of C_4 has six operations associated with it and so on.

We can see that the second row of table 6.3 leads to

$$\Gamma_2: \quad \psi^2 = \tfrac{1}{48}[1 + 3 - 6 - 6 + 8 + 1 + 3 - 6 - 6 + 8] = 0 \tag{6.46}$$

and although we are not going to test *every* row of table 6.3, we can appreciate that *only* ψ^1 is non-zero.

As we proceed from the lowest eigenvalue along the line Δ the free-electron eigenfunction is $e^{2\pi i\alpha x/a}$. That this function is also the correct basis function for irreducible representation of Δ_1 can be shown as follows. From the character table of Δ, i.e., table 6.5, we can see that:

$$\Delta_1: \quad \psi^1 = \tfrac{1}{8} e^{2\pi i\alpha x/a}[1 + 1 + 2 + 2 + 2] = e^{2\pi i\alpha x/a} \tag{6.47}$$

$$\Delta_2: \quad \psi^2 = \tfrac{1}{8} e^{2\pi i\alpha x/a}[1 + 1 - 2 + 2 - 2] = 0 \tag{6.48}$$

$$\Delta_2': \quad \psi^{2'} = \tfrac{1}{8} e^{2\pi i x/a}[1 + 1 - 2 - 2 + 2] = 0 \tag{6.49}$$

and so on, showing that only ψ^1 is non-zero.

Table 6.1 shows that, at an X point ($\alpha = 0{\cdot}5$), the appropriate arbitrary function can be taken to be $\psi(x) = e^{\pi i x/a}$. Also table 6.2 shows which operations, of the group D_{4h}, leave the function $e^{\pi i x/a}$ invariant and those which convert it to $e^{-\pi i x/a}$. The application of (6.44) therefore gives

$$X_1: \quad \psi^1 = \tfrac{1}{16}[e^{\pi i x/a} + 2\,e^{-\pi i x/a} + e^{\pi i x/a} + 2\,e^{\pi i x/a} + 2\,e^{-\pi i x/a}$$
$$+ e^{-\pi i x/a} + 2\,e^{\pi i x/a} + e^{-\pi i x/a} + 2\,e^{-\pi i x/a} + 2\,e^{\pi i x/a}]$$
$$= \tfrac{1}{2}[e^{\pi i x/a} + e^{-\pi i x/a}] = \cos\frac{\pi x}{a} \tag{6.50}$$

$$X_4': \quad \psi^4 = \tfrac{1}{16}[e^{\pi i x/a} - 2\,e^{-\pi i x/a} + e^{\pi i x/a} + 2\,e^{\pi i x/a} - 2\,e^{-\pi i x/a} - e^{-\pi i x/a}$$
$$+ 2\,e^{\pi i x/a} - e^{-\pi i x/a} - 2\,e^{-\pi i x/a} + 2\,e^{\pi i x/a}]$$
$$= \tfrac{1}{2}[e^{\pi i x/a} - e^{-\pi i x/a}] = i\sin\frac{\pi x}{a} \tag{6.51}$$

The use of the other rows of table 6.4 reveals that all other eigenfunctions vanish. Naturally this implies that we have chosen a good arbitrary function $\psi(x)$ from which our two functions can be projected. A bad choice could well have led to a null result. Thus, the symmetrized eigenfunctions for the states X_1 and X_4' are essentially $\cos \pi x/a$ and $\sin \pi x/a$ respectively. We actually expect this from our knowledge of elementary solid state physics which tells us that the electron eigenfunction assumes the form of a standing wave at a band-edge. The appearance of these *non-propagating* states can of course be described physically in terms of Bragg reflections at zone boundaries.

6.9 Problems

6.1 If a space-group operation $\{\boldsymbol{\alpha}|\mathbf{t}\}$ is a point group operation followed by a translation $\mathbf{t}$ show that

$$\{E|\mathbf{R}\}\{\boldsymbol{\alpha}|\mathbf{t}\} = \{\boldsymbol{\alpha}|\mathbf{t}\}\{E|\mathbf{R}'\}$$

where $\mathbf{R}'$ and $\mathbf{R}$ are lattice vectors.

Hence show that

$$\langle k|R'\rangle = \langle k|\boldsymbol{\alpha}^{-1}|R\rangle = \langle k'|R\rangle$$

where $\mathbf{k}'$ and $\mathbf{k}$ label the irreducible representations of the group of pure translations.

6.2 Using the results of 6.1 prove that the eigenvalues $E(\mathbf{k})$ of the time-independent Schrödinger equation satisfy

$$E(\mathbf{k}') = E(\mathbf{k})$$

which shows that energy bands are invariant with respect to point-group operations.

6.3 If the time-dependent Schrödinger equation is

$$H\psi = -i\hbar\frac{\partial\psi}{\partial t}$$

show from a consideration of the representation of a translational operation that energy bands in solids possess time-inversion symmetry.

6.4 In section 6.8 only the case $\mathbf{K}_0 = 2\pi/a\ (0, 0, 0)$ was considered. Set $(p, q, r) = (\mp 1, 0, 0), (0, \mp 1, 0)$, or $(0, 0, \mp 1)$, assume that $F = e^{(2\pi i/a)(\alpha x + z)}$, and then show from eq. (6.44) that

$$\Delta_1:\ \phi_1 = \frac{1}{2}\left\{\cos\left(\frac{2\pi z}{a}\right) + \cos\left(\frac{2\pi y}{a}\right)\right\} e^{(2\pi i\alpha x)/a}$$

Similarly, starting with an appropriate F, show that

$$\Delta_5:\ \phi_5 = i\,e^{2\pi i x/a}\sin\left(\frac{2\pi y}{a}\right)$$

$$\text{or}\quad i\,e^{2\pi i x/a}\sin\left(\frac{2\pi z}{a}\right)$$

$$\Gamma_1:\ \psi_1 = \frac{1}{3}\left\{\cos\left(\frac{2\pi x}{a}\right) + \cos\left(\frac{2\pi y}{a}\right) + \cos\left(\frac{2\pi z}{a}\right)\right\}$$

6.5 Derive the free electron energy bands for a face-centred cubic lattice. Show that the reducible representation of O_h for the eight-fold degenerate state at $\mathbf{k} = 0$ is

$$\Gamma^{\text{red}}(R) = \Gamma_1 \oplus \Gamma_2' \oplus \Gamma_{15} \oplus \Gamma_{25}'$$

Show that the group of the point $(p, q, r) = (0, 1, \frac{1}{2})$ contains eight operations divided into five classes namely E, C_4^2, C_2, JC_4^2, and JC_4. Determine, without reference to a character table, the probable dimensions of the irreducible representations of this group.

6.6 Show the group of the Σ axis of Fig. 6.3 is E, C_2, JC_4^2, JC_2 and establish that this group is simply C_{2v}.

Plot the energy bands in the Γ–M direction and show that three bands emerge from Γ for $\xi = 1$ and two emerge from M for $\xi = 0{\cdot}5$. Determine the nature of the energy levels at M for $\xi = 0{\cdot}5$.

7 Molecular vibrations and normal modes

7.0 Introduction

If a system is capable of sustaining vibrations then it is clearly important to be able to discover their nature because they govern many physical properties such as the emission and absorption spectra of molecules. We do this by determining the number of independent frequencies the system can possess and the configuration assumed by the system when it is vibrating.

Vibrating systems are studied in many branches of physics and engineering; however there is an essential similarity between the way in which we analyse a vibrating molecule, say, and a vibrating macroscopic mechanical system such as a double pendulum. Thus although this chapter is devoted to an analysis of the vibrational modes of some simple molecules we shall (since we are only seeking the allowed frequencies of vibration) use classical models, i.e., a molecule will be treated as a set of mass points which interact as if connected by springs.

The determination of the allowed frequencies of a system is clearly an eigenvalue problem. In fact the allowed frequencies are eigenfrequencies which are associated with eigenvectors, which are particular displacements of the vibrating unit. We encountered an eigenvalue problem in the last chapter where we discovered that, because of the symmetry of the problem, the application of group theory enabled us to decide, at the outset, exactly how many distinct eigenvalues there were together with their degree of degeneracy. Obviously a vibrating system must possess some symmetry so that once again group theory will allow us to determine, before the task of solving the problem in any detail even begins, *exactly* how many eigenfrequencies there are together with their degeneracy. However, before we come to the application of group theory to vibrating systems there are a number of concepts we wish to clarify, and it is of value to examine, to some extent, the more long-winded approach. Actually the analysis of the vibrations of a system reduces to the problem of finding the *normal modes* of vibration so we will therefore begin by considering what is meant by this term.

7.1 Normal modes

Let us consider a spring hanging vertically with a weight attached at the bottom. If we pull the weight downwards through a small distance and release it, then (provided Hooke's law obtains, i.e., restoring force proportional to displacement) it will oscillate in simple harmonic motion with a frequency characteristic of the spring. We can also imagine performing the less usual procedure of pulling the weight sideways, without increasing the length of the spring, in which case it will oscillate as a simple pendulum. Each of these two oscillations can exist independently of the other and in both cases the restoring force is directed towards the equilibrium position; but if the weight is displaced simultaneously outwards and downwards, then the restoring force will not be directed towards the equilibrium position and the weight will describe a path in space, known as a Lissajous' figure, that is a combination of the two simple harmonic motions in the vertical and horizontal directions. The general motion of the vibrating system is *not* a simple harmonic motion but is resolvable into two independent simple harmonic motions known as the *normal modes* of vibration, and their frequencies are called the *normal frequencies*.

In this particular example the resolution into normal modes is obvious by inspection, but in the case of more complicated vibrations we must devise a rigorous way of analysing a given motion into a set of normal modes in each of which the system is performing a simple harmonic motion.

7.2 Vibrating systems of point masses

Suppose a system, consisting of N interconnected particles without *external* constraints, possesses an equilibrium configuration specified by the $3N$ Cartesian coordinates X_i^0 $(1 \leqslant i \leqslant 3N)$ of the particle positions, where the origin is considered to be at the centre of mass and $X_1^0 X_2^0 X_3^0$ are the $X^0 Y^0 Z^0$ coordinates of particle number 1, $X_4^0 X_5^0 X_6^0$ are the $X^0 Y^0 Z^0$ coordinates of particle number 2, and so on. In a general displacement all the particles move in such a way that a particle occupies a new position with coordinates $X_i = X_i^0 + x_i$ where x_i is a *small* displacement of the particle from its equilibrium position. Such a displacement can be written as a $3N$-dimensional vector $\mathbf{r}$ defined in terms of $3N$ basis vectors $\mathbf{x}_i$ pointing along the directions of the three Cartesian coordinate axes where x_i, are, of course, measured from an origin centred on the ith particle. This $3N$-dimensional vector is

$$\mathbf{r} = \sum_i x_i \mathbf{x}_i \tag{7.1}$$

The total energy E in the non-equilibrium configuration of the particles depends on the coordinates x_i and, written in terms of the kinetic energy T and potential energy V, is

$$\begin{aligned} E &= T + V \\ &= \sum_i \tfrac{1}{2} m_i \dot{x}_i^2 + V \end{aligned} \tag{7.2}$$

where $\dot{x}_i = \mathrm{d}x_i/\mathrm{d}t$, t is the time, and m_i is the mass of the particle. N.B. although the sum is from 1 to $3N$ there are, of course, only N masses.

The potential energy V of the system, in terms of x_i, can always be expressed as a Maclaurin expansion about the equilibrium position of the particles. Now, since we are only going to be interested in small vibrations, we can neglect all terms of this expansion above the second power in x_i. In so doing we adopt what is known as the harmonic approximation. The expansion to second order is

$$V = V_0 + \sum_i (\partial V/\partial x_i)_0 x_i + \tfrac{1}{2} \sum_i \sum_j (\partial^2 V/\partial x_i\, \partial x_j)_0 x_i x_j \tag{7.3}$$

where the suffix 0 indicates the equilibrium position. Obviously since V_0 is the potential energy of the equilibrium configuration it may be arbitrarily taken to be zero. Also because V is by definition a minimum for an equilibrium configuration all the derivatives $(\partial V/\partial x_i)_0$ must be zero, hence

$$\begin{aligned} V &= \tfrac{1}{2} \sum_i \sum_j (\partial^2 V/\partial x_i\, \partial x_j)_0 x_i x_j \\ &= \tfrac{1}{2} \sum_i \sum_j k_{ij} x_i x_j \end{aligned} \tag{7.4}$$

where

$$k_{ij} = (\partial^2 V/\partial x_i \partial x_j)_0 = k_{ji} \tag{7.5}$$

We should observe that V is independent of the origin so that we are at liberty to place the origin at any convenient position.

The expressions for the kinetic and potential energies can now be written as

$$2T = \langle \dot{x} | \mathbf{M} | \dot{x} \rangle \tag{7.6}$$

and

$$2V = \langle x | \mathbf{K} | x \rangle \tag{7.7}$$

where

$$\langle x| = \langle x_1, x_2, \ldots, x_{3N}| \tag{7.8a}$$

and

$$\langle \dot{x}| = \langle \dot{x}_1, \dot{x}_2, \ldots, \dot{x}_{3N}| \tag{7.8b}$$

$\mathbf{M}$ is a diagonal matrix with m_i along the leading diagonal and $\mathbf{K}$ is the *symmetric* matrix of the elements k_{ij} (cf. eq. (7.5)).

The procedure for finding the normal modes and vibrational frequencies of the system basically involves making a change from the Cartesian basis $\mathbf{x}_i$ to a new basis $\mathbf{q}_i$ via a real transformation matrix $\tilde{\boldsymbol{\alpha}}$, i.e.,

$$\begin{bmatrix} \mathbf{q}_1 \\ \mathbf{q}_2 \\ \vdots \\ \mathbf{q}_{3N} \end{bmatrix} = \tilde{\boldsymbol{\alpha}} \begin{bmatrix} \mathbf{x}_1 \\ \mathbf{x}_2 \\ \vdots \\ \mathbf{x}_{3N} \end{bmatrix} \tag{7.9}$$

The *coordinates* q_i in the new basis therefore transform as

$$|q\rangle = \boldsymbol{\alpha}^{-1}|x\rangle \tag{7.10}$$

(cf. Appendix A.11) so that

$$|x\rangle = \boldsymbol{\alpha}|q\rangle \tag{7.11}$$

In this new basis eq. (7.1) becomes $\mathbf{r} = \sum_i q_i \mathbf{q}_i$ and the expressions (7.6) and (7.7) for the kinetic and potential energies become

$$2T = \langle \dot{q}|\tilde{\boldsymbol{\alpha}}\mathbf{M}\boldsymbol{\alpha}|\dot{q}\rangle \tag{7.12}$$

$$2V = \langle q|\tilde{\boldsymbol{\alpha}}\mathbf{K}\boldsymbol{\alpha}|q\rangle \tag{7.13}$$

The essential point now is that, as shown in the Appendix A.20, it is always possible to find a non-singular transformation matrix $\boldsymbol{\alpha}$ which simultaneously transforms $\mathbf{M}$ to a unit matrix and $\mathbf{K}$ to a diagonal form, i.e., we can always find a matrix $\boldsymbol{\alpha}$ such that

$$\tilde{\boldsymbol{\alpha}}\mathbf{M}\boldsymbol{\alpha} = \mathbf{I} \tag{7.14}$$

$$\tilde{\boldsymbol{\alpha}}\mathbf{K}\boldsymbol{\alpha} = \mathbf{L} \tag{7.15}$$

where $\mathbf{I}$ is the unit matrix and $\mathbf{L}$ is a diagonal matrix with diagonal elements equal to λ_i. Thus using eq. (7.14) and (7.15) eqs. (7.6) and (7.7) become

$$2T = \langle \dot{q}|\dot{q}\rangle \tag{7.16}$$

$$2V = \langle q|\mathbf{L}|q\rangle \tag{7.17}$$

which if written out in full are

$$T = \tfrac{1}{2}\sum_i \dot{q}_i^2 \tag{7.18}$$

$$V = \tfrac{1}{2}\sum_i \lambda_i q_i^2 \tag{7.19}$$

The important feature of these equations is that we have transformed away the cross product terms of V, showing that the total energy E can be written as

$$E = \tfrac{1}{2}\sum_i (\dot{q}_i^2 + \lambda_i q_i^2) \tag{7.20}$$

The total energy is however a constant which allows us to write

$$\frac{dE}{dt} = \sum_i \dot{q}_i(\ddot{q}_i + \lambda_i q_i) = 0 \tag{7.21}$$

and hence

$$\ddot{q}_i + \lambda_i q_i = 0 \tag{7.22}$$

Equation (7.22) is, of course, the differential equation describing a simple harmonic motion with an oscillation frequency $\nu_i = \lambda_i^{1/2}/2\pi$. The advantage of the above procedure really rests on the removal of the cross terms from the expression for V and in doing this we have broken the actual system down into a set of $3N$ elementary modes of vibration which are all independent of each other. These elementary, independent, simple harmonic motions with frequencies ν_i are called the *normal modes*, whilst the coordinates q_i are called the *normal coordinates*. Two or more modes with the *same* frequency are still independent. However, the normal coordinates q_i of such modes can no longer be uniquely constructed. For example if a system has frequencies $\nu_1 = \nu_2$ and normal coordinates q_1 and q_2 we could also have configurations $q_1 + q_2$ and $q_1 - q_2$; indeed any linearly independent combinations are acceptable. The vectors $\mathbf{q}_i$ form the *normal basis* and each basis vector will yield the configuration of the particles when vibrating in the ith normal mode. The word normal arises because we are in fact, when using the basis vectors $\mathbf{q}_i$, working in a multi-dimensional space in which each dimension is associated with one of the set $\mathbf{q}_i$; these basis vectors are *normal* to each other, just as are the directions $\mathbf{x}$, $\mathbf{y}$, and $\mathbf{z}$ in an ordinary Cartesian three-dimensional space. In general a system executing small amplitude vibrations will not be in a normal mode but its motion can always be regarded as some linear combination of normal modes, and also since these modes *are independent* a system initially excited into a normal mode must always remain in that mode.

Although we have, in principle, solved the problem of determining the normal modes we have still to calculate $\boldsymbol{\alpha}$ and hence λ_i. However, as we shall now see, we can develop an equation for λ_i without a knowledge of $\boldsymbol{\alpha}$ although naturally we will still need $\boldsymbol{\alpha}$ in order to find $\mathbf{q}_i$. We can in fact determine λ_i from eqs. (7.14) and (7.15) by noting that they can be re-written as

$$\tilde{\boldsymbol{\alpha}}\mathbf{M}\boldsymbol{\alpha} = \mathrm{diag}\,(1) \tag{7.23}$$

$$\tilde{\boldsymbol{\alpha}}\mathbf{K}\boldsymbol{\alpha} = \mathrm{diag}\,(\lambda_i) \tag{7.24}$$

so that by introducing an arbitrary λ we obtain

$$\tilde{\boldsymbol{\alpha}}(\mathbf{K} - \lambda\mathbf{M})\boldsymbol{\alpha} = \text{diag}\,(\lambda_i - \lambda) \tag{7.25}$$

Now the right-hand side of eq. (7.25) has a determinant which vanishes when $\lambda = \lambda_i$ in which case we obtain (cf. eq. (A.14))

$$|\tilde{\boldsymbol{\alpha}}|\,|\mathbf{K} - \lambda\mathbf{M}|\,|\boldsymbol{\alpha}| = 0 \tag{7.26}$$

which shows that the λ_i are simply the roots of the determinantal equation (known as the secular equation)

$$|\mathbf{K} - \lambda\mathbf{M}| = 0 \tag{7.27}$$

From eq. (7.27) we can see, as anticipated, that the eigenfrequencies can be determined without having to carry out a diagonalization process involving $\boldsymbol{\alpha}$; however the determination of the normal coordinates q does require the construction of $\boldsymbol{\alpha}$ which can of course be obtained by following the procedures of the Appendix A.20. When the latter has been done we can find the new basis vectors $\mathbf{q}_i$ and, from eq. (7.10), the normal coordinates q_i. Actually we can get a picture of the relative motion of the particles from the relationship of $\mathbf{q}_i$ to $\mathbf{x}_i$ and the amplitudes of the vibration of a particle executing a normal mode can also be obtained from eq. (7.11) by setting all *except one* of the q_i equal to zero. For example for the first normal mode we have, from eq. (7.11)

$$\begin{bmatrix} x_1 \\ x_2 \\ \vdots \\ x_{3N} \end{bmatrix} = \boldsymbol{\alpha} \begin{bmatrix} q_1 \\ 0 \\ \vdots \\ 0 \end{bmatrix} \tag{7.28}$$

where $q_1 = A_1 \cos \sqrt{\lambda_1} t$, A_1 being the amplitude.

Finally, we should point out that although there are $3N$ normal modes, (one for each 'degree of freedom' of the system), we can, in general, identify six which have zero frequency; this is because they are not vibrational modes at all but are made up of three translational and three rotational motions. Such motions correspond to translations in the X, Y and Z directions, and rotations about these directions. Hence there are only $3N - 6$ vibrational normal modes (i.e., non-zero frequency normal modes). An exception to this occurs when all the particles lie on a straight line, in which case the rotational state is specified by only two variables. We can see this by first remembering that a rotating body requires two angles to fix the axis of rotation and a further angle to specify the angle of rotation about this axis. In the case of a linear system of point masses we can therefore see that the rotation about the line joining the masses produces no displacement of the masses and that such a rotation ceases to be a degree of freedom. Thus linear systems have $3N - 5$ vibrational normal modes.

7.3 Normal modes of molecules

Although the preceding sections contain material which is relevant to any type of vibrating system we will for the rest of this chapter restrict our attention to molecular vibrations, in which the molecules will be regarded as a set of point masses, held together with Hooke's law restoring forces.

(a) The diatomic molecule

In order to simplify this problem let us place atoms of mass m_1 and m_2 on the X-axis as shown in Fig. 7.1. In this case there are $N = 2$ particles,

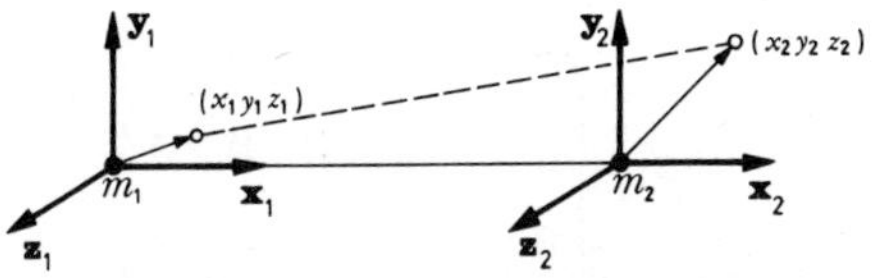

Fig. 7.1 *Displacement of the atoms in a diatomic molecule. ● equilibrium position of atoms. ○ position of displaced atoms.*

so that the system is linear and there is one vibrational normal mode (i.e., $(3N - 5) = 1$). The displacement coordinates x_i will have the values $x_1 y_1 z_1, x_2 y_2 z_2$ corresponding respectively to the two atoms. In a general displacement, involving translation and rotation, the molecule changes its equilibrium length which is

$$S_0 = (X_2^0 - X_1^0) \tag{7.29a}$$

to

$$S = [\{(X_2^0 + x_2) - (X_1^0 + x_1)\}^2 + (y_2 - y_1)^2 + (z_2 - z_1)^2]^{1/2} \tag{7.29b}$$

The length of the molecule, if we neglect squares of small quantities, therefore changes by an amount

$$S - S^0 = x_2 - x_1 \tag{7.30}$$

Now if the force constant of the force tending to restore the atoms to equilibrium is k, then the potential energy V is given by

$$2V = k(x_2 - x_1)^2 \tag{7.31a}$$

and the kinetic energy T is given by

$$2T = m_1(\dot{x}_1^2 + \dot{y}_1^2 + \dot{z}_1^2) + m_2(\dot{x}_2^2 + \dot{y}_2^2 + \dot{z}_2^2) \tag{7.31b}$$

Thus the matrices **K** and **M** of eqs. (7.6) and (7.7) are

$$\mathbf{K} = \begin{bmatrix} k & -k & 0 & 0 & 0 & 0 \\ -k & k & 0 & 0 & 0 & 0 \\ 0 & 0 & 0 & 0 & 0 & 0 \\ 0 & 0 & 0 & 0 & 0 & 0 \\ 0 & 0 & 0 & 0 & 0 & 0 \\ 0 & 0 & 0 & 0 & 0 & 0 \end{bmatrix} \tag{7.32}$$

and

$$\mathbf{M} = \begin{bmatrix} m_1 & 0 & 0 & 0 & 0 & 0 \\ 0 & m_2 & 0 & 0 & 0 & 0 \\ 0 & 0 & m_1 & 0 & 0 & 0 \\ 0 & 0 & 0 & m_2 & 0 & 0 \\ 0 & 0 & 0 & 0 & m_1 & 0 \\ 0 & 0 & 0 & 0 & 0 & m_2 \end{bmatrix} \tag{7.33}$$

where the variables are taken in the order x_1, x_2, y_1, y_2, z_1, and z_2. Hence the solution of eq. (7.27), namely the secular equation, is

$$\lambda = 0, 0, 0, 0, 0, \frac{k(m_1 + m_2)}{m_1 m_2} \tag{7.34}$$

Thus, as we expected, there are six normal modes with only one non-zero vibrational frequency whose value is

$$\nu = (1/2\pi)(k/\mu)^{1/2} \tag{7.35}$$

where μ is the reduced mass given by

$$\frac{1}{\mu} = \frac{1}{m_1} + \frac{1}{m_2}$$

Finally, as shown in the Appendix A.20, the diagonalizing matrix $\boldsymbol{\alpha}$ is given by

$$\boldsymbol{\alpha} = \begin{bmatrix} \frac{1}{(m_1+m_2)^{1/2}}\begin{bmatrix} \sqrt{\frac{m_2}{m_1}} & 1 \\ -\sqrt{\frac{m_1}{m_2}} & 1 \end{bmatrix} & \begin{matrix} 0 & 0 & 0 & 0 \\ 0 & 0 & 0 & 0 \end{matrix} \\ \begin{matrix} 0 & 0 \\ 0 & 0 \\ 0 & 0 \\ 0 & 0 \end{matrix} & \begin{matrix} 1/\sqrt{m_1} & 0 & 0 & 0 \\ 0 & 1/\sqrt{m_2} & 0 & 0 \\ 0 & 0 & 1/\sqrt{m_1} & 0 \\ 0 & 0 & 0 & 1/\sqrt{m_2} \end{matrix} \end{bmatrix} \tag{7.36}$$

so that from eq. (7.10) the normal coordinates which satisfy eqs. (7.18) and (7.19) are given by

$$\begin{aligned} q_1 &= \left(\frac{m_1 m_2}{m_1+m_2}\right)^{1/2}(x_1 - x_2) & \lambda_1 &= k/\mu \\ q_2 &= \frac{1}{(m_1+m_2)^{1/2}}(m_1 x_1 + m_2 x_2) & \lambda_2 &= 0 \\ q_3 &= (m_1)^{1/2} y_1 & \lambda_3 &= 0 \\ q_4 &= (m_2)^{1/2} y_2 & \lambda_4 &= 0 \\ q_5 &= (m_1)^{1/2} z_1 & \lambda_5 &= 0 \\ q_6 &= (m_2)^{1/2} z_2 & \lambda_6 &= 0 \end{aligned} \tag{7.37}$$

The actual motion of the atoms in the first of these normal modes can be found by setting $q_1 = a\cos(\sqrt{\lambda_1}t)$ (i.e., a solution of eq. (7.22)) and all the other $q_i = 0$ (cf. eq. (7.28)), and then solving for the nine values of x, y, and z. The result is $x_1 = Fm_2 a\cos(\sqrt{\lambda_1}t)$, $x_2 = -Fm_1 a\cos(\sqrt{\lambda_1}t)$ where

$$F = \left\{\frac{1}{m_1 m_2(m_1+m_2)}\right\}^{1/2}$$

and obviously the other coordinates are zero. We can obtain the motion of the atoms in the other normal modes by a similar method and the first three modes are shown in Fig. 7.2. We should point out that the mode q_3 can be thought of as a linear combination of a translation and a rotation and is in fact a linear combination of two *equal* frequency normal modes. Indeed we have a five-fold degenerate zero eigenfrequency; the degeneracy tells us that the choice of q_i is not unique and the fact that the frequency is zero tells us that the system is *not* vibrating in these modes, hence they

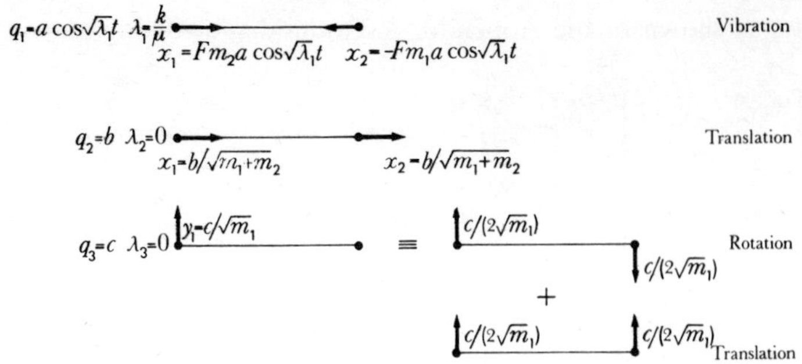

Fig. 7.2 *The first three normal modes of a diatomic molecule. a, b, and c are the normal mode amplitudes. The mode q_3 is a linear combination of translation and rotation.*

must correspond to translations, rotations, or linear combinations of them.

(b) The triatomic molecule of the water type

This molecule, illustrated in Fig. 7.3 consists of an atom of mass M at the apex (semi-angle ϕ) of an isosceles triangle and two atoms of equal masses m at the other corners. The force constant between the unlike atoms will

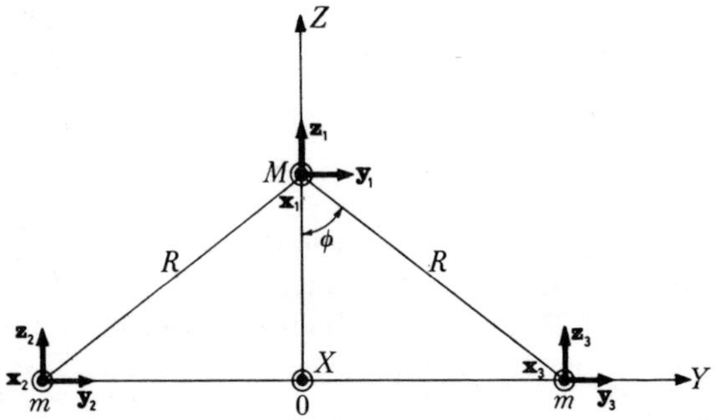

Fig. 7.3 *A triatomic molecule of the water type. The coordinates of the atoms are measured in the XYZ coordinate system centred on O. x_i, y_i, z_i are measured from the equilibrium positions of the atoms. The X and x_i axes are perpendicular to the plane of the triangle and are denoted by ⊙.*

be taken as k_1 and that between the like atoms as k_2. From Fig. 7.3 we can see that the coordinates $X^0Y^0Z^0$ for the equilibrium situation are:

	X^0	Y^0	Z^0
Atom 1	0	0	$R\cos\phi$
Atom 2	0	$-R\sin\phi$	0
Atom 3	0	$R\sin\phi$	0

After a general displacement the new coordinates will be:

	X	Y	Z
Atom 1	x_1	y_1	$R\cos\phi + z_1$
Atom 2	x_2	$-R\sin\phi + y_2$	z_2
Atom 3	x_3	$R\sin\phi + y_3$	z_3

Hence the changes in length of the bonds can now be calculated. For example, the distance R_{12} between atoms 1 and 2 in the perturbed position is

$$R_{12} = [(x_1 - x_2)^2 + (-R\sin\phi + y_2 - y_1)^2 + (z_2 - R\cos\phi - z_1)^2]^{1/2}$$

and when we ignore terms of second order, we find that

$$R_{12} = [R^2 + 2R\sin\phi(y_1 - y_2) + 2R\cos\phi(z_1 - z_2)]^{1/2}$$

Thus, by putting

$$R_{12} = R_{12}^0 + \Delta R_{12}$$

where R_{12}^0 is the equilibrium interatomic bond length and ΔR_{12} is the change in this length, we find that

$$\Delta R_{12} = (y_1 - y_2)\sin\phi + (z_1 - z_2)\cos\phi \tag{7.38}$$

and similarly

$$\Delta R_{23} = (y_2 - y_3) \tag{7.39}$$

$$\Delta R_{31} = (y_3 - y_1)\sin\phi - (z_3 - z_1)\cos\phi \tag{7.40}$$

The potential energy V is therefore

$$2V = k_1 \Delta R_{12}^2 + k_1 \Delta R_{31}^2 + k_2 \Delta R_{23}^2 \tag{7.41}$$

so that using eqs. (7.38), (7.39), and (7.40) we can write,

$$2V = \langle x|\mathbf{K}|x\rangle$$

where the matrix $\mathbf{K}$ is

$$\mathbf{K} = k_1 \begin{bmatrix} 0 & 0 & 0 & 0 & 0 & 0 & 0 & 0 & 0 \\ 0 & 0 & 0 & 0 & 0 & 0 & 0 & 0 & 0 \\ 0 & 0 & 0 & 0 & 0 & 0 & 0 & 0 & 0 \\ 0 & 0 & 0 & 2s^2 & -s^2 & -s^2 & 0 & -sc & sc \\ 0 & 0 & 0 & -s^2 & s^2+\kappa & -\kappa & -sc & sc & 0 \\ 0 & 0 & 0 & -s^2 & -\kappa & s^2+\kappa & sc & 0 & -sc \\ 0 & 0 & 0 & 0 & -sc & sc & 2c^2 & -c^2 & -c^2 \\ 0 & 0 & 0 & -sc & sc & 0 & -c^2 & c^2 & 0 \\ 0 & 0 & 0 & sc & 0 & -sc & -c^2 & 0 & c^2 \end{bmatrix} \tag{7.42}$$

in which $s = \sin\phi$, $c = \cos\phi$, $\kappa = k_2/k_1$, and the coordinates are taken in the order $x_1, x_2, x_3, y_1, y_2, y_3, z_1, z_2, z_3$. The kinetic energy of the system is

$$2T = \langle \dot{x}|\mathbf{M}|\dot{x}\rangle \tag{7.43}$$

where obviously the matrix $\mathbf{M}$ is

$$\mathbf{M} = \begin{bmatrix} M & 0 & 0 & 0 & 0 & 0 & 0 & 0 & 0 \\ 0 & m & 0 & 0 & 0 & 0 & 0 & 0 & 0 \\ 0 & 0 & m & 0 & 0 & 0 & 0 & 0 & 0 \\ 0 & 0 & 0 & M & 0 & 0 & 0 & 0 & 0 \\ 0 & 0 & 0 & 0 & m & 0 & 0 & 0 & 0 \\ 0 & 0 & 0 & 0 & 0 & m & 0 & 0 & 0 \\ 0 & 0 & 0 & 0 & 0 & 0 & M & 0 & 0 \\ 0 & 0 & 0 & 0 & 0 & 0 & 0 & m & 0 \\ 0 & 0 & 0 & 0 & 0 & 0 & 0 & 0 & m \end{bmatrix} \tag{7.44}$$

In order to determine the eigenfrequencies we must solve the secular eq. (7.27) whose solution is a straightforward but laborious matter. We therefore simply quote the results, which are

$$\lambda = 0, 0, 0, 0, 0, 0, k_1(M + 2ms^2)/mM \tag{7.45}$$

together with the roots of the quadratic equation:

$$m^2\lambda^2 - k_1 m(1 + 2k_2/k_1 + 2mc^2/M)\lambda + 2k_1k_2c^2(1 + 2m/M) = 0 \tag{7.46}$$

The diagonalizing matrix $\boldsymbol{\alpha}$ and the normal coordinates can be obtained by the methods discussed earlier. However, we shall not pursue this any further because we are primarily interested in the role of group theory for which, as we shall see, we only require $\mathbf{K}$ and $\mathbf{M}$.

7.4 Exploitation of symmetry

In the above examples we have determined eigenfrequencies and have indicated how to determine normal mode configurations. However, the labour involved in a normal mode calculation is, for many systems, almost prohibitive. On the other hand most molecules possess enough symmetry to enable us to exploit the machinery of group theory which not only reduces the amount of work involved in the calculation but also furnishes us with an insight into the nature of the vibrations.

The symmetry we refer to above is that of the equilibrium configuration of the molecule, but at any moment in time in the vibrating state, the molecule is in a distorted configuration. A symmetry operation is one which leaves the *undistorted* molecule indistinguishable from its previous orientation. Such an operation interchanges *equivalent* atoms. However, when the symmetry operation is performed on the distorted molecule the effect will be the same as that obtained by interchanging displacement vectors amongst equivalent atoms. We will therefore define the action of a symmetry operation in the following way: each atom in a distorted molecule is considered to be displaced through vectors $\mathbf{x}_i$ from its equilibrium position and when a symmetry operation is applied we imagine the atoms to remain where they are and the vector $\mathbf{x}_3$ of atom A_3, in Fig. 7.4

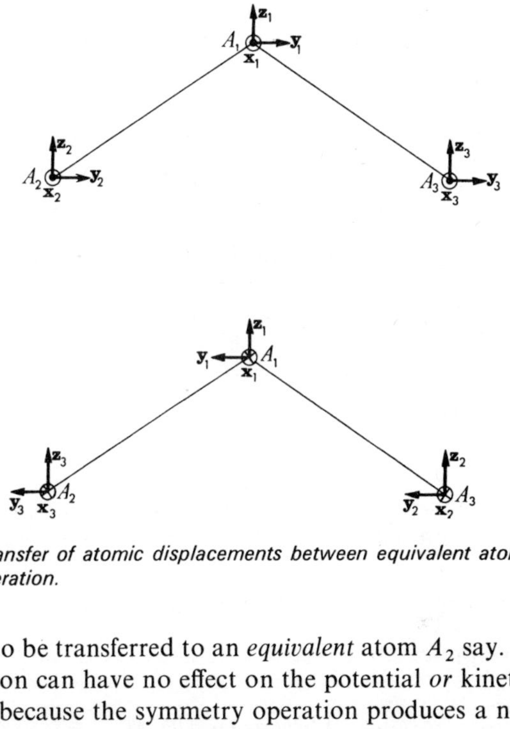

Fig. 7.4 *Transfer of atomic displacements between equivalent atoms due to a symmetry operation.*

for example, to be transferred to an *equivalent* atom A_2 say. Now a symmetry operation can have no effect on the potential *or* kinetic energy of the molecule because the symmetry operation produces a new situation which is physically equivalent to the old one, i.e., the distances between the atoms and the angles between the bonds are preserved and the quadratic forms of T *and* V will remain unchanged.

Now that we have seen that the symmetry of the equilibrium form of the molecule can be exploited we will consider exactly how group theory will help us to determine and classify the normal modes. We begin by asserting that the $3N$ Cartesian basis vectors $\mathbf{x}_i$, with reference to which the matrices $\mathbf{K}$ and $\mathbf{M}$ are determined, form a basis for a reducible $3N$-dimensional representation of the group of symmetry operations of the undistorted molecule; this will be referred to as the Cartesian representation. Hence by reducing this representation in the manner of section 2.8 we can determine the irreducible representations to which the $3N$ translation, rotational, and vibrational modes belong. Also we can immediately find the degeneracy of each normal mode. In addition to the above we can by using the projection operators (section 3.3) find those linear combinations of the coordinate bases $\mathbf{x}_i$ which will form bases for the irreducible representations; in fact the projection operators will project out of the Cartesian basis $\mathbf{x}$ a symmetrized basis $\mathbf{s}$ which will bring the energy matrices $\mathbf{K}$ and $\mathbf{M}$ into block form and thus greatly simplify the solution of the secular equation.

If we begin with the unsymmetrized basis $\mathbf{x}$ then a symmetry operation R whose representation is $\mathbf{\Gamma}(R)$ transforms the *coordinates* x_i to new coordinates x_i' by the equations (cf. eq. (A.131))

$$|x\rangle = \mathbf{\Gamma}(R)|x'\rangle \tag{7.47}$$

or, remembering that $\mathbf{\Gamma}(R)$ can be complex (cf. section (2.4))

$$\langle x| = \langle x'|\mathbf{\Gamma}^{\dagger}(R) \tag{7.48}$$

The potential energy of the system, for the operations we are considering, is, of course, invariant. This invariance of the potential energy is expressed by

$$2V = \langle x|\mathbf{K}|x\rangle = \langle x'|\mathbf{K}'|x'\rangle \tag{7.49}$$

where $\mathbf{K}'$ is the potential energy matrix in the new basis. Thus, using eqs. (7.47) and (7.48) we obtain

$$\mathbf{K}' = \mathbf{\Gamma}^{\dagger}(R)\mathbf{K}\mathbf{\Gamma}(R) \tag{7.50}$$

However, $\mathbf{K}$ possesses the symmetry of the equilibrium state of the molecule so that we can *for a symmetry operation* assert that $\mathbf{K}' = \mathbf{K}$ and therefore

$$\mathbf{K} = \mathbf{\Gamma}^{\dagger}(R)\mathbf{K}\mathbf{\Gamma}(R) \tag{7.51}$$

Now since all representations can be assumed to be unitary (cf. section 2.2) we can write

$$\mathbf{\Gamma}^{\dagger}(R) = \mathbf{\Gamma}^{-1}(R)$$

and hence eq. (7.51) implies that

$$\mathbf{K}\mathbf{\Gamma}(R) = \mathbf{\Gamma}(R)\mathbf{K} \tag{7.52a}$$

Similar arguments apply to the kinetic energy matrix so that

$$\mathbf{M}\mathbf{\Gamma}(R) = \mathbf{\Gamma}(R)\mathbf{M} \tag{7.52b}$$

If we now produce the symmetrized basis $\mathbf{s}$ from the Cartesian basis $\mathbf{x}$ by means of the projection operators O^j we, in effect, perform the basis transformation

$$[\mathbf{s}_1, \mathbf{s}_2, \ldots, \mathbf{s}_{3N}] = [\mathbf{x}_1, \mathbf{x}_2, \ldots, \mathbf{x}_{3N}]\boldsymbol{\beta} \tag{7.53}$$

and the coordinate transformation

$$|s\rangle = \boldsymbol{\beta}^{-1}|x\rangle \tag{7.54}$$

Hence the potential energy (eq. (7.49)), when referred to the symmetrized basis **s**, is given by the expression

$$2V = \langle s|\boldsymbol{\beta}^{\dagger}\mathbf{K}\boldsymbol{\beta}|s\rangle \tag{7.55}$$

which we can write as

$$2V = \langle s|\mathbf{K}''|s\rangle \tag{7.56}$$

where

$$\mathbf{K}'' = \boldsymbol{\beta}^{\dagger}\mathbf{K}\boldsymbol{\beta} \tag{7.57}$$

Now in this basis, although we cannot say that $\mathbf{K}'' = \mathbf{K}$, we can say that

$$\mathbf{K}''\boldsymbol{\Gamma}^{\text{red}}(R) = \boldsymbol{\Gamma}^{\text{red}}(R)\mathbf{K}'' \tag{7.58}$$

because, firstly eq. (7.52) is true in any basis, and secondly the representations $\boldsymbol{\Gamma}(R)$ in the symmetrized basis **s** must be in the block form $\boldsymbol{\Gamma}^{\text{red}}(R)$ (cf. section 3.5).

A direct consequence of eq. (7.58) is that $\mathbf{K}''$ *must also be in block form.* Although we are not going to prove this statement the essential steps are as follows. Suppose $\boldsymbol{\Gamma}^{\text{red}}(R)$ contains two irreducible representations $\boldsymbol{\Gamma}^1(R)$ and $\boldsymbol{\Gamma}^2(R)$, the former once and the latter twice. Then

$$\boldsymbol{\Gamma}^{\text{red}}(R) = \begin{bmatrix} \boldsymbol{\Gamma}^1 & 0 & 0 \\ 0 & \boldsymbol{\Gamma}^2 & 0 \\ 0 & 0 & \boldsymbol{\Gamma}^2 \end{bmatrix} \tag{7.59}$$

If we now partition the matrix $\mathbf{K}''$ into submatrices $\mathbf{K}''_{ij}$ so that

$$\mathbf{K}'' = \begin{bmatrix} \mathbf{K}''_{11} & \mathbf{K}''_{12} & \mathbf{K}''_{13} \\ \mathbf{K}''_{21} & \mathbf{K}''_{22} & \mathbf{K}''_{23} \\ \mathbf{K}''_{31} & \mathbf{K}''_{32} & \mathbf{K}''_{33} \end{bmatrix} \tag{7.60}$$

where $\mathbf{K}''_{11}$ is of the same order as $\boldsymbol{\Gamma}^1(R)$ and $\mathbf{K}''_{22}, \mathbf{K}''_{33}$ are of the same order as $\boldsymbol{\Gamma}^2(R)$, then eq. (7.58) gives us equations such as

$$\boldsymbol{\Gamma}^m(R)\mathbf{K}''_{ij} = \mathbf{K}''_{ij}\boldsymbol{\Gamma}^n(R) \tag{7.61}$$

If $\boldsymbol{\Gamma}^m(R)$ and $\boldsymbol{\Gamma}^n(R)$ are non-equivalent irreducible representations then it can be shown that $\mathbf{K}''_{ij}$ is a null matrix, whilst if $\boldsymbol{\Gamma}^m(R)$ and $\boldsymbol{\Gamma}^n(R)$ are equivalent irreducible representations then $\mathbf{K}''_{ij}$ is a constant times the unit matrix. As a consequence $\mathbf{K}''$ can be partitioned into a set of matrices which are either null matrices or constant matrices, for example, if $\boldsymbol{\Gamma}^1(R)$ were one-dimensional and $\boldsymbol{\Gamma}^2(R)$ were two-dimensional, then $\mathbf{K}''$ would have the block form:

$$\mathbf{K}'' = \left[\begin{array}{c|cc|cc} a & 0 & 0 & 0 & 0 \\ \hline 0 & b & 0 & c & 0 \\ 0 & 0 & b & 0 & c \\ \hline 0 & d & 0 & e & 0 \\ 0 & 0 & d & 0 & e \end{array}\right] \tag{7.62}$$

It can then be seen that $\mathbf{K}''$ can be brought to simpler form corresponding to each irreducible representation by suitably transposing the rows and columns by a unitary similarity transform. In fact in the above example, interchange of the third and fourth rows, followed by interchange of the third and fourth columns brings it to the form:

$$\mathbf{K}'' = \left[\begin{array}{c|cc|cc} a & 0 & 0 & 0 & 0 \\ \hline 0 & b & c & 0 & 0 \\ 0 & d & e & 0 & 0 \\ \hline 0 & 0 & 0 & b & c \\ 0 & 0 & 0 & d & e \end{array}\right] \tag{7.63}$$

Hence $\mathbf{K}''$ is in block form. Thus any matrix such as $\boldsymbol{\beta}$ which symmetrizes the basis must bring the matrix $\mathbf{K}$ into block form. Similar arguments apply to the matrix $\mathbf{M}$. We should observe that the use of the *symmetrized* basis $\mathbf{s}$ only brings $\mathbf{K}$ and $\mathbf{M}$ into *block* form, and that a further transformation is necessary to obtain the *normal* basis $\mathbf{q}$ in which $\mathbf{K}$ and $\mathbf{M}$ are in diagonal form.

7.5 Molecules of the water type

The results of the above section will now be illustrated by considering once again the vibrations of a non-linear triatomic molecule of the water type. We will therefore consider:

(a) The reduction of the Cartesian representation

We can see from Fig. 7.5 that this system belongs to the group C_{2v} which contains the symmetry operations E, C_2, σ'_v, and σ''_v. By applying these operations in turn to the nine Cartesian basis vectors $\mathbf{x}_1, \mathbf{x}_2, \mathbf{x}_3, \mathbf{y}_1, \mathbf{y}_2,$

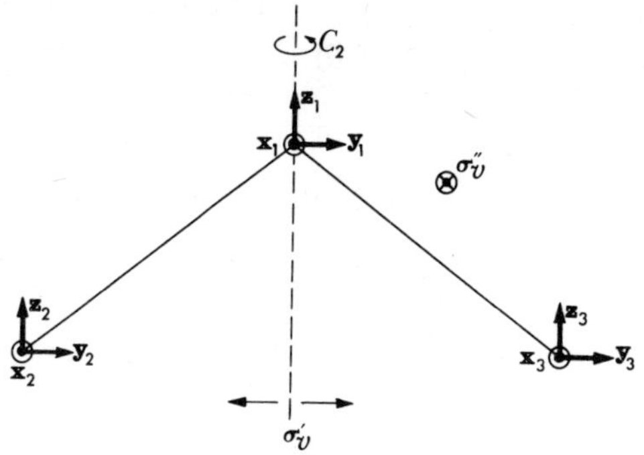

Fig. 7.5 *Symmetry elements of the group C_{2v}, i.e., the symmetry group of the water molecule.*

$\mathbf{y}_3, \mathbf{z}_1, \mathbf{z}_2$, and $\mathbf{z}_3$ (where we now identify the displacement of the atoms with coordinates $x_i y_i z_i$ respectively) we obtain their nine-dimensional reducible representations. For example under the operation C_2 we see (Fig. 7.5) that

$$(\mathbf{x}_1 \quad \mathbf{y}_1 \quad \mathbf{z}_1) \text{ becomes } (\mathbf{x}_1' \quad \mathbf{y}_1' \quad \mathbf{z}_1') \equiv (-\mathbf{x}_1 \quad -\mathbf{y}_1 \quad \mathbf{z}_1)$$
$$(\mathbf{x}_2 \quad \mathbf{y}_2 \quad \mathbf{z}_2) \text{ becomes } (\mathbf{x}_2' \quad \mathbf{y}_2' \quad \mathbf{z}_2') \equiv (-\mathbf{x}_3 \quad -\mathbf{y}_3 \quad \mathbf{z}_3)$$
$$(\mathbf{x}_3 \quad \mathbf{y}_3 \quad \mathbf{z}_3) \text{ becomes } (\mathbf{x}_3' \quad \mathbf{y}_3' \quad \mathbf{z}_3') \equiv (-\mathbf{x}_2 \quad -\mathbf{y}_2 \quad \mathbf{z}_2)$$

This transformation can therefore be written as

$$\begin{bmatrix} \mathbf{x}_1' \\ \mathbf{x}_2' \\ \mathbf{x}_3' \\ \mathbf{y}_1' \\ \mathbf{y}_2' \\ \mathbf{y}_3' \\ \mathbf{z}_1' \\ \mathbf{z}_2' \\ \mathbf{z}_3' \end{bmatrix} = \begin{bmatrix} -1 & 0 & 0 & 0 & 0 & 0 & 0 & 0 & 0 \\ 0 & 0 & -1 & 0 & 0 & 0 & 0 & 0 & 0 \\ 0 & -1 & 0 & 0 & 0 & 0 & 0 & 0 & 0 \\ 0 & 0 & 0 & -1 & 0 & 0 & 0 & 0 & 0 \\ 0 & 0 & 0 & 0 & 0 & -1 & 0 & 0 & 0 \\ 0 & 0 & 0 & 0 & -1 & 0 & 0 & 0 & 0 \\ 0 & 0 & 0 & 0 & 0 & 0 & 1 & 0 & 0 \\ 0 & 0 & 0 & 0 & 0 & 0 & 0 & 0 & 1 \\ 0 & 0 & 0 & 0 & 0 & 0 & 0 & 1 & 0 \end{bmatrix} \begin{bmatrix} \mathbf{x}_1 \\ \mathbf{x}_2 \\ \mathbf{x}_3 \\ \mathbf{y}_1 \\ \mathbf{y}_2 \\ \mathbf{y}_3 \\ \mathbf{z}_1 \\ \mathbf{z}_2 \\ \mathbf{z}_3 \end{bmatrix} \tag{7.64}$$

The above equation can therefore be written as

$$(\mathbf{x}') = \mathbf{\Gamma}^c(C_2)(\mathbf{x}) \tag{7.65}$$

where $\mathbf{\Gamma}^c(C_2)$ is the matrix representing the operation C_2 in the Cartesian representation. The character of this representation is therefore

$$\chi^c(C_2) = -1 \tag{7.66}$$

Similarly we can find

$$\chi^c(E) = 9 \tag{7.67}$$

$$\chi^c(\sigma_v') = 1 \tag{7.68}$$

$$\chi^c(\sigma_v'') = 3 \tag{7.69}$$

We can now use the character table of C_{2v} (table 7.1) and eq. (2.37) to determine the irreducible components of $\mathbf{\Gamma}^c(R)$. The equation we require is

$$a_i = \frac{1}{g}\sum_R \chi^{i*}(R)\chi^c(R) \tag{7.70}$$

Table 7.1 Character table of C_{2v} together with the characters of the reducible Cartesian representation and its translational, rotational and vibrational components N.B. we have dropped the superscript red). The right hand column shows the irreducible representations to which the translational vectors $\mathbf{T}_x$, $\mathbf{T}_y$, $\mathbf{T}_z$ and the rotational vectors $\mathbf{R}_x$, $\mathbf{R}_y$, $\mathbf{R}_z$ belong

	E	C_2	σ_v'	σ_v''		
A_1	1	1	1	1	T_z	
A_2	1	1	−1	−1		R_z
B_1	1	−1	1	−1	T_x	R_y
B_2	1	−1	−1	1	T_y	R_x
$\Gamma^c(R)$	9	−1	1	3		
$\Gamma^t(R)$	3	−1	1	1		
$\Gamma^r(R)$	3	−1	−1	−1		
$\Gamma^v(R)$	3	1	1	3		

where a_i is the number of times the ith irreducible representation appears in the reducible Cartesian representation, g is the order of the group, and $\chi^i(R)$ may be obtained from the character table. Thus we have

$$\begin{aligned} a_{A_1} &= \tfrac{1}{4}(9 - 1 + 1 + 3) = 3 \\ a_{A_2} &= \tfrac{1}{4}(9 - 1 - 1 - 3) = 1 \\ a_{B_1} &= \tfrac{1}{4}(9 + 1 + 1 - 3) = 2 \\ a_{B_2} &= \tfrac{1}{4}(9 + 1 - 1 + 3) = 3 \end{aligned} \tag{7.71}$$

The reduced Cartesian representation is therefore given by

$$\mathbf{\Gamma}^{c\,\mathrm{red}}(R) = 3\mathbf{A}_1 \oplus \mathbf{A}_2 \oplus 2\mathbf{B}_1 \oplus 3\mathbf{B}_2 \tag{7.72}$$

where we have labelled the irreducible representations in accordance with the notation introduced in section 2.7.

As we have discussed before, six of the nine normal modes are zero frequency modes and correspond to pure translations and pure rotations.

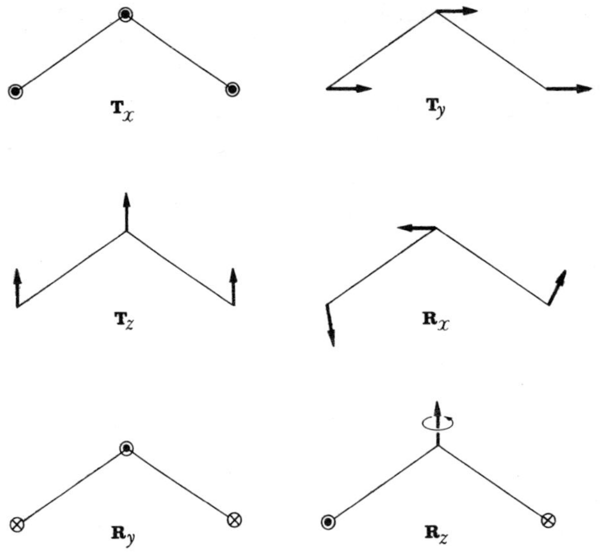

Fig. 7.6 *The three translational* ($\mathbf{T}_x$, $\mathbf{T}_y$, $\mathbf{T}_z$) *and the three rotational* ($\mathbf{R}_x$, $\mathbf{R}_y$, $\mathbf{R}_z$) *modes of the water molecule. The x-axis is perpendicular to the page.* ⊙ *denotes a vector coming out of the page.* ⊗ *denotes a vector going into the page.*

These are illustrated in Fig. 7.6. Since we are primarily interested in the vibrational modes of the molecule it is clearly of the utmost importance to identify which irreducible representations of eq. (7.72) are associated with translations and rotations. We will therefore write

$$\boldsymbol{\Gamma}^{c\,\mathrm{red}}(R) = \boldsymbol{\Gamma}^{t\,\mathrm{red}}(R) \oplus \boldsymbol{\Gamma}^{r\,\mathrm{red}}(R) \oplus \boldsymbol{\Gamma}^{v\,\mathrm{red}}(R) \tag{7.73}$$

where $\boldsymbol{\Gamma}^{t\,\mathrm{red}}(R)$, $\boldsymbol{\Gamma}^{r\,\mathrm{red}}(R)$, and $\boldsymbol{\Gamma}^{v\,\mathrm{red}}(R)$ are the reduced representations associated with pure translations, pure rotations, and pure vibrations respectively.

We can achieve this decomposition by first noting that an arbitrary translation can be written as a linear combination of independent translations $\mathbf{T}_x$, $\mathbf{T}_y$, and $\mathbf{T}_z$ along the X, Y, and Z axes. These linearly independent translations will transform in the same way as the basis vectors (which are polar vectors, cf. appendix A.24) of a three-dimensional space. Also if we choose the origin to be at the centre of mass then the rotations can be represented by axial vectors in this three-dimensional space which we must remember behave in the opposite manner to polar vectors under reflections or improper rotations (cf. appendix A.24 and section 5.0).

Now, we could find the irreducible representations to which the translations and rotations belong by looking them up in the character table (cf. appendix C) in which we would find that the translations and rotations make up $\mathbf{A}_1 \oplus \mathbf{A}_2 \oplus 2\mathbf{B}_1 \oplus 2\mathbf{B}_2$, and hence the reduced representation associated with pure vibrations is

$$\Gamma^{v\,\text{red}}(R) = 2\mathbf{A}_1 \oplus \mathbf{B}_2 \tag{7.74}$$

Alternatively we may obtain this result from first principles by erecting a coordinate system $X\ Y\ Z$ at the centre of mass in which we can represent the translations and rotations as shown in Fig. 7.7. The representation

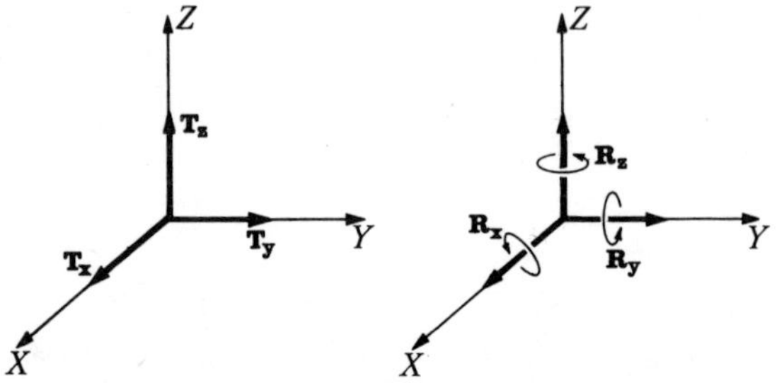

Fig. 7.7 *The representation of the three translations and the three rotations as vectors in an X Y Z coordinate system.*

of the operation C_2, for example, using the unit translation vectors $\mathbf{T}_x, \mathbf{T}_y, \mathbf{T}_z$ as a basis is $\Gamma^t(C_2)$ given by

$$C_2(\mathbf{T}_x, \mathbf{T}_y, \mathbf{T}_z) = [\mathbf{T}_x, \mathbf{T}_y, \mathbf{T}_z]\begin{bmatrix} -1 & 0 & 0 \\ 0 & -1 & 0 \\ 0 & 0 & 1 \end{bmatrix} = [\mathbf{T}_x, \mathbf{T}_y, \mathbf{T}_z]\Gamma^t(C_2) \tag{7.75}$$

The character of the representation is

$$\chi^t(C_2) = -1 \tag{7.76}$$

The rotations $\mathbf{R}_x$, $\mathbf{R}_y$, $\mathbf{R}_z$ will transform in the same way because C_2 does not involve improper rotations or reflections and will have the same character

$$\chi^r(C_2) = -1 \tag{7.77}$$

However, whereas under the operation σ'_v we have

$$\sigma'_v(\mathbf{T}_x, \mathbf{T}_y, \mathbf{T}_z) = [\mathbf{T}_x, \mathbf{T}_y, \mathbf{T}_z]\begin{bmatrix} 1 & 0 & 0 \\ 0 & -1 & 0 \\ 0 & 0 & 1 \end{bmatrix} \tag{7.78}$$

so that

$$\chi^t(\sigma'_v) = 1 \tag{7.79}$$

we find for the axial vectors that

$$\sigma_v'(\mathbf{R}_x, \mathbf{R}_y, \mathbf{R}_z) = [\mathbf{R}_x, \mathbf{R}_y, \mathbf{R}_z] \begin{bmatrix} -1 & 0 & 0 \\ 0 & 1 & 0 \\ 0 & 0 & -1 \end{bmatrix} \tag{7.80}$$

so that

$$\chi^r(\sigma_v') = -1 \tag{7.81}$$

Similarly for the operations, E and σ_v'' we find

$$X^t(E) = 3, \qquad X^r(E) = 3 \tag{7.82}$$

$$X^t(\sigma_v'') = 1, \qquad X^r(\sigma_v'') = -1 \tag{7.83}$$

These characters are shown in table 7.1 and by direct inspection, or by the application of eq. (2.37), it is clear that

$$\mathbf{\Gamma}^{t\,\mathrm{red}}(R) = \mathbf{A}_1 \oplus \mathbf{B}_1 \oplus \mathbf{B}_2 \tag{7.84}$$

$$\mathbf{\Gamma}^{r\,\mathrm{red}}(R) = \mathbf{A}_2 \oplus \mathbf{B}_1 \oplus \mathbf{B}_2 \tag{7.85}$$

Hence from eqs. (7.72), (7.84), (7.85) we reproduce the equation given earlier, viz.:

$$\mathbf{\Gamma}^{v\,\mathrm{red}}(R) = 2\mathbf{A}_1 \oplus \mathbf{B}_2$$

The importance of this result is that we can see that there are three vibrational modes which, because A_1 and B_2 are one-dimensional representations, are non-degenerate. From a quantum mechanical point of view, the molecule is a simple harmonic oscillator with an energy E given by

$$E = (v_1 + \tfrac{1}{2})h\nu_1 + (v_2 + \tfrac{1}{2})h\nu_2 + (v_3 + \tfrac{1}{2})h\nu_3 \tag{7.86}$$

where ν_1, ν_2, and ν_3 are the classical vibrational frequencies and v_1, v_2, and v_3 are integers, i.e., three quantum numbers. The knowledge of the irreducible representations to which the vibrations belong will enable us to classify the eigenfunctions and hence will allow us to determine the selection rules for transitions between these states; this will be discussed in chapter 8.

(b) Construction of a symmetrized basis

A symmetric basis consists of vectors which span sub-spaces corresponding to each irreducible component of the Cartesian reducible representation. This symmetric basis will contain for the water molecule nine linearly independent vectors, three of which belong to A_1, one to A_2, two to B_1, and three to B_2 (eq. (7.72)). We could find these vectors by using the projection operators O^j which project, from an arbitrary vector, that part belonging to its irreducible representation. The basis so produced is not in general what we would regard as a fully symmetrized basis (cf. section 3.5) because although the functions produced by O^j are orthogonal to each other when they belong to *different* irreducible representations, they are not necessarily orthogonal to each other when two or more belong to the same irreducible representation. However it is possible in the case of the water molecule to produce a fully symmetrized set of orthogonal functions for each representation from the set produced with the aid of O^j, because as all the irreducible representations are one-dimensional, O^j is the full projection operator.

We could, to begin with, without even using the projection operators, write down the three translation vectors:

$$\begin{aligned} \mathbf{s}_x &= \mathbf{x}_1 + \mathbf{x}_2 + \mathbf{x}_3 \\ \mathbf{s}_y &= \mathbf{y}_1 + \mathbf{y}_2 + \mathbf{y}_3 \\ \mathbf{s}_z &= \mathbf{z}_1 + \mathbf{z}_2 + \mathbf{z}_3 \end{aligned} \tag{7.87}$$

knowing that these must form a symmetrized basis for $\mathbf{\Gamma}^t(R)$. The complete symmetrized basis can however be obtained in a systematic way by successively applying the projection operators O^j of section 3.3 to *all* the basis vectors of the original unsymmetrized set. In the case of C_{2v} the general expression for the projection operator is

$$O^j = \tfrac{1}{4}[\chi^j(E)P_E + \chi^j(C_2)P_{C_2} + \chi^j(\sigma_v')P_{\sigma_v'} + \chi^j(\sigma_v'')P_{\sigma_v''}] \tag{7.88}$$

where $\chi^j(R)$ is the character of the operation R (i.e., E, C_2, σ_v', σ_v'') of the group in the representation $\mathbf{\Gamma}^j(R)$. If we apply each of the projection operators O^{A_1}, O^{A_2}, O^{B_1}, and O^{B_2} in turn to each of the Cartesian basis vectors $\mathbf{x}_i\mathbf{y}_i\mathbf{z}_i$ then we obtain the results given in table 7.2. For example, we can see that

$$\begin{aligned} O^{B_2}(\mathbf{z}_2) &= \tfrac{1}{4}[(1 \times \mathbf{z}_2) + (-1 \times \mathbf{z}_3) + (-1 \times \mathbf{z}_3) + (1 \times \mathbf{z}_2)] \\ &= \tfrac{1}{2}(\mathbf{z}_2 - \mathbf{z}_3) \end{aligned} \tag{7.89}$$

Table 7.2 The action of the projection operators of the group C_{2v} on the Cartesian basis vectors $\mathbf{x}_i, \mathbf{y}_i, \mathbf{z}_i$

	O^{A_1}	O^{A_2}	O^{B_1}	O^{B_2}
$\mathbf{x}_1$	0	0	$\mathbf{x}_1$	0
$\mathbf{x}_2$	0	$\mathbf{x}_2 - \mathbf{x}_3$	$\mathbf{x}_2 + \mathbf{x}_3$	0
$\mathbf{x}_3$	0	$\mathbf{x}_2 - \mathbf{x}_3$	$\mathbf{x}_2 + \mathbf{x}_3$	0
$\mathbf{y}_1$	0	0	0	$\mathbf{y}_1$
$\mathbf{y}_2$	$\mathbf{y}_2 - \mathbf{y}_3$	0	0	$\mathbf{y}_2 + \mathbf{y}_3$
$\mathbf{y}_3$	$\mathbf{y}_2 - \mathbf{y}_3$	0	0	$\mathbf{y}_2 + \mathbf{y}_3$
$\mathbf{z}_1$	$\mathbf{z}_1$	0	0	0
$\mathbf{z}_2$	$\mathbf{z}_2 + \mathbf{z}_3$	0	0	$\mathbf{z}_2 - \mathbf{z}_3$
$\mathbf{z}_3$	$\mathbf{z}_2 + \mathbf{z}_3$	0	0	$\mathbf{z}_2 - \mathbf{z}_3$

An examination of table 7.2 reveals that there are nine linearly independent vectors distributed amongst the representations A_1, A_2, B_1, and B_2 in accordance with eq. (7.72). These nine vectors could be used immediately as a symmetrized basis. However, as we pointed out earlier, simple linear combinations provide us with the translation vectors and it is convenient to take account of this fact. Hence for a symmetrized basis we take:

$$
\begin{aligned}
A_1:&\quad \mathbf{s}_1 = \mathbf{z}_1 + \mathbf{z}_2 + \mathbf{z}_3 \\
A_1:&\quad \mathbf{s}_2 = \mathbf{y}_2 - \mathbf{y}_3 \\
A_1:&\quad \mathbf{s}_3 = \mathbf{z}_2 + \mathbf{z}_3 \\
A_2:&\quad \mathbf{s}_4 = \mathbf{x}_2 - \mathbf{x}_3 \\
B_1:&\quad \mathbf{s}_5 = \mathbf{x}_1 + \mathbf{x}_2 + \mathbf{x}_3 \\
B_1:&\quad \mathbf{s}_6 = \mathbf{x}_2 + \mathbf{x}_3 \\
B_2:&\quad \mathbf{s}_7 = \mathbf{y}_1 + \mathbf{y}_2 + \mathbf{y}_3 \\
B_2:&\quad \mathbf{s}_8 = \mathbf{y}_2 + \mathbf{y}_3 \\
B_2:&\quad \mathbf{s}_9 = \mathbf{z}_2 - \mathbf{z}_3
\end{aligned}
\tag{7.90}
$$

These vectors are shown in Fig. 7.8 where we can easily recognize $\mathbf{s}_1, \mathbf{s}_5, \mathbf{s}_7$ as pure translations and we can see evidence of rotation in $\mathbf{s}_4$, $\mathbf{s}_6$, $\mathbf{s}_8$, and $\mathbf{s}_9$, and of vibration in $\mathbf{s}_2$, $\mathbf{s}_3$, and $\mathbf{s}_9$. Only $\mathbf{s}_4$ is a pure rotation whilst the others are hybrid motions. We should emphasize that the $\mathbf{s}_i$ are the symmetrized basis vectors and *not* the normal mode basis vectors because

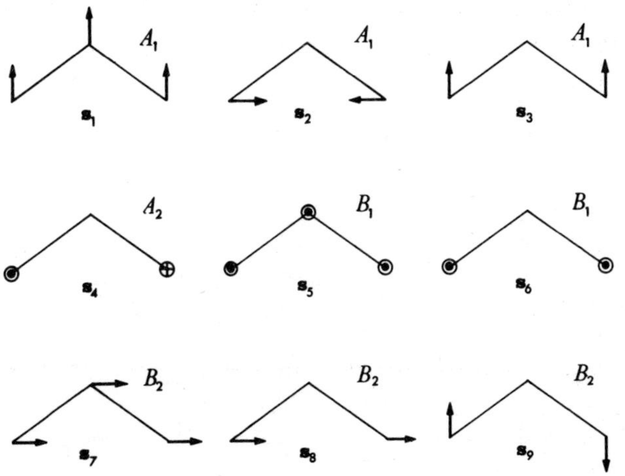

Fig. 7.8 *Illustration of the nine symmetrized basis vectors. The notation is the same as in Fig. 7.6.*

the symmetrized basis will only, as we shall now see, reduce **K** and **M** to block form.

From eqs. (7.90) we can now relate the symmetrized basis $\mathbf{s}_i$ to the Cartesian basis $\mathbf{x}_i$ ($\mathbf{y}_i$ and $\mathbf{z}_i$) by the transformation matrix $\boldsymbol{\beta}$ of eq. (7.53), written in *column* form as

$$(\mathbf{s}) = \boldsymbol{\beta}^{\dagger}(\mathbf{x}) \tag{7.91}$$

or

$$\begin{bmatrix} \mathbf{s}_1 \\ \mathbf{s}_2 \\ \mathbf{s}_3 \\ \mathbf{s}_4 \\ \mathbf{s}_5 \\ \mathbf{s}_6 \\ \mathbf{s}_7 \\ \mathbf{s}_8 \\ \mathbf{s}_9 \end{bmatrix} = \begin{bmatrix} 0 & 0 & 0 & 0 & 0 & 0 & 1 & 1 & 1 \\ 0 & 0 & 0 & 0 & 1 & -1 & 0 & 0 & 0 \\ 0 & 0 & 0 & 0 & 0 & 0 & 0 & 1 & 1 \\ 0 & 1 & -1 & 0 & 0 & 0 & 0 & 0 & 0 \\ 1 & 1 & 1 & 0 & 0 & 0 & 0 & 0 & 0 \\ 0 & 1 & 1 & 0 & 0 & 0 & 0 & 0 & 0 \\ 0 & 0 & 0 & 1 & 1 & 1 & 0 & 0 & 0 \\ 0 & 0 & 0 & 0 & 0 & 1 & 1 & 0 & 0 \\ 0 & 0 & 0 & 0 & 0 & 0 & 0 & 1 & -1 \end{bmatrix} \begin{bmatrix} \mathbf{x}_1 \\ \mathbf{x}_2 \\ \mathbf{x}_3 \\ \mathbf{y}_1 \\ \mathbf{y}_2 \\ \mathbf{y}_3 \\ \mathbf{z}_1 \\ \mathbf{z}_2 \\ \mathbf{z}_3 \end{bmatrix} \tag{7.92}$$

Now, as pointed out previously, the advantage of using a symmetrized basis lies in the fact that the potential and kinetic energy matrices **K** and **M** are transformed into block form (eq. (7.63)). Indeed, using eq. (7.57), namely

$$\mathbf{K}'' = \boldsymbol{\beta}^{\dagger}\mathbf{K}\boldsymbol{\beta}$$

and the expressions for **K** and $\boldsymbol{\beta}^{\dagger}$ given in eqs. (7.42) and (7.92) respectively we can show that

$$\mathbf{K}'' = 2k_1 \left[\begin{array}{ccc|c|cc|ccc} 0 & 0 & 0 & & & & & & \\ 0 & s^2 + 2\kappa & sc & & & & & 0 & \\ 0 & sc & c^2 & & & & & & \\ \hline & & & 0 & & & & & \\ \hline & & & & 0 & 0 & & & \\ & & & & 0 & 0 & & & \\ \hline & 0 & & & & & 0 & 0 & 0 \\ & & & & & & 0 & s^2 & sc \\ & & & & & & 0 & sc & c^2 \end{array}\right] \begin{array}{l} \\ A_1 \\ \\ A_2 \\ \\ B_1 \\ \\ B_2 \\ \\ \end{array} \tag{7.93}$$

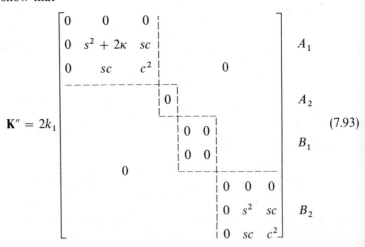

and similarly using eq. (7.44) for **M** and

$$\mathbf{M}'' = \boldsymbol{\beta}^{\dagger}\mathbf{M}\boldsymbol{\beta} \tag{7.94}$$

we can show that

$$\mathbf{M}'' = \left[\begin{array}{ccc|c|cc|ccc} M+2m & 0 & 2m & & & & & & \\ 0 & 2m & 0 & & & & & & \\ 2m & 0 & 2m & & & & & & \\ \hline & & & 2m & & & & & \\ \hline & & & & M+2m & 2m & & & \\ & & & & 2m & 2m & & & \\ \hline & & & & & & M+2m & 2m & 0 \\ & & & & & & 2m & 2m & 0 \\ & & & & & & 0 & 0 & 2m \end{array}\right] \begin{array}{l} \\ A_1 \\ \\ A_2 \\ \\ B_1 \\ \\ B_2 \\ \\ \end{array} \tag{7.95}$$

We now notice that the matrices **K** and **M** are, as we anticipated, in block form and these have been labelled, on the right-hand side, to indicate the irreducible representation to which each block belongs. Notice also that the dimension of each block is equal to the dimension of the representation multiplied by the number of times it appears in the reduced representation (e.g., A_1 is one-dimensional and appears three times in the representation, cf. eq. (7.72)). The fact that the blocks of **K** contain a considerable number of zeros is a consequence of choosing of $\mathbf{s}_1$, $\mathbf{s}_5$, and $\mathbf{s}_7$ to be translation vectors. Indeed we could produce more zeros in both **M** and **K** by trying various linear combinations of the vectors $\mathbf{s}_i$, within a given representation, especially those that we could pick out as being pure rotations.

The eigenfrequencies of the normal modes can now be found by solving the secular eq. (7.27) in the form

$$|\mathbf{K}'' - \lambda\mathbf{M}''| = 0 \tag{7.96}$$

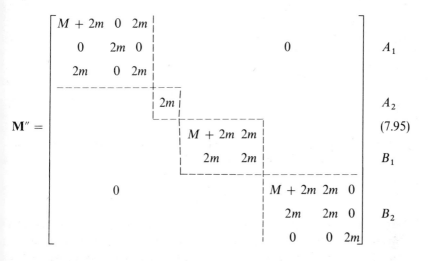

which on account of its block structure, is simpler to solve than using eqs. (7.42) and (7.44). In fact eq. (7.96) reduces to the four equations (cf. eq. (A.62))

$$A_1: \begin{vmatrix} -(M+2m)\lambda & 0 & -2m\lambda \\ 0 & k_1(2s^2+4\kappa)-2m\lambda & 2sck_1 \\ -2m\lambda & 2sck_1 & 2k_1c^2-2m\lambda \end{vmatrix} = 0 \tag{7.97}$$

$$A_2: \ |2m\lambda| = 0 \tag{7.98}$$

$$B_1: \ \lambda^2\begin{vmatrix} M+2m & 2m \\ 2m & 2m \end{vmatrix} = 0 \tag{7.99}$$

$$B_2: \begin{vmatrix} -(M+2m)\lambda & -2m\lambda & 0 \\ -2m\lambda & 2k_1s^2-2m\lambda & 2sck_1 \\ 0 & 2sck_1 & 2k_1c^2-2m\lambda \end{vmatrix} = 0 \tag{7.100}$$

Equations (7.97)–(7.100) yield six zero roots, corresponding to the translational and rotational modes. Equation (7.100) yields the root

$$\lambda = k_1\frac{(M+2ms^2)}{Mm} \tag{7.101}$$

in agreement with eq. (7.45) and, finally, eq. (7.97) yields eq. (7.46). We have thus obtained the same result as before. However although we have obtained the normal frequencies, we have not produced the normal coordinates. In fact we have gone as far as possible on the grounds of symmetry alone and in order to determine the normal modes we must reduce **K** and **M** to diagonal form which means we must find a further transformation matrix γ such that

$$(\mathbf{q}) = \gamma^\dagger(\mathbf{s}) \tag{7.102}$$

where $\mathbf{q}$ is the normal basis. We could find γ analytically by diagonalizing the individual blocks, using the methods given in the Appendix A.20, or alternatively we can recognize that $\mathbf{q}_i$ is simply some linear combination of the vectors $\mathbf{s}_i$ within a given representation. It is therefore considerably easier to determine the normal basis $\mathbf{q}_i$ from the symmetrized basis $\mathbf{s}_i$, rather than from the Cartesian basis $\mathbf{x}_i$. We should emphasize that the construction of the normal basis is not a matter for group theory. However an intuitive decomposition of the symmetric basis can often be achieved by inspection of the figures showing the atomic displacements given by the symmetrized basis.

In such a decomposition we inspect the configuration given by the vectors $\mathbf{s}_i$ and try to recognize the presence of pure rotation and translation that belong to the same irreducible representation. If we are successful, we then express the configuration given by $\mathbf{s}_i$ as a combination of a figure demonstrating the rotational, translational part and a figure showing the residual, vibrational, part. This type of resolution is shown in Fig. 7.9 for the case of the water molecule we have analysed above. Naturally the residual vibrational parts do not necessarily possess vectors showing the directions of motion which the atoms possess in a normal mode vibration;

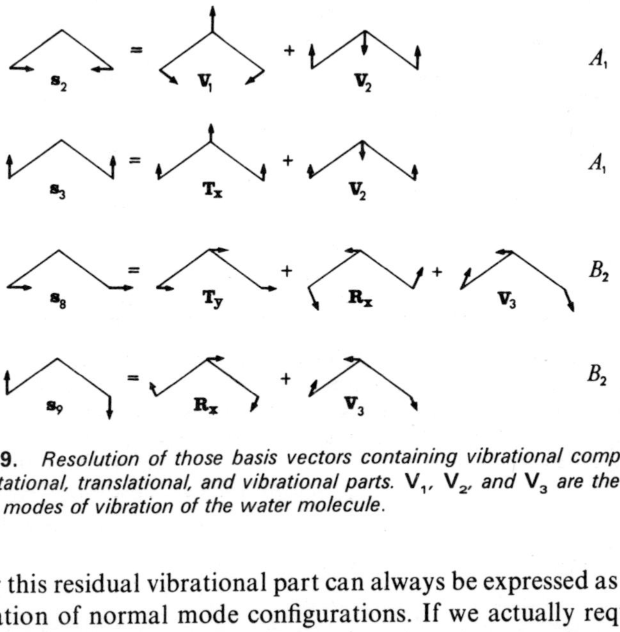

Fig. 7.9. *Resolution of those basis vectors containing vibrational components into rotational, translational, and vibrational parts.* $\mathbf{V}_1$, $\mathbf{V}_2$, *and* $\mathbf{V}_3$ *are the actual normal modes of vibration of the water molecule.*

however this residual vibrational part can always be expressed as a linear combination of normal mode configurations. If we actually require the *amplitudes* of the motion then we have to produce the normal coordinates which then requires us to find a transformation matrix which will simultaneously bring $\mathbf{K}$ to diag (λ_i) and $\mathbf{M}$ to $\mathbf{I}$.

7.6 Problems

7.1 Show that the expressions for q_i given in eq. (7.37), when substituted into eqs. (7.18) and (7.19), yield eqs. (7.31).

7.2 Write down the expressions (7.42) and (7.44) for $\mathbf{K}$ and $\mathbf{M}$ respectively for the case of three equal masses at the corners of an equilateral triangle (i.e., $M = m$, $k_1 = k_2 = k$) and solve the resulting secular equation, showing that the results agree with eqs. (7.45) and (7.46).

7.3 Obtain the normal modes of the diatomic molecule, considering motion in the x-direction only, by using the symmetry group S_2. If the masses are equal show that the symmetrized basis is also the normal basis.

7.4 Consider a molecule consisting of three equal masses at the corners of an equilateral triangle. Use the projection operators of the group D_3 (assume z is perpendicular to the plane) to obtain the symmetric basis, and by taking account of the translational and rotational modes show that these can be written as

		λ
$A_1: \mathbf{s}_1 = -\frac{\sqrt{3}}{2}\mathbf{x}_2 + \frac{\sqrt{3}}{2}\mathbf{x}_3 + \mathbf{y}_1 - \frac{1}{2}\mathbf{y}_2 - \frac{1}{2}\mathbf{y}_3$	V_1	$3k/m$
$A_2: \mathbf{s}_2 = \mathbf{x}_1 - \frac{1}{2}\mathbf{x}_2 - \frac{1}{2}\mathbf{x}_3 + \frac{\sqrt{3}}{2}\mathbf{y}_2 - \frac{\sqrt{3}}{2}\mathbf{y}_3$	R_z	0
$A_2: \mathbf{s}_3 = \mathbf{z}_1 + \mathbf{z}_2 + \mathbf{z}_3$	T_z	0
$E: \mathbf{s}_4 = \mathbf{x}_1 + \mathbf{x}_2 + \mathbf{x}_3$	T_x	0
$E: \mathbf{s}_5 = \mathbf{y}_1 + \mathbf{y}_2 + \mathbf{y}_3$	T_y	0
$E: \mathbf{s}_6 = 2\mathbf{z}_1 - \mathbf{z}_2 - \mathbf{z}_3$	R_x	0
$E: \mathbf{s}_7 = -\mathbf{z}_2 + \mathbf{z}_3$	R_y	0
$E: \mathbf{s}_8 = \frac{\sqrt{3}}{2}\mathbf{x}_2 - \frac{\sqrt{3}}{2}\mathbf{x}_3 + \mathbf{y}_1 - \frac{1}{2}\mathbf{y}_2 - \frac{1}{2}\mathbf{y}_3$	V_2	$3k/2m$
$E: \mathbf{s}_9 = \mathbf{x}_1 - \frac{1}{2}\mathbf{x}_2 - \frac{1}{2}\mathbf{x}_3 - \frac{\sqrt{3}}{2}\mathbf{y}_2 + \frac{\sqrt{3}}{2}\mathbf{y}_3$	V_3	$3k/2m$

Hence show that the matrix $\boldsymbol{\beta}$ constructed from the symmetrized basis completely diagonalizes the matrices **K** and **M** so that, in this example of very high symmetry, the symmetrized basis is also the normal basis. Check that the action on $\mathbf{x}_1$ and $\mathbf{y}_1$, of the selection of projection operators shown in the figure are correct.

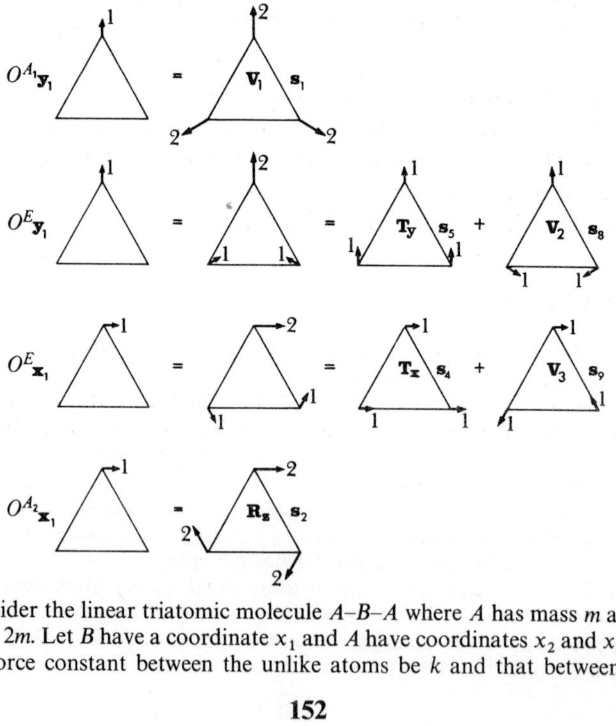

7.5 Consider the linear triatomic molecule A–B–A where A has mass m and B has mass $2m$. Let B have a coordinate x_1 and A have coordinates x_2 and x_3, and let the force constant between the unlike atoms be k and that between the like atoms be zero. Use the methods of section A.20 to find the diagonalizing matrix $\boldsymbol{\alpha}$ and hence the normal coordinates and normal modes (ignore y and z displacements). Use the symmetry group S_2 to obtain the symmetrized basis and show that this reduces **K** and **M** to block form.

7.6 In some normal mode analyses it is convenient to work in *internal coordinates* which means that we remove the rotational and translational modes *ab initio* by setting the total angular and linear momenta equal to zero. In one dimension the latter condition becomes

$$\sum m_i x_i = 0$$

Use this condition to eliminate the coordinate x_1 of the atom B of problem 7.8 to obtain

$$\mathbf{K} = \frac{k}{2}\begin{bmatrix} 5 & 3 \\ 3 & 5 \end{bmatrix}; \qquad \mathbf{M} = \frac{m}{2}\begin{bmatrix} 3 & 1 \\ 1 & 3 \end{bmatrix}$$

where k is the force constant and m is the mass. Hence solve the vibrational problem in internal coordinates to obtain λ_i, $\boldsymbol{\alpha}$, $\mathbf{q}_i$, and q_i.

8 Molecular orbitals

8.0 Orbitals

In this chapter we will briefly discuss some of the basic ideas involved in the theory of molecular orbitals and the study of molecular spectroscopy. We will see that many of the group theoretical techniques we have previously developed can be used to simplify the problems involved.

A molecule can be imagined to be formed by bringing together a number of atoms so that in order to understand this process we must first consider the nature of the free atoms. These consist of a nucleus, normally surrounded by closed shells of electrons which are in turn surrounded by a number of valence electrons that determine the chemical properties of the atom. The eigenfunctions ψ describing a single valence electron are derived from the Schrödinger equation

$$H\psi = \varepsilon\psi \tag{8.1}$$

where ε is the eigenvalue and H contains a potential energy term which we can obtain by assuming that the electron moves in an average field due to the influence of the nucleus and all the other electrons, that is the valence electrons are imagined to move independently of each other. The eigenfunctions ψ are called the *atomic orbitals.*

As a simple example let us consider the hydrogen atom which has a single valence electron orbiting a proton nucleus. We can write the atomic orbital in this case, using the spherical polar coordinates of Fig. 8.1, as

$$\psi_{nlm}r, \theta, \phi) = R_{ln}(r)Y_l^m(\theta, \phi) \tag{8.2}$$

where n is the principal quantum number, l is the orbital angular momentum quantum number which takes on the values 0 to $n-1$, and m is the magnetic quantum number, such that $-l \leqslant m \leqslant l$. The spherical harmonics $Y_l^m(\theta, \phi)$ describe the angular dependence of the wave-function. In accordance with standard spectroscopic notation the orbitals ψ are known as

s-orbitals when $l = 0$

p-orbitals when $l = 1$

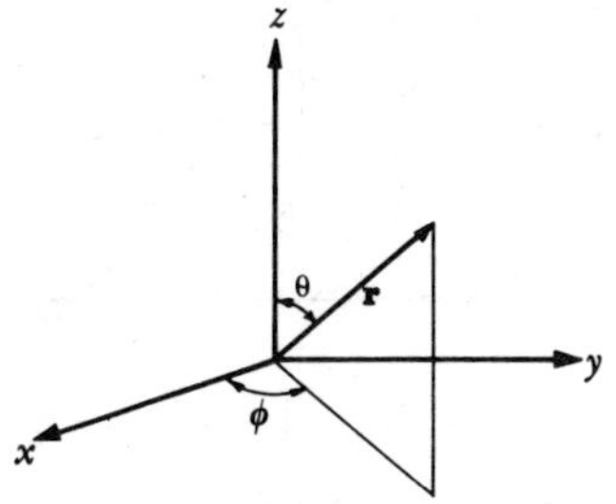

Fig. 8.1 *Spherical polar coordinates.*

and

$$d\text{-orbitals when} \quad l = 2 \quad \text{and so on.}$$

In most discussions of orbitals it is common practice to give a pictorial representation of the angular part $Y_l^m(\theta, \phi)$. Thus, when $l = 1$ we have three *p*-orbitals which can be associated with the *x*, *y*, and *z* axes, i.e.,

$$p_x = b \sin\theta \cos\phi \propto Y_1^1 + Y_1^{-1} \tag{8.3}$$

$$p_y = b \sin\theta \sin\phi \propto Y_1^1 - Y_1^{-1} \tag{8.4}$$

$$p_z = a \cos\theta \propto Y_1^0 \tag{8.5}$$

and these can be plotted as a locus in *x*, *y*, *z* space. For example the locus of

$$\cos\theta = p_z/a \tag{8.6}$$

is plotted in Fig. 8.2(a) and if we imagine rotating this locus about the *z* axis we would obtain a three-dimensional figure consisting of two spheres in contact at the origin. As an alternative to this procedure we could plot contours of constant $|\psi(r, \theta, \phi)|^2$ (i.e., the electron density). Finally, the quantity $a^2 \cos^2\theta$ is often plotted, where a is chosen so that a surface enclosing 90 per cent of the charge of the electron is obtained; this is shown in Fig. 8.2(b). The p_x and p_y orbitals can be represented in a similar manner and will be found to be identical to the p_z orbital except for being stretched out along the *x* and *y* axes respectively.

Now when atoms combine to form molecules the atomic orbitals overlap and their electrons will belong to *molecular orbitals*, i.e., orbitals associated with the molecule. Such molecular orbitals can be determined in various ways but a very popular one is to assume that a molecular orbital can be written as a suitable linear combination of the atomic orbitals; a procedure which can be given a very simple justification. For example let us consider two hydrogen atoms combining to form a hydrogen molecule (cf. Fig. 8.3), where we have two electrons going into an orbital around both nuclei. In this case it is clear that when an electron is in the neighbourhood of *A* it is practically screened from the atom at *O′*

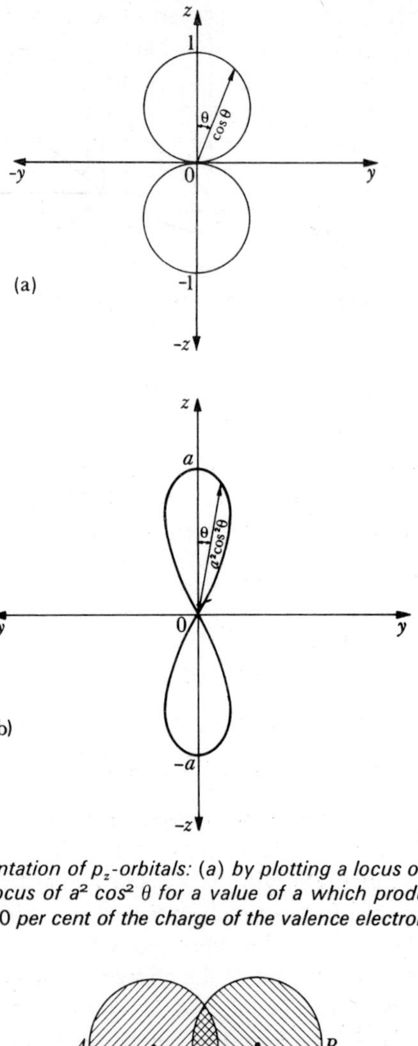

Fig. 8.2 *Representation of p_z-orbitals: (a) by plotting a locus of cos θ for a = 1; (b) by plotting a locus of $a^2 \cos^2 \theta$ for a value of a which produces a boundary enclosing at least 90 per cent of the charge of the valence electrons.*

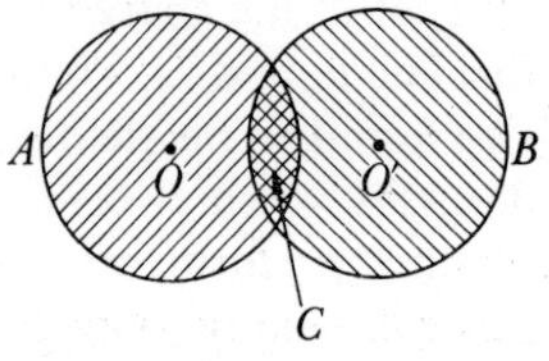

Fig. 8.3 *The molecular orbital of a hydrogen molecule with atoms centred at O and O′.*

by the atom at O. In this position the electron must be described by a molecular orbital that looks very much like the atomic orbital associated with the atom at O. Similarly at B an electron is best described by a molecular orbital resembling an atomic orbital of the atom at O'. In the overlap region C an electron must be described in terms of a molecular orbital which is some linear combination of atomic orbitals belonging to the atoms at O and O'. It is necessary in forming a description based on linear combinations of atomic orbitals to use orbitals that spatially overlap each other and have comparable energies.

Having now established some of the language of molecular orbital theory we will briefly consider some of the types of bonding that can occur by examining the nature of the bonding in the aromatic hydrocarbon molecule benzene C_6H_6.

This molecule consists of six carbon atoms arranged at the apices of a regular hexagon (which will be assumed to lie in the x-y plane), and to each of these atoms is attached a single hydrogen atom. Each carbon atom has six electrons two of which are in closed s-shells whilst the remainder are in s and p orbitals. In the formation of the molecule, as shown in Fig. 8.4, *one* valence electron of carbon is required to bond a hydrogen atom to a carbon atom and *two* valence electrons are required to form the so-called σ-bonds between the carbon atoms; the fourth electron is in a p_z-orbital which will be discussed later.

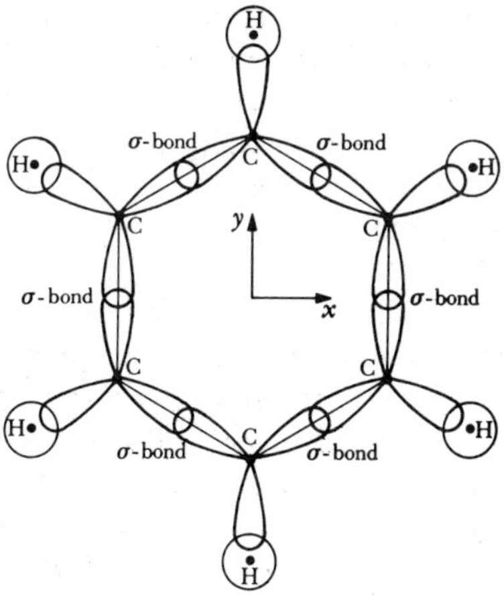

Fig. 8.4 *The benzene molecule lying in the x-y plane illustrating the formation of σ-bonds.*

The σ-bonds are very strong and are really those responsible for the existence of the molecule; they lie in the plane of the hexagon and are created by a judicious mixing (hybridization) of the s, p_x, and p_y atomic orbitals of carbon. The p_z-orbital has a node on the x-y plane and cannot contribute to σ-bonding.

This type of bonding is due therefore to the overlap of orbitals which extend over the x-y plane containing the carbon atoms. It is, as we have said, very strong, implying that the electrons involved will be difficult to excite by an external influence, and in fact the electrons in σ-bonds only contribute to molecular spectra at very short wavelengths.

Let us now consider the fourth valence electron which is in a p_z-orbital. This orbital only has a non-zero value above or below the plane of the hexagon and overlaps with those on neighbouring carbon atoms to form the π-bonding orbital. This type of bonding is weak and the electrons in the π-orbitals are free to travel around the molecule, i.e., each electron is shared equally by all six carbon atoms. These π-electrons, being only loosely bound to the molecule, can therefore be excited by external influences and largely govern many of the spectroscopic properties of benzene.

The remarks we have made, using the example of benzene, are equally applicable to other hydrocarbons.

8.1 Symmetry and the molecular orbitals of benzene

We are now going to use group theory to assist us to determine and classify the energy levels of the π-electrons of benzene. The group we shall use is the point group D_{6h}, i.e., the group of operations under which the hexagon remains invariant. Physically this means that the electrons in the π-orbitals have a potential energy that possesses the symmetry of D_{6h}. Hence this group is the group of the Hamiltonian of the π-electrons (cf. section 3.7) and the individual π-orbitals must form basis functions for the irreducible representations of this group.

The point group D_{6h} is the direct product group $D_6 \otimes C_{1h}$, where C_{1h} is the group containing E and σ_h alone. However C_{1h} has only two irreducible representations whose basis functions are either symmetric or antisymmetric with respect to σ_h so that when we have determined the basis functions of D_6 it is a trivial matter to produce those of D_{6h}. In view of these observations we shall restrict our attention to an examination of benzene using only the group D_6 a step which reduces the amount of labour involved.

To begin with we consider the six p_z-orbitals centred on the carbon atoms of benzene, which form basis functions for a six-dimensional reducible representation of the group D_6 which contains, as shown in Fig. 8.5, the elements of the cyclic group C_6, together with the six two-fold rotations perpendicular to the principal (z) axis.

The six p_z-orbitals will be denoted by ϕ_1, ϕ_2, ϕ_3, ϕ_4, ϕ_5, and ϕ_6, where

$$\phi_j \equiv \phi(\mathbf{r} - \mathbf{R}_j) \tag{8.7}$$

(cf. Fig. 8.6), which we can write as a row vector

$$(\tilde{\boldsymbol{\phi}}) = (\boldsymbol{\phi}_1, \boldsymbol{\phi}_2, \boldsymbol{\phi}_3, \boldsymbol{\phi}_4, \boldsymbol{\phi}_5, \boldsymbol{\phi}_6) \tag{8.8}$$

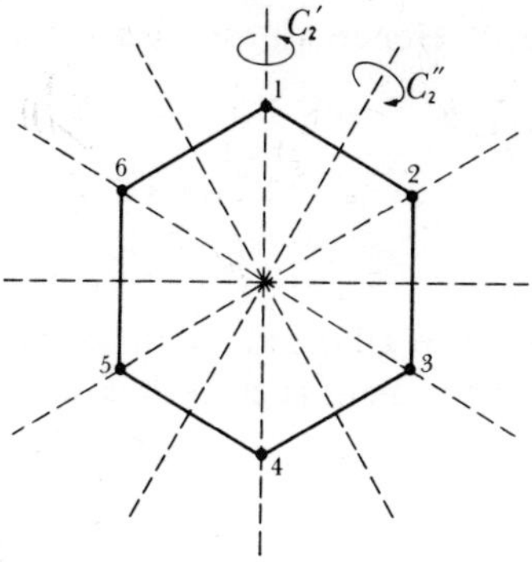

Fig. 8.5 *The symmetry elements of the group D_6, a subgroup of the symmetry group of the benzene molecule. The six-fold axis giving elements $2C_6$, $2C_3$, and C_2 is perpendicular to the page.*

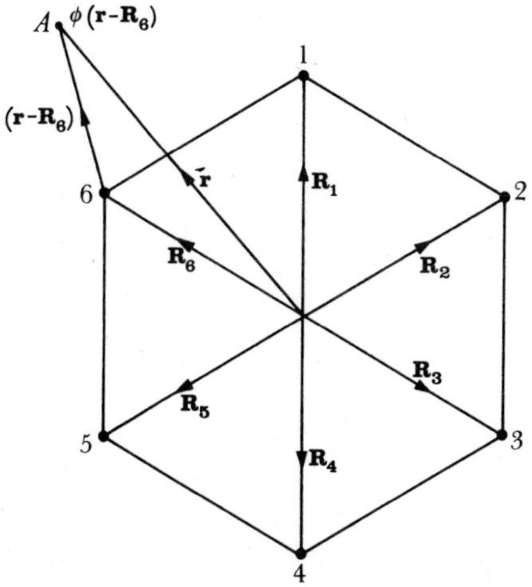

Fig. 8.6 *Vector positions of the carbon atoms in the benzene molecule. $\phi(\mathbf{r} - \mathbf{R}_6)$ for an arbitrary vector $\mathbf{r}$ is evaluated at the point A.*

We shall now consider what representations of the operators of the group D_6 we can obtain by using this basis. These representations must be reducible since D_6 has no six-dimensional irreducible representations.

We know (eq. (3.21)) that

$$\begin{aligned} P_R\phi_j &= P_R\phi(\mathbf{r} - \mathbf{R}_j) \\ &= \phi(\mathbf{\Gamma}^{-1}(R)(\mathbf{r} - \mathbf{R}_j)) \end{aligned} \tag{8.9a}$$

where R is a group element. Hence, for example,

$$\begin{aligned} P_{C_6}\phi_1 &= \phi(\mathbf{\Gamma}^{-1}(C_6)(\mathbf{r} - \mathbf{R}_1)) \\ &= \phi(\mathbf{\Gamma}(C_6^5)(\mathbf{r} - \mathbf{R}_1)) \\ &= \phi(\mathbf{r} - \mathbf{R}_2) = \phi_2 \end{aligned} \tag{8.9b}$$

Proceeding in this manner for all the functions ϕ_i and the elements of the group D_6 we find that for the identity operation we have

$$P_E(\tilde{\mathbf{\phi}}) = (\tilde{\mathbf{\phi}})\begin{bmatrix} 1 & 0 & 0 & 0 & 0 & 0 \\ 0 & 1 & 0 & 0 & 0 & 0 \\ 0 & 0 & 1 & 0 & 0 & 0 \\ 0 & 0 & 0 & 1 & 0 & 0 \\ 0 & 0 & 0 & 0 & 1 & 0 \\ 0 & 0 & 0 & 0 & 0 & 1 \end{bmatrix} \tag{8.10}$$

and the character $\chi(E)$ of this representation is equal to 6. Similarly a rotation of 180° about an axis through atoms 1 and 4 yields

$$P_{C_2}(\tilde{\mathbf{\phi}}) = (\tilde{\mathbf{\phi}})\begin{bmatrix} 1 & 0 & 0 & 0 & 0 & 0 \\ 0 & 0 & 0 & 0 & 0 & 1 \\ 0 & 0 & 0 & 0 & 1 & 0 \\ 0 & 0 & 0 & 1 & 0 & 0 \\ 0 & 0 & 1 & 0 & 0 & 0 \\ 0 & 1 & 0 & 0 & 0 & 0 \end{bmatrix} \tag{8.11}$$

and the character $\chi(C_2)$ is equal to 2. The same result is true for the other C_2 operations. None of the remaining operations of the group leaves any atom in the same position, hence their characters must be zero. Now that we are in possession of the character of the reducible representation based on the p_z-orbitals we can find its irreducible components by using eq. (2.37), namely

$$a_i = \frac{1}{g}\sum \chi^{\mathrm{red}}(R)\chi^{i*}(R) \tag{8.12}$$

where $\chi^{\mathrm{red}}(R)$ is the character of the reducible representation, $\chi^i(R)$ is the character of the ith irreducible representation, g is the order of the group, and a_i is the number of times the ith irreducible representation appears in the reducible representation.

In order to use eq. (8.12) we require the character table of D_6 which is given in table 8.1, together with the characters of the reducible representa-

Table 8.1 Character table of D_6

D_6	E	$2C_6$	$2C_3$	C_2	$3C_2'$	$3C_2''$
A_1	1	1	1	1	1	1
A_2	1	1	1	1	−1	−1
B_1	1	−1	1	−1	1	−1
B_2	1	−1	1	−1	−1	1
E_1	2	1	−1	−2	0	0
E_2	2	−1	−1	2	0	0
Γ^{red}	6	0	0	0	2	0

tion that we have just calculated. Using this table we obtain

$$\begin{aligned} a_1 &= \tfrac{1}{12}[6 + 0 + 0 + 0 + 6 + 0] = 1 \\ a_2 &= \tfrac{1}{12}[6 + 0 + 0 + 0 - 6 + 0] = 0 \\ a_3 &= \tfrac{1}{12}[6 + 0 + 0 + 0 + 6 + 0] = 1 \\ a_4 &= \tfrac{1}{12}[6 + 0 + 0 + 0 - 6 + 0] = 0 \\ a_5 &= \tfrac{1}{12}[12 + 0 + 0 + 0 + 0 + 0] = 1 \\ a_6 &= \tfrac{1}{12}[12 + 0 + 0 + 0 + 0 + 0] = 1 \end{aligned} \tag{8.13}$$

which shows that

$$\Gamma^{\text{red}}(R) = \mathbf{A}_1 \oplus \mathbf{B}_1 \oplus \mathbf{E}_1 \oplus \mathbf{E}_2 \tag{8.14}$$

Equation (8.14) thus shows us that there are four π-electron energy levels; two of which (A_1 and B_1) are singly degenerate and two are doubly degenerate (E_1 and E_2).

We can now determine by using the projection operators those linear combinations of the six p_z-orbitals ϕ_j which act as basis functions for the irreducible representations of eq. (8.14). This can be done by using eq. (3.48), namely

$$\Phi^i = O^i\phi = \frac{1}{g}\sum_R \chi^{i*}(R)P_R\phi \tag{8.15}$$

where g is the order of the group, $\chi^i(R)$ is the character of the ith irreducible representation of R, ϕ is chosen to be one of the p_z-orbitals, and Φ^i is a symmetrized function which belongs to the ith irreducible representation. These symmetrized functions will be used to bring the secular equation, for determining the energy eigenvalues of the π-electrons, into block form.

Thus, if we choose ϕ_1, and the irreducible representation A_1, we obtain from eq. (8.15),

$$\begin{aligned} \Phi^1 &= \tfrac{1}{12}(\phi_1 + \phi_2 + \phi_6 + \phi_3 + \phi_5 + \phi_4 \\ &\quad + \phi_1 + \phi_3 + \phi_5 + \phi_2 + \phi_4 + \phi_6) \\ &= \tfrac{1}{6}(\phi_1 + \phi_2 + \phi_3 + \phi_4 + \phi_5 + \phi_6) \end{aligned} \tag{8.16}$$

Similarly for the irreducible representation B_1 we obtain

$$\Phi^3 = \tfrac{1}{6}(\phi_1 - \phi_2 + \phi_3 - \phi_4 + \phi_5 - \phi_6) \tag{8.17}$$

The determination of the symmetrized functions belonging to E_1 and E_2 is a little more complicated because these are two-dimensional representations and obviously require two basis functions each. Thus in order to generate two independent functions belonging to E_1 or E_2 we must select two arbitrary functions in a suitable manner, and this clearly allows an element of choice, because for example we could use ϕ_1 and ϕ_2 or alternatively make other selections from the set ϕ_i. *Any* selection which leads to two independent functions for each is, of course, acceptable.

If we use ϕ_1 and ϕ_2 we find that for E_1 we have

$$\begin{aligned} \Phi^5 &= \tfrac{1}{12}(2\phi_1 + \phi_2 - \phi_3 - 2\phi_4 - \phi_5 + \phi_6) \\ \Phi^{5'} &= \tfrac{1}{12}(\phi_1 + 2\phi_2 + \phi_3 - \phi_4 - 2\phi_5 - \phi_6) \end{aligned} \tag{8.18}$$

whilst for E_2 we have

$$\begin{aligned} \Phi^6 &= \tfrac{1}{12}(2\phi_1 - \phi_2 - \phi_3 + 2\phi_4 - \phi_3 - \phi_6) \\ \Phi^{6'} &= \tfrac{1}{12}(-\phi_1 + 2\phi_2 - \phi_3 - \phi_4 + 2\phi_5 - \phi_6) \end{aligned} \tag{8.19}$$

Now A_1 is a one-dimensional representation so that the function generated by eq. (8.15) must of necessity be a basis function of that representation, because since the character of a one-dimensional representation matrix is the *same* as the matrix we are in this case using the *full projection operator* of eq. (3.41). However, for a two-dimensional irreducible representation it is clear from eqs. (3.45) and (3.48) that the functions Φ^5, $\Phi^{5'}$ are *not* basis functions of this representation but merely belong to the representation in the sense discussed in chapter 3. The only problem that this creates is that the two functions such as Φ^5 and $\Phi^{5'}$ are not orthogonal as they would be if they were the proper basis functions of the two-dimensional representation. (N.B. A proper basis function of an irreducible representation has two labels; one to denote the irreducible representation and one to label the column to which it belongs and as is shown in section 3.4, they are orthogonal between different representations and different columns of that representation.) However, as we shall see later, we can exploit orthogonality in determining the eigenvalues of the π-electrons in the molecule so it is of some value to find the linear combinations of Φ^5 and $\Phi^{5'}$ and of Φ^6 and $\Phi^{6'}$ which are orthogonal to each other, i.e., the basis functions of E_1 and E_2. This can be done by observing that the original functions (cf. eqs. (8.18) and (8.19)), thought of as vectors, are of equal length so that the sums and differences are mutually orthogonal. We must therefore form

$$
\begin{aligned}
\Phi_1^5 &= \Phi^5 + \Phi^{5'} = \tfrac{1}{4}(\phi_1 + \phi_2 - \phi_4 - \phi_5) \\
\Phi_2^5 &= \Phi^5 - \Phi^{5'} = \tfrac{1}{12}(\phi_1 - \phi_2 - 2\phi_3 - \phi_4 + \phi_5 + 2\phi_6) \\
\Phi_1^6 &= \Phi^6 + \Phi^{6'} = \tfrac{1}{12}(\phi_1 + \phi_2 - 2\phi_3 + \phi_4 + \phi_5 - 2\phi_6) \\
\Phi_2^6 &= \Phi^6 - \Phi^{6'} = \tfrac{1}{4}(\phi_1 - \phi_2 + \phi_4 - \phi_5)
\end{aligned}
\tag{8.20}
$$

The functions $\Phi^1, \Phi^3, \Phi_1^5, \Phi_2^5, \Phi_1^6$, and Φ_2^6 are then the mutually orthogonal π-orbital eigenfunctions which can now be used to determine the eigenvalues.

8.2 Energy eigenvalues of the π-electrons of benzene

The eigenvalues of the π-electrons can be obtained directly from the Schrödinger equation

$$H\psi = \varepsilon\psi \tag{8.21}$$

where H is the Hamiltonian of the electrons, and has the symmetry D_{6h}. We could of course solve eq. (8.21) directly using the orthogonal functions Φ_k^j because they are clearly the functions which produce a diagonal representation of H. Hence

$$\varepsilon = \frac{\langle \Phi_k^j | H | \Phi_k^j \rangle}{\langle \Phi_k^j | \Phi_k^j \rangle} \tag{8.22}$$

However for most molecules we would not be able to get as far as the eigenfunctions, *using group theory alone*. An instance of such a case would occur if the reduced representation based on the p_z-orbitals contained $3A_1$ (say), in which case we would be able to obtain three independent basis functions, but as pointed out earlier, we could not guarantee that they would be mutually orthogonal or that they would be the functions from which ε could be calculated using eq. (8.22).

Thus in order to illustrate how the general case would be solved we will first of all produce an equation for the eigenvalues without using symmetrized basis function and we will then show how the use of symmetrized functions leads to simplifications.

The solution of eq. (8.21) is obtained (cf. section 3.9) by expanding the eigenfunction ψ into a *finite* set of atomic orbitals, i.e.,

$$\psi(\mathbf{r}) = \sum_{k=1}^{6} a_k \phi_k(\mathbf{r} - \mathbf{R}_k) \tag{8.23}$$

where $\phi_k(\mathbf{r} - \mathbf{R}_k)$ is a p_z-orbital centred on the kth carbon atom at position defined by the vector $\mathbf{R}_k$; and a_k are arbitrary constants. On substituting into the Schrödinger eq. (8.21) we obtain

$$H \sum_k a_k \phi_k = \varepsilon \sum_k a_k \phi_k \tag{8.24}$$

Hence, by multiplying by ϕ_i^* and integrating over all the space, we obtain an equation that can be written in the Dirac notation (cf. Appendix A.6) as,

$$\sum_k a_k [\langle \phi_i | H | \phi_k \rangle - \varepsilon \langle \phi_i | \phi_k \rangle] = 0 \tag{8.25}$$

This equation must be true for all a_k, so the eigenvalues are the solutions of the so-called *secular equation*:

$$|\langle \phi_i | H | \phi_k \rangle - \varepsilon \langle \phi_i | \phi_k \rangle| = 0 \tag{8.26}$$

or

$$|H_{ik} - \varepsilon S_{ik}| = 0 \tag{8.27}$$

where

$$H_{ik} = \langle \phi_i | H | \phi_k \rangle \tag{8.28}$$

$$= \int \phi^*(\mathbf{r} - \mathbf{r}_i) H \phi(\mathbf{r} - \mathbf{r}_k) \, \mathrm{d}r \tag{8.29}$$

and

$$S_{ik} = \langle \phi_i | \phi_k \rangle \tag{8.30}$$

$$= \int \phi^*(\mathbf{r} - \mathbf{r}_i) \phi(\mathbf{r} - \mathbf{r}_k) \, \mathbf{dr} \tag{8.31}$$

The integrals S_{ik} are known as the overlap integrals, between the same atomic orbitals centred on different atoms, and can be calculated exactly; in practice the functions ϕ_i are normalized so that

$$S_{ii} = 1 \tag{8.32}$$

also the overlap between nearest neighbour atoms will be the same for all atoms, so we shall put

$$S_{ik} = S \ (i = k + 1, \text{or } k - 1) \tag{8.33}$$

The overlap between other than nearest neighbours can be ignored, and if we do this we are using what is known as the Hückel approximation.

The integrals H_{ik} can in principle be calculated provided we have an expression for the Hamiltonian H, which in the Hückel approximation is taken as that of a single electron moving in the potential field of all the nuclei and the other electrons. If we once again introduce only self and nearest neighbour interactions we can write

$$H_{ii} = \langle \phi_i | H | \phi_i \rangle = \alpha \tag{8.34}$$

and

$$H_{ik} = \langle \phi_i | H | \phi_k \rangle = \beta \ (i = k + 1, i = k - 1) \tag{8.35}$$

By this means we reduce the number of parameters which can be compared with experiment and used to ascertain the correctness of the assumed Hamiltonian and atomic orbitals.

With these approximations, eq. (8.27) becomes

$$\begin{vmatrix} \alpha-\varepsilon & \beta-S\varepsilon & 0 & 0 & 0 & \beta-S\varepsilon \\ \beta-S\varepsilon & \alpha-\varepsilon & \beta-S\varepsilon & 0 & 0 & 0 \\ 0 & \beta-S\varepsilon & \alpha-\varepsilon & \beta-S\varepsilon & 0 & 0 \\ 0 & 0 & \beta-S\varepsilon & \alpha-\varepsilon & \beta-S\varepsilon & 0 \\ 0 & 0 & 0 & \beta-S\varepsilon & \alpha-\varepsilon & \beta-S\varepsilon \\ \beta-S\varepsilon & 0 & 0 & 0 & \beta-S\varepsilon & \alpha-\varepsilon \end{vmatrix} = 0 \quad (8.36)$$

which can be solved for the eigenvalues ε. (Note that as the molecule is cyclic atom 1 has atoms 6 and 2 as neighbours.)

If we use the symmetrized orbitals of eqs. (8.16), (8.17), (8.18), and (8.19) then eq. (8.27) is reduced to the block form:

$$\begin{vmatrix} (\alpha+2\beta)-(1+2S)\varepsilon & 0 & 0 & 0 & 0 & 0 \\ 0 & (\alpha-2\beta)-(1-2S)\varepsilon & 0 & 0 & 0 & 0 \\ 0 & 0 & 2(\alpha+\beta)-2(1+S)\varepsilon & (\alpha+\beta)-(1+S)\varepsilon & 0 & 0 \\ 0 & 0 & (\alpha+\beta)-(1+S)\varepsilon & 2(\alpha+\beta)-2(1+S)\varepsilon & 0 & 0 \\ 0 & 0 & 0 & 0 & 2(\alpha-\beta)-2(1-S)\varepsilon & (\alpha-\beta)-(1-S)\varepsilon \\ 0 & 0 & 0 & 0 & (\alpha-\beta)-(1-S)\varepsilon & 2(\alpha-\beta)-2(1-S)\varepsilon \end{vmatrix} = 0$$

(8.37)

The matrix elements in eq. (8.37) are evaluated in the same manner as those in eq. (8.36); for example the element H_{43} is given by

$$\begin{aligned} H_{43} &= \langle \Phi^{5'} | H | \Phi^{5} \rangle \\ &= \tfrac{1}{144} \langle \phi_1 + 2\phi_2 + \phi_3 - \phi_4 - 2\phi_5 - \phi_6 | H | 2\phi_1 + \phi_2 - \phi_3 \\ &\quad - 2\phi_4 - \phi_5 + \phi_6 \rangle \\ &= \tfrac{1}{144}(6\alpha + 6\beta) \end{aligned} \quad (8.38)$$

whilst

$$S_{43} = \tfrac{1}{144}(6 + 6S) \tag{8.39}$$

Common factors have been removed from the secular eq. (8.37), which can also be solved for the (same) eigenvalues ε.

Finally if we use the orthogonal functions $\Phi^j_{(k)}$ of eqs. (8.16), (8.17), and (8.20) then the secular equation (8.27) becomes completely diagonal, i.e.,

$$\begin{array}{c} A_1 \\ B_1 \\ \\ E_1 \\ \\ \\ E_2 \\ \end{array}\begin{vmatrix} (\alpha+2\beta)-(1+2S)\varepsilon & 0 & 0 & 0 & 0 & 0 \\ 0 & (\alpha-2\beta)-(1-2S)\varepsilon & 0 & 0 & 0 & 0 \\ 0 & 0 & (\alpha+\beta)-(1+S)\varepsilon & 0 & 0 & 0 \\ 0 & 0 & 0 & (\alpha+\beta)-(1+S)\varepsilon & 0 & 0 \\ 0 & 0 & 0 & 0 & (\alpha-\beta)-(1-S)\varepsilon & 0 \\ 0 & 0 & 0 & 0 & 0 & (\alpha-\beta)-(1-S)\varepsilon \end{vmatrix} = 0 \tag{8.40}$$

where we have indicated at the left-hand side the distribution of irreducible representations.

The solutions of eq. (8.40) (and of eqs. (8.36) and (8.37)) are

$$\begin{aligned} A_1:\quad \varepsilon_1 &= \frac{\alpha + 2\beta}{1 + 2S} \\ B_1:\quad \varepsilon_2 &= \frac{\alpha - 2\beta}{1 - 2S} \\ E_1:\quad \varepsilon_3 &= \frac{\alpha + \beta}{1 + S} \\ E_2:\quad \varepsilon_4 &= \frac{\alpha - \beta}{1 - S} \end{aligned} \tag{8.41}$$

where, since E_1 and E_2 are two-dimensional irreducible representations, it is clear that the energies ε_3 and ε_4 are doubly degenerate. Note that if the full group D_{6h} were used we should find that these levels are labelled A_{2u}, B_{2g}, E_{1g}, and E_{2u} respectively. The energy level scheme of the benzene molecule given by eq. (8.41) is shown in Fig. 8.7 where we indicate sche-

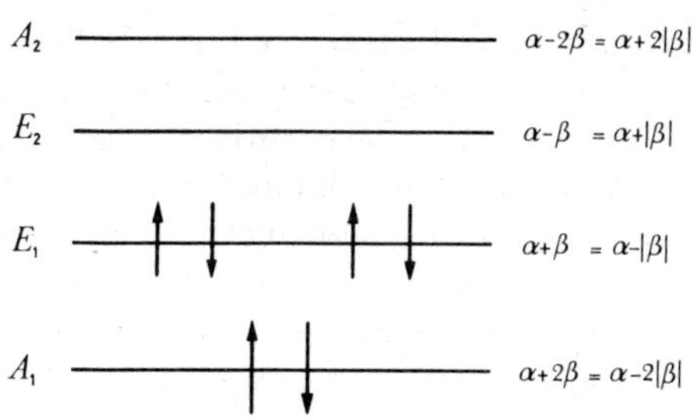

Fig. 8.7 *The one-electron energy states of the π-electrons of benzene. The six electrons fill the levels as shown, the arrows indicating the directions of the spin.*

matically that the effect of reducing the symmetry of the environment of a carbon atom by arranging the carbon atoms on a hexagonal ring is to split the energy level of an electron in a p_z-orbital of a free carbon atom into four levels, two of which are doubly degenerate. The levels have been placed in ascending order of energy by noting that the totally symmetric state, which belongs to A_1, must have the lowest energy; this implies that β must be negative. We can do this by analogy with the problem of a classical particle in a box in which we know that the lowest energy state is a totally symmetric state. Further confirmation is provided by the fact that the π-orbital belonging to A_1 (eq. (8.16)) does not change sign as we move from atom to atom, the lowest energy state of the particle in a box also has a wave function that does not change sign within its range.

Now that we have the energy levels of the π-orbitals we can find the ground and excited states of the benzene molecule by placing the six p-electrons into these levels, taking account of the Pauli exclusion principle that allows two electrons of oppositely directed spin in each level. These have been indicated by the arrows in Fig. 8.7 which shows the ground state of the molecule. The lowest energy transition corresponding to an energy change of 2β is obtained by promoting one electron to the state E_2, and is found by experiment to occur at approximately 6 eV. The simple Hückel theory using p_z-orbitals yields a value for 2β of 1·7 eV but the calculation can be improved by using empirical orbitals.

8.3 Electronic selection rules

In section 3.11 we saw that the possibility of an electron making a transition from some initial state $\psi_i(\mathbf{r})$ to some final state $\psi_f(\mathbf{r})$ under the action of an external agency depends upon the representations to which these states belong. We are now going to illustrate this restriction by considering the example of electronic transitions in molecules for the particular case in which an electron is induced to go from one eigenstate to another by the action of an electromagnetic wave. We should point out that (so called spontaneous) transitions *can* take place from higher to lower energy levels in the absence of any external agency, and that transitions may also be induced by, for example, the incidence of fast electrons. In the former case, however, it can be shown that the selection rule for the spontaneous transition is the same as that for the one induced by an electromagnetic field.

Let us denote the initial eigenfunction ψ_i by $|i\rangle$ and the final eigenfunction ψ_f by $|f\rangle$. Then a transition is only possible if the matrix element $\langle i|H'|f\rangle$ is non-zero where H' is the *additional* energy the electron acquires in the presence of the external agency.

We will now consider the nature of H' in rather more detail for the case of an electromagnetic wave. The classical Hamiltonian of an electron (mass m and charge e) in the presence of an electromagnetic wave that is determined by a vector potential $\mathbf{A}$ is given by

$$H = \frac{1}{2m}(\mathbf{p} - e\mathbf{A})^2 \tag{8.42}$$

where $\mathbf{p}$ is the particle momentum and we assume that all static fields are absent so that the electric field $\mathbf{E}$ and magnetic field $\mathbf{B}$ of the wave are determined by

$$\begin{aligned} \operatorname{div} \mathbf{A} &= 0 \\ \mathbf{E} &= -\frac{\partial \mathbf{A}}{\partial t} \\ \mathbf{B} &= \operatorname{curl} \mathbf{A} \end{aligned} \tag{8.43}$$

Equation (8.42) can be re-written as

$$H = \frac{p^2}{2m} - \frac{e}{m}(\mathbf{p} \cdot \mathbf{A}) + \frac{e^2 A^2}{2m} \tag{8.44}$$

hence, if the term $e^2A^2/2m$ can be neglected, we can see that in the presence of the external field the electron acquires an additional energy

$$H' = -\frac{e}{m}(\mathbf{p} \cdot \mathbf{A}) \tag{8.45}$$

Now if τ is a typical time period of the wave we can write (cf. eq. (8.43))

$$\mathbf{E} \sim -\frac{\mathbf{A}}{\tau} \tag{8.46}$$

and

$$\mathbf{p} \sim \frac{m\mathbf{r}}{\tau} \tag{8.47}$$

so that the additional energy H' is

$$H' = e(\mathbf{r} \cdot \mathbf{E}) \tag{8.48}$$

This is now simply the energy of an electric dipole of moment $e\mathbf{r}$ in an electric field $\mathbf{E}$. The matrix element $\langle i|H'|f\rangle$ of the transition has the form:

$$\langle i|H'|f\rangle = e\langle i|\mathbf{r}\,.\,\mathbf{E}|f\rangle = e\{E_x\langle i|x|f\rangle + E_y\langle i|y|f\rangle + E_z\langle i|z|f\rangle\} \quad (8.49)$$

and such transitions are therefore referred to as *dipole* transitions. Equation (8.49) is still true for the quantum mechanical case in which $\mathbf{p}$ must be replaced by $-i\hbar(\partial/\partial\mathbf{r})$ and we must remember that components of $\mathbf{p}$ and $\mathbf{A}$ do not commute.

In order to decide then, whether a dipole transition is forbidden or not, we must determine whether the matrix element of the dipole moment, given by eq. (8.49), is zero or not.

As we discovered in section 3.11 a matrix element of the form $\langle i|x|f\rangle$ *will* be zero unless the direct product $\mathbf{\Gamma}^{i*}(R) \otimes \mathbf{\Gamma}^{x}(R) \otimes \mathbf{\Gamma}^{f}(R)$ contains the totally symmetric representation (usually denoted by $\mathbf{\Gamma}^{1}(R)$) where $\mathbf{\Gamma}^{i}(R)$, $\mathbf{\Gamma}^{x}(R)$, and $\mathbf{\Gamma}^{f}(R)$ are the representations (not necessarily reducible) to which the functions $\psi_i(\mathbf{r})$, x, and $\psi_f(\mathbf{r})$ respectively belong. Whether or not the direct product $\mathbf{\Gamma}^{i*}(R) \otimes \mathbf{\Gamma}^{d}(R) \otimes \mathbf{\Gamma}^{f}(R)$ (where d is x, y, or z) contains $\mathbf{\Gamma}^{1}(R)$ can be assessed by making use of the character table of the group concerned. In fact the character of the direct product representation is simply the product of the characters for each representation, i.e.,

$$\chi^{i*\otimes d\otimes f}(R) = \chi^{i*}(R)\chi^{d}(R)\chi^{f}(R) \quad (8.50)$$

and we can use eq. (8.50) to determine a_1 the number of times $\mathbf{\Gamma}^{1}(R)$ appears in the direct product representation from

$$a_1 = \frac{1}{g}\sum_{R} \chi^{1*}(R)\chi^{i*}(R)\chi^{d}(R)\chi^{f}(R) \quad (8.51)$$

where g is the order of the group.

As an example we will consider again the π-orbitals of benzene for which we shall use the group D_{6h} (as this is the full group of the benzene molecule) whose character table is in appendix C. Now there are six electrons to be distributed among the π-orbitals and there are obviously many ways in which this can be done. In the ground state, for example, there will be two electrons in the A_{2u} level and four in the E_{1g} level so that the state function of the ground state of the *whole molecule* is therefore some product function of these individual π-orbitals. Such a product function must be a basis function of the direct product representation

$$\begin{aligned} &\mathbf{A}_{2u} \otimes \mathbf{A}_{2u} \otimes \mathbf{E}_{1g} \otimes \mathbf{E}_{1g} \otimes \mathbf{E}_{1g} \otimes \mathbf{E}_{1g} \\ &\quad = 3\mathbf{A}_{1g} \oplus 3\mathbf{A}_{2g} \oplus 5\mathbf{E}_{2g} \end{aligned} \quad (8.52)$$

However, the product function is a closed shell configuration and must therefore belong to A_{1g}. We need therefore only consider electronic transitions from a state with symmetry A_{1g} to other possible symmetry states of the molecule.

We shall begin by observing (from appendix C) that z belongs to A_{2u}, whereas both x and y belong to E_{1u}. Hence for an allowed transition we must find those representations $\mathbf{\Gamma}^f(R)$ of the final state for which

$$z: \quad \mathbf{A}_{1g}^* \otimes \mathbf{A}_{2u} \otimes \mathbf{\Gamma}^f(R) = \mathbf{A}_{1g} \oplus \cdots \tag{8.53}$$

$$x, y: \quad \mathbf{A}_{1g}^* \otimes \mathbf{E}_{1u} \otimes \mathbf{\Gamma}^f(R) = \mathbf{A}_{1g} \oplus \cdots \tag{8.54}$$

where, for clarity, we have used the labels A_{1g}, A_{2u}, E_{1u} to indicate the known representations. Naturally if the direct products on the left-hand sides of eqs. (8.53) and (8.54) happen to be irreducible representations then the right-hand sides are simply $\mathbf{A}_{1g}$.

The search for the solutions $\mathbf{\Gamma}^f(R)$ is started by calculating the character of the direct product representation $\mathbf{\Gamma}^i(R) \otimes \mathbf{\Gamma}^d(R)$. For example, the character $\chi^{A_{1g} \otimes A_{2u}}(R)$ obtained from the characters of A_{1g} and A_{2u} namely:

$$\begin{aligned} \chi^{A_{1g}}(R) &= 1 \quad 1 \quad 1 \quad 1 \quad 1 \quad 1 \quad 1 \quad 1 \quad 1 \quad 1 \quad 1 \quad 1 \\ \chi^{A_{2u}}(R) &= 1 \quad 1 \quad 1 \quad 1 \quad -1 \quad -1 \quad -1 \quad -1 \quad -1 \quad -1 \quad 1 \quad 1 \end{aligned} \tag{8.55}$$

is

$$\chi^{A_{1g} \otimes A_{2u}}(R) = 1 \quad 1 \quad 1 \quad 1 \quad -1 \quad -1 \quad -1 \quad -1 \quad -1 \quad -1 \quad 1 \quad 1 \tag{8.56}$$

which is of course the character of A_{2u}. It should now be obvious that the only solution for $\mathbf{\Gamma}^f(R)$ is $\mathbf{A}_{2u}$ because this is the only character row by which we can multiply eq. (8.56) to give the characters of A_{1g}. In this case the direct product representation is irreducible. Thus the only allowed electronic transition from A_{1g} due to an interaction polarized along the z-axis is to A_{2u}.

Similarly $\mathbf{A}_{1g} \otimes \mathbf{E}_{1u}$ of eq. (8.53) can easily be shown to be $\mathbf{E}_{1u}$ and the unknown $\mathbf{\Gamma}^f(R)$ can be found by multiplying the row of characters of E_{1u} by other rows of the character table. However, we soon find that E_{1u} is the only row which produces a representation containing A_{1g}. Now the representation $\mathbf{E}_{1u} \otimes \mathbf{E}_{1u}$ is reducible and has a character given by

$$\chi^{E_{1u} \otimes E_{1u}}(R) = 4 \quad 1 \quad 1 \quad 4 \quad 0 \quad 0 \quad 4 \quad 1 \quad 1 \quad 4 \quad 0 \quad 0 \tag{8.57}$$

and the number of times A_{1g} appears in this irreducible representation (cf. eq. (8.51)) is

$$a_{A_{1g}} = \tfrac{1}{24}\{4 + 2 + 2 + 4 + 0 + 0 + 4 + 2 + 2 + 4 + 0 + 0\} = 1 \tag{8.58}$$

Finally, therefore, the only electronic transitions, by means of dipole interactions, from the ground state are

$$z: \quad A_{1g} \rightarrow A_{2u} \tag{8.59}$$

$$x, y: \quad A_{1g} \rightarrow E_{1u} \tag{8.60}$$

The first excited state of benzene has a configuration with two electrons in the state A_{2u}, three electrons in E_{1g} and one electron in E_{2u}. The state function of this configuration is therefore a basis function for the reducible direct product representation

$$\mathbf{A}_{2u} \otimes \mathbf{A}_{2u} \otimes \mathbf{E}_{1g} \otimes \mathbf{E}_{1g} \otimes \mathbf{E}_{1g} \otimes \mathbf{E}_{2u} = \mathbf{E}_{1g} \otimes \mathbf{E}_{1g} \otimes \mathbf{E}_{1g} \otimes \mathbf{E}_{2u} \qquad (8.61)$$

The character $\chi(R)$ of this representation can be obtained by means of eq. (3.79) and the character table of D_{6h} (cf. appendix C). The answer is

$$\chi: \quad 16 \quad -1 \quad 1 \quad -16 \quad 0 \quad 0 \quad -16 \quad 1 \quad -1 \quad 16 \quad 0 \quad 0$$

Thus using eq. (2.37) we can show that the only non-zero coefficients are

$$a_{B_{1u}} = 3; \qquad a_{B_{2u}} = 3; \qquad a_{E_{1u}} = 5 \qquad (8.62)$$

Hence

$$\mathbf{E}_{1g} \otimes \mathbf{E}_{1g} \otimes \mathbf{E}_{1g} \otimes \mathbf{E}_{2u} = 3\mathbf{B}_{1u} \oplus 3\mathbf{B}_{2u} \oplus 5\mathbf{E}_{1u} \qquad (8.63)$$

The first excited state therefore consists of eleven possible energy levels. (N.B. the one-electron energy level diagram of Fig. 8.7 would appear to give only one level instead of eleven; however in a many-electron system there must always be some interactions which lift degeneracies.) Finally then, the transition rules of eqs. (8.59) and (8.60) tell us that transitions can only occur to the E_{1u} components of the excited state.

8.4 Vibrational spectra

In chapter 7 we considered the normal modes of vibration of a molecule. In fact we saw that, in general, a molecule has $3N - 6$ vibrational modes, three rotational modes, and three translational modes. Thus when we consider the spectra of molecules we must, in principle, not only take electronic transitions into account but also transitions which arise from rotational and vibrational levels. (N.B. the translational modes are not quantized.)

The vibrational levels arise from the quantization of the vibrational modes of the molecule and the states of vibration, neglecting anharmonic effects, are some linear combination of independent normal modes which can either be degenerate or non-degenerate. We regard a normal mode as a simple harmonic oscillator with a frequency ν, say, whose energy levels are given by

$$E = (v + \tfrac{1}{2})h\nu \qquad (8.64)$$

However, if two normal modes are degenerate then we have a situation where two normal modes have the *same* frequency so that the energy is

$$E = (v_1 + \tfrac{1}{2} + v_2 + \tfrac{1}{2})h\nu = (v_1 + v_2 + 1)h\nu \qquad (8.65)$$

where v_1 and v_2 are quantum numbers associated with each mode and we can simply write $v_1 + v_2$ as another integer v. The extension of the argument to higher degrees of degeneracy is obvious.

The electronic energy levels are, in terms of wave-numbers, typically 10^4–10^5 cm^{-1} apart whereas the vibrational levels are typically 10^2–10^4 cm^{-1} apart. Also the rotational motion of the molecule imposes a fine structure on the vibrational levels giving rise to rotational levels which are typically 10^{-1}–10^{-2} cm^{-1} apart. Thus while it is clear that a careful account must be taken of the role of the various levels of a molecule during a transition it is also clear that in certain regimes we can study the various types of levels quite separately. We will therefore now concentrate entirely on the vibrational levels which are studied in the infra-red region, the spectrum, i.e., in the wave-number range 10^2–10^4 cm^{-1}. In this regime electronic transitions do not occur because there is not enough energy to transfer electrons from one level to another. We will not discuss the rotational levels because we are not interested here in the rotational fine structure.

Each normal mode is regarded as a one-dimensional independent oscillator whose wave-function ψ satisfies

$$\frac{d^2\psi}{d^2x^2} - \frac{2m}{\hbar^2}(2\pi^2 m\nu^2 x^2)\psi = -\frac{2m}{\hbar^2}E\psi \qquad (8.66)$$

where x is the displacement coordinate, m is the reduced mass of the system, and ν is the vibrational frequency. If we now define the normal coordinate $q = \sqrt{m}x$ and write $\lambda = 4\pi^2\nu^2$ we can transform eq. (8.66) to

$$\frac{d^2\psi}{dq^2} + \frac{2}{\hbar^2}\left(E - \frac{\lambda}{2}q^2\right)\psi = 0 \qquad (8.67)$$

whose solution is

$$\psi_v(q) = NH_v(q)\exp\left\{-\frac{\lambda^{1/2}q^2}{2\hbar}\right\} \qquad (8.68)$$

where N is a normalization constant, $H_v(q)$ is a Hermite polynomial, and the total energy of the oscillator is

$$E = (v + \tfrac{1}{2})h\nu \qquad (8.69)$$

The first three Hermite polynomials are

$$H_0(q) = 1; \qquad H_1(q) = \frac{2q\lambda^{1/4}}{\hbar^{1/2}}; \qquad H_2(q) = \frac{4q^2\lambda^{1/2}}{\hbar} - 2 \qquad (8.70)$$

The total vibrational wave-function of a system possessing $3N - 6$ normal coordinates q_i is

$$\psi_v = N \exp\left\{-\frac{1}{2\hbar}\sum_{i=1}^{3N-6} \lambda_i^{1/2} q_i^2\right\} \prod_{i=1}^{3N-6} H_{v_i}(q_i) \tag{8.71}$$

where N is again a normalization constant.

The ground state function in which $v_i = 0$ for all the i modes is simply

$$\psi_v = N \exp\left\{-\frac{1}{2\hbar}\sum_{i=1}^{3N-6} \lambda_i^{1/2} q_i^2\right\} = N\psi_v^G \tag{8.72}$$

which we can see is a totally symmetric function because the argument of the exponential has the same form as the potential energy. Hence the symmetry of any vibrational state is entirely determined by the Hermite polynomials.

Let us consider first of all a single-quantum state in which $v_1 = 1$ and $v_i = 0$ for $i \neq 1$. The wave-function for such a state is then

$$\psi_v = N\psi_v^G q_1 \tag{8.73}$$

where we have absorbed the other factors of $H_1(q_1)$ into the normalization constant. The symmetry of ψ_v is determined by q_1 and hence will possess the symmetry of the normal mode, i.e., ψ_v acts as basis function for the same irreducible representation as q_1. A higher energy state can be obtained by putting two quanta into the system. We can do this by putting the two quanta into the same normal mode giving rise to an excited level known as an *overtone* level. Alternatively we can put the two quanta into different normal modes; thus producing a so-called *combination* level.

For an overtone level the wave-function is defined by setting $v_1 = 2$, $v_i = 0$ for $i \neq 1$ and can be written as

$$\psi_v^O = N\psi_v^G H_2(q_1) = N\psi_v^G\left(\frac{4\lambda^{1/2}}{\hbar} q_1^2 - 2\right) \tag{8.74}$$

If q_1 is non-degenerate q_1^2 will belong to the totally symmetric representation and hence ψ_v^O will be totally symmetric.

For a combination level we have, for instance, $v_1 = 1, v_2 = 1$ and $v_i = 0$, $i \neq 1, i \neq 2$ and the wave-function becomes

$$\begin{aligned}\psi_v^C &= N\psi_v^G H_1(q_1) H_1(q_2) \\ &= \frac{4\lambda_1^{1/4}\lambda_2^{1/4}}{\hbar} N\psi_v^G q_1 q_2\end{aligned} \tag{8.75}$$

In this case ψ_v^C transforms as the product function $q_1 q_2$ and hence will belong to the direct product representation associated with q_1 and q_2.

As an example let us consider the vibrational modes of ammonia. The symmetry operations which leave this molecule invariant constitute the group C_{3v}. These operations are E, two C_3 rotations, and three σ_v reflections (cf. Fig. 8.8). This molecule contains four atoms and hence, in the manner of section 7.5, gives rise to a twelve-dimensional reducible representation of C_{3v}.

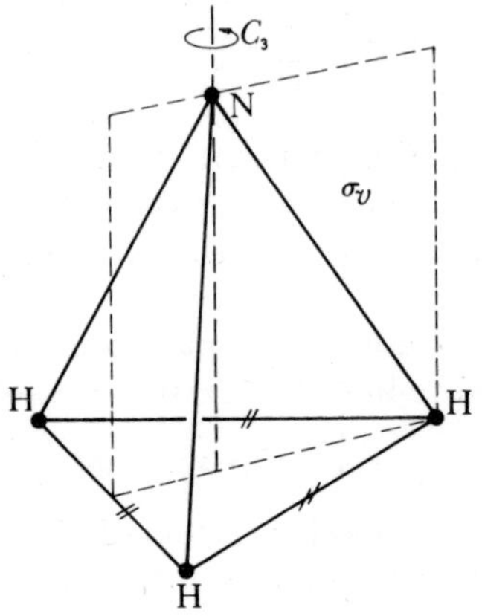

Fig. 8.8 *Symmetry elements of the group C_{3v} the symmetry group of the ammonia molecule.*

The character of the representation of the identity element is obviously 12. The character of the representations of C_3 and σ_v can be worked out by assigning coordinates to each atom and performing the symmetry operations (cf. section 7.5(a)). The results are 0 for C_3 and 2 for σ_v. Hence by using the character table of C_{3v} given in appendix C we obtain

$$\mathbf{\Gamma}^c(R) = 3\mathbf{A}_1 \oplus \mathbf{A}_2 \oplus 4\mathbf{E} \tag{8.76}$$

where $\mathbf{\Gamma}^c(R)$ is the reduced Cartesian representation and A_1, A_2, and E now label the irreducible representations of C_{3v} (cf. section 2.7). As we pointed out in chapter 7, $\mathbf{\Gamma}^c(R)$ actually contains $\mathbf{\Gamma}^t(R)$ and $\mathbf{\Gamma}^r(R)$, i.e., reducible representations of translational and rotational operations; hence

$$\mathbf{\Gamma}^c(R) = \mathbf{\Gamma}^v(R) \oplus \mathbf{\Gamma}^t(R) \oplus \mathbf{\Gamma}^r(R) \tag{8.77}$$

where $\mathbf{\Gamma}^v(R)$ is the reducible representation of the six normal vibrational modes. $\mathbf{\Gamma}^t(R)$ and $\mathbf{\Gamma}^r(R)$ are respectively $\mathbf{A}_1 \oplus \mathbf{E}$ and $\mathbf{A}_2 \oplus \mathbf{E}$ (these can be obtained from the character table of C_{3v} or from first principles as discussed in section 7.5(a)). Hence

$$\mathbf{\Gamma}^v(R) = 2\mathbf{A}_1 \oplus 2\mathbf{E} \tag{8.78}$$

showing that there are two singly degenerate and two doubly degenerate vibrational modes. The total energy is therefore

$$E = (v_1 + \tfrac{1}{2})h\nu_1 + (v_2 + \tfrac{1}{2})h\nu_2 + (v_3 + 1)h\nu_3 + (v_4 + 1)h\nu_4 \quad (8.79)$$

where ν_1 and ν_2 are the frequencies of the normal modes with A_1 symmetry and ν_3 and ν_4 are the frequencies of the normal modes with E symmetry. These energy levels are illustrated in Fig. 8.9 for certain one quantum and two quantum states.

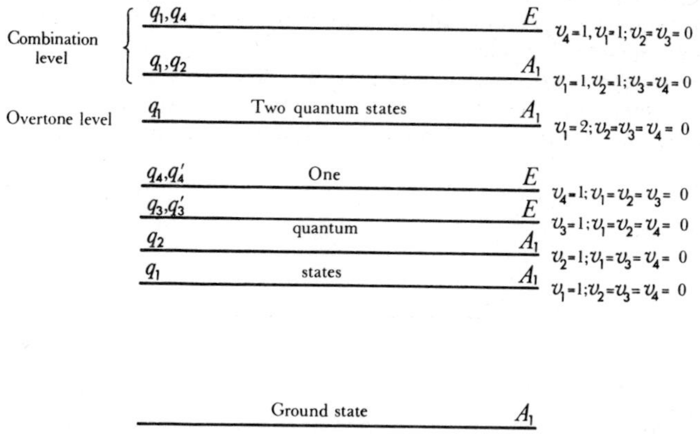

Fig. 8.9 *The vibrational energy levels of ammonia showing the four one quantum states and some of the two quantum states. The values of the quantum numbers are shown beside the levels. $q_1, \ldots, q_4'$ label the normal modes.*

The probability of a transition between vibrational levels of a particular normal mode is proportional to the square of the matrix element $\int \psi_v^{i*}(q) \times \mathbf{M}\psi_v^f(q)\,dq$ where i and f denote initial and final vibrational states and $\mathbf{M}$ a dipole moment which has components M_x, M_y, and M_z (N.B. the transition is a dipole transition in the sense discussed in section 8.3). The components of the dipole moment can in this case be expanded into the form

$$M_x = (M_x)_0 + \left(\frac{\partial M_x}{\partial q}\right)_0 q \quad (8.80)$$

and so on where we must interpret $(M_x)_0$ as the *permanent* dipole possessed by the molecule in the ground state. Obviously since $(M_x)_0$ is independent of q it cannot give any contribution to the transition matrix element. Hence the transition matrix element is essentially $\int \psi_v^i(q)\{q\}\psi_v^f(q)\,dq$. This matrix will vanish unless its direct product representation contains the totally symmetric representation. In fact by using the Hermite polynomials it can be readily shown that the selection rule is $\Delta v = \pm 1$.

8.5 Problems

8.1 Naphthalene ($C_{10}H_8$) consists of two hexagonal rings joined along one edge with a carbon atom at each intersection. Show that the symmetry operations of the molecule belong to D_{2h}. Using the ten p_z-orbitals of the carbon atoms set up a reducible Cartesian representation of D_{2h} and determine its irreducible components. Hence comment on the energy level diagram of naphthalene.

8.2 Construct the projection operators of the group D_{2h} and produce the symmetrized basis using the p_z-orbitals. Having done this derive and solve the secular equation for naphthalene (cf. eq. (8.26)) using the Hückel approximation.

8.3 Determine the possible electronic transitions that can take place from the ground state of naphthalene (which, of course, has symmetry A_{1g}).

8.4 The symmetry operations of the water molecule belong to C_{2v}. Determine the electronic selection rules.

8.5 Find the symmetry operations of ethylene (C_2H_4) given in the figure. Determine the symmetry of the normal modes of vibration. Select some of these normal modes and construct and label some overtone and combination levels.

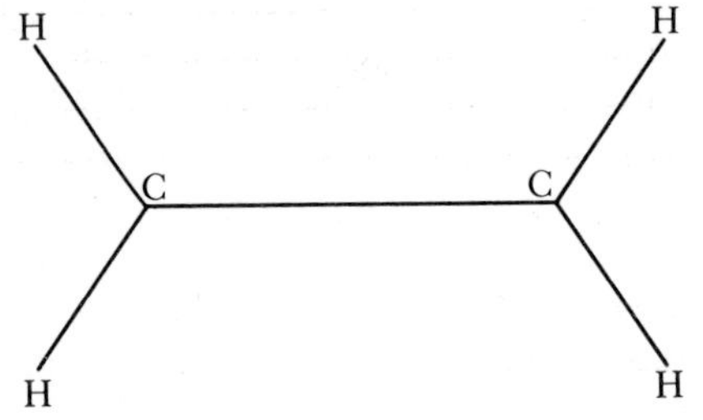

9 Symmetry of atoms

9.0 Introduction

In this chapter we shall establish and exploit the symmetry of an atom by considering the behaviour of its electrons in the attractive field of the nucleus. An understanding of this behaviour, as is well-known from atomic spectroscopy, is founded upon the use of certain quantum numbers which are used to classify the discrete electronic energy states. It is therefore important to develop a theoretical framework which, while unable to provide quantitative results, will enable us to understand the origin of the labels that we refer to as quantum numbers. Group theory is ideally suited to achieve this aim because, as we shall see, by exploiting the symmetry of the atom (or precisely the symmetry of the Hamiltonian of the atom) we can develop a theoretical model which is complete, except for its inability to predict the actual magnitudes of the energy levels.

In chapter 3 we discussed the way in which eigenvalue and eigenfunction calculations can be considerably simplified with the aid of group theoretical techniques. Such simplifications centred on the use of a set of basis functions, which were not necessarily eigenfunctions of the Hamiltonian, to construct a matrix representation of the Hamiltonian. Projection operators and character tables then allowed us to produce, from the initial set of basis functions, another set of so-called *symmetrized* functions which could then be used to bring the matrix representation to block form.

Such an approach is, of course, very valuable but it was developed only with *finite* groups in mind. It is therefore well suited to a molecular system such as benzene, whose energy levels were enumerated in section 8.2 but is rather more complicated to set up for a spherically symmetric system, such as a free atom or ion for which we have to use the *continuous* full rotation group R_3 (cf. section 1.9).

The relevance of R_3 can be seen by appreciating the fact that, in the study of atoms, we usually consider the valence electrons to be moving in a central field. This is an approximation which amounts to assuming that the electrons have a spherically symmetric potential energy so that the Hamiltonian is invariant under the group of symmetry operations of a sphere. This group is actually the direct product group $R_3 \otimes S_2$ where R_3 is the continuous full rotation group and S_2 is the finite group that contains only the identity and the inversion operator.

There are several ways of formulating the Hamiltonian of an atom within a central field approximation; but these need not concern us here because, provided we are satisfied that the Hamiltonian *is* invariant under the group R_3, this is virtually all that we need to know. From this knowledge we will be able to draw many important and exact conclusions.

9.1 Basis functions and irreducible representations of R_3

The Hamiltonian of an atom always contains kinetic energy terms which in turn contain the operator ∇^2. This operator in the spherical polar coordinates (r, θ, ϕ) is

$$\nabla^2 \equiv \frac{1}{r^2}\left\{\frac{\partial}{\partial r}\left(r^2\frac{\partial}{\partial r}\right) + \frac{1}{\sin\theta}\frac{\partial}{\partial\theta}\left(\sin\theta\frac{\partial}{\partial\theta}\right) + \frac{1}{\sin^2\theta}\frac{\partial^2}{\partial\phi^2}\right\} \tag{9.1}$$

Now, even if we add to ∇^2 a term corresponding to the potential energy of the electrons, it will not affect the angular part of the Hamiltonian because we are making a central field approximation, which requires the potential energy to depend only on $|\mathbf{r}|$. Let us therefore consider the angular part L^2 of ∇^2 which is given by

$$L^2 \equiv -\left\{\frac{1}{\sin\theta}\frac{\partial}{\partial\theta}\left(\sin\theta\frac{\partial}{\partial\theta}\right) + \frac{1}{\sin^2\theta}\frac{\partial^2}{\partial\phi^2}\right\} \tag{9.2}$$

where we should now recognize that L^2 is a familiar operator that possesses as eigenfunctions the spherical harmonics $Y_l^m(\theta, \phi)$. The label l is a positive integer (including zero) and the label m is another integer lying in the range $-l \leqslant m \leqslant l$. The definition of $Y_l^m(\theta, \phi)$ that we shall use is

$$Y_l^m(\theta, \phi) = (-1)^{1/2(m+|m|)}\left[\frac{(2l+1)(l-|m|)!}{4\pi(l+|m|)!}\right]^{1/2} P_l^{|m|}(\cos\theta)\,\mathrm{e}^{im\phi} \tag{9.3}$$

where the functions $P_l^{|m|}(\cos\theta)$ are the associated Legendre functions.

The eigenvalues of L^2 are $l(l+1)$, i.e.,

$$L^2 Y_l^m(\theta, \phi) = l(l+1)Y_l^m(\theta, \phi) \tag{9.4}$$

Also since m can, for a particular value of l, take on the $(2l+1)$ values from $-l$ to l it is clear that (9.4) is satisfied by the $(2l+1)$ spherical harmonics. There is thus a $(2l+1)$-fold degeneracy associated with each eigenvalue of (9.4) because each function is labelled by m.

An operator P_R that belongs to the group R_3 (a subgroup of the group of the Hamiltonian), where the suffix R indicates an arbitrary rotation, commutes with the Hamiltonian (cf. section 3.7) so that

$$\begin{aligned} P_R L^2 Y_l^m(\theta, \phi) &= P_R l(l+1) Y_l^m(\theta, \phi) \\ L^2 P_R Y_l^m(\theta, \phi) &= l(l+1) P_R Y_l^m(\theta, \phi) \end{aligned} \tag{9.5}$$

which shows that $P_R Y_l^m(\theta, \phi)$ is either $Y_l^m(\theta, \phi)$, another spherical harmonic $Y_l^{m'}(\theta, \phi)$ or a linear combination of harmonics from the set of $(2l + 1)$ functions. We can therefore conclude that the spherical harmonics $Y_l^m(\theta, \phi)$ can act as basis functions for a $(2l + 1)$-dimensional representation of P_R. Such a representation, which we will call $\mathbf{D}^l(R)$ is defined by

$$P_R\widetilde{(\mathbf{Y}_l^m(\theta, \phi))} = \widetilde{(\mathbf{Y}_l^m(\theta, \phi))}\mathbf{D}^l(R) \tag{9.6}$$

We must at this early stage point out that since l is an integer the dimension of $\mathbf{D}^l(R)$ is necessarily an odd number. Even-dimensional representations can be constructed and are interpreted using the concept of electron *spin*; however, these are not ordinary representations in the sense of eq. (3.8) and they will be discussed later.

As we have seen in the earlier chapters, a lot of important results can be obtained from a knowledge of the character of a representation rather than the representation itself. We will therefore direct our attention to the determination of $\chi^l(R)$, the character of $\mathbf{D}^l(R)$. This objective could be achieved by first of all determining $\mathbf{D}^l(R)$ and then subsequently calculating the trace of $\mathbf{D}^l(R)$. Now this can prove to be rather a difficult exercise; however, fortunately, we can make use of the fact that the characters of all elements in the same class, are identical. In order to be able to do this we must recognize that a rotation through an angle α, say, about a particular axis through the centre of a sphere is related to an identical rotation about any other axis passing through the centre of the sphere by a *similarity* transformation (if S indicates some other rotation and R indicates a rotation of α about a certain axis then $S^{-1}RS$ is now merely a rotation of α about some new axis). Thus all rotations of α about all possible axes are in the same class, and hence have the same character.

We need, therefore, only consider rotations about the z axis, say, with the assurance that the character we obtain from $\mathbf{D}^l(\alpha)$ which represents a rotation of α about the z axis, is numerically equal to the character we would obtain from $\mathbf{D}^l(R)$ where R is a rotation of α about an arbitrary axis.

The set of rotations about the z-axis constitutes the *axial* group C_∞ which is an Abelian subgroup of R_3. (The Abelian nature of C_∞ can be understood from the fact that successive rotations α_1 and α_2 must be the same as the successive rotations α_2 and α_1.)

Now if the basis function $Y_l^m(\theta, \phi)$ is subjected to an operation P_α, which produces a rotation of α about the z axis, then we can make use of the relationship (cf. eq. (9.3)) that

$$Y_l^m(\theta, \phi \pm \alpha) = \mathrm{e}^{\pm im\alpha} Y_l^m(\theta, \phi) \tag{9.7}$$

to produce $\mathbf{D}^l(\alpha)$ a matrix representation of P_α. Thus it follows from eq. (9.7) that the representation is obtained from

$$P_\alpha(Y_l^l(\theta,\phi),\ldots,Y_l^{-l}(\theta,\phi)) = (Y_l^l(\theta,\phi),\ldots,Y_l^{-l}(\theta,\phi))\begin{bmatrix} e^{-il\alpha} & & & \\ & e^{-i(l-1)\alpha} & & 0 \\ & & \ddots & \\ 0 & & & e^{il\alpha} \end{bmatrix} \tag{9.8}$$

so that

$$\mathbf{D}^l(\alpha) = \begin{bmatrix} e^{-il\alpha} & & & \\ & e^{-i(l-1)\alpha} & & 0 \\ & & \ddots & \\ 0 & & & e^{il\alpha} \end{bmatrix} \tag{9.9}$$

The trace of $\mathbf{D}^l(\alpha)$ is the character $\chi^l(\alpha)$ of the representation of P_α and is given by

$$\begin{aligned}\chi^l(\alpha) &= \sum_{m=-l}^{l} e^{im\alpha} \\ &= e^{-il\alpha}[1 + e^{i\alpha} + \ldots + e^{2il\alpha}] \\ &= e^{-il\alpha}\frac{[1 - e^{i(2l+1)\alpha}]}{1 - e^{i\alpha}} \\ &= e^{i\alpha/2}[e^{i(l+1/2)\alpha} - e^{-i(l+1/2)\alpha}]/(e^{i\alpha} - 1) = \frac{\sin(l+\frac{1}{2})\alpha}{\sin \alpha/2}\end{aligned} \tag{9.10}$$

Although we now possess the character $\chi^l(\alpha)$ we have not yet discussed the reducibility of $\mathbf{D}^l(\alpha) \equiv \mathbf{D}^l(R)$. We shall therefore clear this point up by asserting that the matrices $\mathbf{D}^l(R)$ are irreducible on the grounds that it is in fact impossible to produce a *single* matrix which will, by a similarity transformation, simultaneously bring *all* the matrices $\mathbf{D}^l(R)$ to block form. We therefore conclude that the basis functions of the full rotation group give rise to $(2l + 1)$-dimensional *irreducible* representations, i.e., odd dimensional representations.

Now R_3, of course, commutes with the Hamiltonian of the atom so that the basis functions of R_3 are eigenfunctions of the Hamiltonian and vice versa. Since this is the case we can now appreciate that the eigenfunctions of the Hamiltonian can be *labelled* with the *quantum numbers* l and m where, group theoretically, l denotes the irreducible representation, and m denotes the column to which the functions belong. Also, as we discussed in section 3.7, we can see that because a distinct eigenvalue of the Hamiltonian can be associated with each irreducible representation $\mathbf{D}^l(R)$

then each such eigenvalue is in fact, $(2l + 1)$-fold degenerate since it is associated with $(2l + 1)$ basis functions. (N.B. these results will be modified when electron spin is incorporated into the development.) Actually although we refer to l and m as labels we can perhaps anticipate here that they are, respectively, the familiar angular momentum and magnetic quantum numbers.

We should point out here that the part of the Hamiltonian which depends only on $|\mathbf{r}|$ is also involved in the determination of the electron eigenvalues and introduces a quantum number n; the principal quantum number where n labels the co-called radial eigenfunction $R_{ln}(r)$ and does not arise from the spherical symmetry. The full eigenfunction of an electron in an atom is therefore $\psi_{nlm}(r, \theta, \phi) = R_{ln}(r)Y_l^m(\theta, \phi)$ where, since $R_{ln}(r)$ is invariant under the group R_3, it is clear that $\psi_{nlm}(r, \theta, \phi)$ can also act as a basis function for the representation $\mathbf{D}^l(R)$. The energy of an electron is therefore a function of both n and l where, as we know from the theory of the hydrogen atom, l has a maximum value of $(n - 1)$. Thus for a particular value of n there are n associated values of l, each value of l corresponding to a distinct energy level (for hydrogen these levels are practically indistinguishable).

9.2 The rotation–inversion group $R_3 \otimes S_2$

Quite apart from being invariant under the operations of the group R_3 the Hamiltonian of an atom, since the potential energy function is spherically symmetric, is also invariant under the operations of S_2, where S_2 contains only the identity and the inversion operations: namely E and i. Each element of S_2 is obviously in a class by itself so that, (since the sum of the squares of the dimensions of their irreducible representations is equal to two, i.e., the number of group elements, cf. eq. (2.23)), their irreducible representations must be one-dimensional. These irreducible representations of S_2 must clearly be either 1 or -1. The character table of S_2, which is also, therefore, the table of its irreducible representations is

	A	
Irreducible representation	E	i
$\Gamma^g(A)$	1	1
$\Gamma^u(A)$	1	-1

where we have adopted the conventional labels for the matrices $\mathbf{\Gamma}(A)$.

The basis functions of $\mathbf{\Gamma}^g(A)$ and $\mathbf{\Gamma}^u(A)$ are, respectively, odd and even because if we introduce the operators P_E and P_i we have

$$\begin{aligned} P_E\psi_g &= \psi_g\mathbf{\Gamma}^g(E) = \psi_g \\ P_i\psi_g &= \psi_g\mathbf{\Gamma}^g(i) = \psi_g \\ P_E\psi_u &= \psi_u\mathbf{\Gamma}^u(E) = \psi_u \\ P_i\psi_u &= \psi_u\mathbf{\Gamma}^u(i) = -\psi_u \end{aligned} \tag{9.11}$$

The eigenvalues of P_E and P_i are obviously ± 1 so that eqs. (9.11) can be taken as a definition of even or odd functions which are said to have even or odd *parity*. The symbols g and u come from the German g: gerade $\equiv$ even, u: ungerade $\equiv$ odd.

We can now conclude that, since the group of the Hamiltonian contains the group S_2, the eigenfunctions of the. Hamiltonian of an atom must also be basis functions for a representation of S_2. Hence the eigenfunctions of such a Hamiltonian must always be either odd or even.

Now the spherical harmonics which we have previously used as basis functions for a $(2l + 1)$-dimensional representation of R_3 are, in view of the above remarks, also basis functions for a representation of S_2. We can therefore from $Y_l^m(\theta, \phi)$ actually construct irreducible representations of S_2. For example

$$P_i Y_l^m(\theta, \phi) = Y_l^m(\pi - \theta, \phi + \pi) = Y_l^m(\theta, \phi)(-1)^l \tag{9.12}$$

where we have used $e^{im(\phi+\pi)} = (-1)^{|m|} e^{im\phi}$ and $P_l^{|m|}(\cos(\pi - \theta)) = (-1)^{l-|m|} P_l^{|m|}(\cos\theta)$. Equation (9.12) shows that the irreducible representation of P_i is 1 if l is even and -1 if l is odd, which fits in with our previous conclusions. However, the development we now have obtained will enable us to produce in a systematic way irreducible representations of the rotation–inversion group.

The rotation–inversion group of the Hamiltonian is simply the group which contains all the elements formed from the product of E and i with all the elements of R_3. It is, in other words, the direct-product group $R_3 \otimes S_2$ $(= S_2 \otimes R_3)$, whose matrix representations are the direct product matrices formed from the irreducible representations of R_3 and S_2 (cf. section 3.6). Thus the irreducible matrix representations of $R_3 \otimes S_2$ are

$$\begin{aligned} \mathbf{D}^{l\otimes g}(AR) &= \mathbf{\Gamma}^g(A) \otimes \mathbf{D}^l(R);\ A = E, i \\ \mathbf{D}^{l\otimes u}(AR) &= \mathbf{\Gamma}^u(A) \otimes \mathbf{D}^l(R);\ A = E, i \end{aligned} \tag{9.13}$$

As we have seen in eq. (9.12), the irreducible representation of P_i *based* on $Y_l^m(\theta, \phi)$ is $\mathbf{\Gamma}(i) = (-1)^l$ which means that $\mathbf{\Gamma}(i) = \mathbf{\Gamma}^g(i)$ for *even* l and

$\Gamma(i) = \Gamma^u(i)$ for *odd* values of l. Thus we can write

$$
\begin{aligned}
&\text{even } l: && \mathbf{D}^{l\otimes g}(ER) = \mathbf{D}^l(R); && \chi^{l\otimes g}(ER) = \chi^l(R) \\
& && \mathbf{D}^{l\otimes g}(iR) = \mathbf{D}^l(R); && \chi^{l\otimes g}(iR) = \chi^l(R) \\
&\text{odd } l: && \mathbf{D}^{l\otimes u}(ER) = \mathbf{D}^l(R); && \chi^{l\otimes u}(ER) = \chi^l(R) \\
& && \mathbf{D}^{l\otimes u}(iR) = -\mathbf{D}^l(R); && \chi^{l\otimes u}(iR) = -\chi^l(R)
\end{aligned}
\tag{9.14}
$$

The inclusion of parity, i.e., inversion is therefore quite a simple operation. Moreover, although l determines the parity of the function, we should recognize that g and u are also perfectly satisfactory quantum labels. This is only true for a one-electron atom. In the case of many-electron atoms the orbital angular momentum quantum number does *not* determine the parity.

9.3 Atoms in crystal fields

Armed with the results we have obtained so far we can digress a little from the main development to examine level splitting in crystals. It is possible to learn a great deal about the electronic structure of atoms or ions by investigating their behaviour in environments which have a lower symmetry than the full rotation group. As we saw in section 3.10 if a system with degenerate eigenvalues is placed in an environment of lower symmetry then the eigenvalues are 'split' into a number of levels. The reason for this is that the irreducible representation associated with a given degenerate level becomes a *reducible* representation when the symmetry is lowered; and the number of irreducible components possessed by the reduced representation is equal to the number of levels into which the original level will ultimately split.

For example in the case of an atom, placed in a cubic crystal, a $(2l + 1)$-dimensional representation of the full rotation group will become a $(2l + 1)$-dimensional reducible representation of the group O_h, which is a subgroup of the rotation–inversion group. We can therefore, by making use of eq. (9.10) and the character table of O_h, determine its irreducible components and hence the degree of splitting which takes place.

Obviously, while this topic is important it is also an extensive one so we will be content here to make a few additional remarks and illustrate the ideas with a particular example. Now a consideration of ions which have closed shells of electrons are of little interest, because, classically, if the total value of the orbital angular momentum is zero there will be no interaction with the crystal field. On the other hand, however, transition metal ions such as chromium or rare earth metal ions, such as cerium have incomplete shells of electrons and are strongly affected by crystalline fields.

Hence, as an example, we shall now consider a situation where an $l = 2$ level of an atom is in a cubic environment; and although we will come to spin later on we are actually anticipating a situation in which any spin degeneracy is not lifted. First of all we shall obtain the character of the now reducible representation from eqs. (9.10) and (9.14). Thus from the character table of O_h, given in appendix C, we can see that its group operations are of the type E, C_3, C_2, C_4, iE, $S_6 = iC_3$, $\sigma_h = iC_2$, and $S_4 = iC_4$ so that only rotations through angles 0, $2\pi/3$, π, and $\pi/2$ are involved. Therefore, since l is even, we obtain from eq. (9.10) the following characters:

	E	$8C_3$	$6C_2'$	$6C_4$	$3C_2$	i	$8S_6$	$6\sigma_d$	$6S_4$	$3\sigma_h$
$\chi^{2\otimes g}(AR)$	5	-1	1	-1	1	5	-1	1	-1	1

From these values and the character table of O_h we can use eq. (2.37) to find the number, and types, of irreducible components of $\mathbf{D}^{2\otimes g}(AR)$. For example, to determine $a_{A_{1g}}$ the number of times A_{1g} appears we use eq. (2.37) to obtain

$$a_{A_{1g}} = \tfrac{1}{48}\{1(5)(1) + 8(-1)(1) + 6(1)(1) + 6(-1)(1) + 3(1)(1) + 1(5)(1) + 8(-1)(1) + 6(1)(1) + 6(-1)(1) + 3(1)(1)\} = 0 \tag{9.15}$$

Thus A_{1g} does not appear at all. It can be shown, in a similar manner, that the only non-zero results we obtain are

$$a_{E_g} = \tfrac{2}{48}\{1(5)(2) + 8(-1)(-1) + 6(1)(0) + 6(-1)(0) + 3(1)(2)\} = 1 \tag{9.16}$$

$$a_{T_{2g}} = \tfrac{2}{48}\{1(5)(3) + 8(-1)(0) + 6(1)(1) + 6(-1)(-1) + 3(1)(-1)\} = 1 \tag{9.17}$$

Thus we conclude, for $l = 2$, that

$$\mathbf{D}^{2\otimes g}(AR) = \mathbf{E}_g \oplus \mathbf{T}_{2g} \tag{9.18}$$

which shows that the five-fold degenerate level of the free atom splits, under the influence of the cubic crystal field, into two levels one of which is two-fold degenerate since E is two-dimensional ($\chi_{E_g}(E) = 2$) and one which is three-fold degenerate ($\chi_{T_{2g}}(E) = 3$). In fact, because the character under i, in this case, is positive we can say that only irreducible representations of O_h with *positive* characters under i can appear in $\mathbf{D}^{2\otimes g}(AR)$. Generalizing we can say that reducible representations with the label g reduce into components with the label g and reducible representations with the label u reduce only into components with a label u. This is a good example of parity conservation, i.e., a function with a given initial parity (even: 'g', or odd: 'u') retains the same parity.

9.4 Angular momentum and the rotation group

We have observed previously that the labels l and m of the basis functions of the full rotation group are quantum numbers and should in fact turn out to be the familiar orbital angular momentum and magnetic quantum numbers. However, it is not absolutely necessary to invoke the classical idea of angular momentum in order to interpret the possible electronic states of an atom. Indeed when we come to the question of electron spin it will immediately become apparent that the absence of a classical analogue actually undermines the usefulness of adhering too rigidly to angular momentum concepts. Nevertheless since it is common practice to talk in terms of angular momentum we will now establish the relationship between angular momentum and the rotation group.

We have up to now discussed R_3 without formally exploiting the fact that it is a continuous group as opposed to the finite groups we have considered in the earlier chapters. We can see that R_3 is a continuous group because all the operators are functions of continuously variable angles. Such a group is technically known as a Lie group.

It now makes sense, therefore, to examine the operation of $P_{\delta\alpha}$ on a function $f(\phi)$ of angle ϕ which transforms it to $f(\phi - \delta\alpha)$ where $\delta\alpha$ is an *infinitesimal* rotation about an arbitrary axis A. This operation is (cf. section 3.1)

$$\begin{aligned} P_{\delta\alpha} f(\phi) &= f(\phi - \delta\alpha) \\ &= f(\phi) - \frac{\partial f}{\partial \phi}\delta\alpha \\ &= \left(1 - \delta\alpha\frac{\partial}{\partial\phi}\right) f(\phi) \end{aligned} \tag{9.19}$$

Thus the action of $P_{\delta\alpha}$ on $f(\phi)$ is almost the identity operation, differing from it only by an *infinitesimal* amount $\delta\alpha(\partial/\partial\phi)$. We can now define an *infinitesimal operator*

$$I_A = -\frac{\partial}{\partial\phi} \tag{9.20}$$

in terms of which $P_{\delta\alpha}$ can be written as

$$P_{\delta\alpha} = 1 + \delta\alpha I_A \tag{9.21}$$

Obviously we can effect a rotation through a *finite* angle α by successive application of eq. (9.21) and indeed if $\delta\alpha$ is defined as α/n such that

$$\alpha = \lim_{\substack{\delta\alpha \to 0 \\ n \to \infty}} (n\delta\alpha) \tag{9.22}$$

then a finite rotation α is effected by

$$\lim_{n\to\infty}\left(1+\frac{\alpha}{n}I_A\right)^n f(\phi) = f(\phi-\alpha) \tag{9.23}$$

The factor $[1 + (\alpha/n)I_A]^n$ can be expanded as

$$\left(1+\frac{\alpha}{n}I_A\right)^n = 1 + n\left(\frac{\alpha I_A}{n}\right) + \frac{n(n-1)}{2!}\left(\frac{\alpha I_A}{n}\right)^2 + \ldots \tag{9.24}$$

showing that

$$\lim_{n\to\infty}\left(1+\frac{\alpha}{n}I_A\right)^n = 1 + \alpha I_A + \frac{\alpha^2 I_A^2}{2!} + \cdots = e^{\alpha I_A} \tag{9.25}$$

We can therefore formally express P_α as

$$P_\alpha = e^{\alpha I_A} \tag{9.26}$$

where the exponential form is merely an economical way of writing P_α, i.e., since I_A is $\partial/\partial\phi$ we can only give substance to P_α in its expanded form.

The connection between I_A and the quantum mechanical angular momentum operator can be established as follows. The classical angular momentum of a particle with a linear momentum $\mathbf{p} = (p_x, p_y, p_z)$ at a position $\mathbf{r} = (x, y, z)$ is the vector

$$\boldsymbol{l} = \mathbf{r} \times \mathbf{p} \tag{9.27}$$

and has Cartesian components

$$\begin{aligned} l_x &= yp_z - zp_y \\ l_y &= zp_x - xp_z \\ l_z &= xp_y - yp_x \end{aligned} \tag{9.28}$$

Now the operator form of $\mathbf{p}$, in the Schrödinger coordinate representation, is

$$\mathbf{p} = -i\hbar\nabla \tag{9.29}$$

where $\hbar = h/2\pi$ (h being Planck's constant) so that, by virtue of the correspondence principle, we obtain from (9.28) the angular momentum operators

$$\mathscr{L}_x = -i\hbar\left(y\frac{\partial}{\partial z} - z\frac{\partial}{\partial y}\right) \tag{9.30}$$

$$\mathscr{L}_y = -i\hbar\left(z\frac{\partial}{\partial x} - x\frac{\partial}{\partial z}\right) \tag{9.31}$$

$$\mathscr{L}_z = -i\hbar\left(x\frac{\partial}{\partial y} - y\frac{\partial}{\partial x}\right) \tag{9.32}$$

Also, in terms of conventional polar coordinates, (r, θ, ϕ), we can easily show that

$$\mathscr{L}_z = -i\hbar\frac{\partial}{\partial\phi} \tag{9.33}$$

The similarity of the infinitesimal operator (9.20) to the angular momentum operator (9.33) is now exposed; indeed if I_A refers to a rotation about the z axis, we obtain

$$\mathscr{L}_z = i\hbar I_A \tag{9.34}$$

Since the rotation about the z axis leads to an infinitesimal operator

$$I_z = -\left(x\frac{\partial}{\partial y} - y\frac{\partial}{\partial x}\right) \tag{9.35}$$

a rotation about the x or y axis must lead, respectively, to

$$\begin{aligned} I_x &= -\left(y\frac{\partial}{\partial z} - z\frac{\partial}{\partial y}\right) \\ I_y &= -\left(z\frac{\partial}{\partial x} - x\frac{\partial}{\partial z}\right) \end{aligned} \tag{9.36}$$

We will now see that these infinitesimal rotations add vectorially. For instance let us consider two vectors $\mathbf{n}$ and $\mathbf{r}$, then obviously the vector $(\mathbf{n} \times \mathbf{r})$ is perpendicular to both $\mathbf{n}$ and $\mathbf{r}$. An infinitesimal rotation of $\mathbf{r}$ to $\mathbf{r} + \delta\mathbf{r}$ about the axis $\mathbf{n}$ is therefore represented by the vector $\delta\mathbf{n}$ such that $\delta\mathbf{r}$ will, to first order, be parallel to the vector $\delta\mathbf{n} \times \mathbf{r}$. Hence two successive rotations $\delta\mathbf{n}$ and $\delta\mathbf{n}'$ about different axes produce

$$\delta\mathbf{r} = \delta\mathbf{n} \times \mathbf{r} + \delta\mathbf{n}' \times (\mathbf{r} + \delta\mathbf{n} \times \mathbf{r}) \approx (\delta\mathbf{n} + \delta\mathbf{n}') \times \mathbf{r}$$

which shows that *infinitesimal* rotations simply add vectorially. We can therefore use the infinitesimal operators I_x, I_y, I_z to express a rotation about an arbitrary axis, defined by the unit vector $\mathbf{u} = (u_x, u_y, u_z)$, as

$$I_u = \mathbf{u} \cdot \mathbf{I} = u_x I_x + u_y I_y + u_z I_z \tag{9.37}$$

which shows that $P_\alpha \equiv e^{-(i/\hbar)\alpha\mathbf{u}.\mathscr{L}}$ where $\mathscr{L} = (\mathscr{L}_x \mathscr{L}_y \mathscr{L}_z)$.

The basis functions $Y_l^m(\theta, \phi)$ are also eigenfunctions of $\mathscr{L}_z$, i.e.,

$$-i\hbar\frac{\partial}{\partial\phi} Y_l^m(\theta, \phi) = \hbar m Y_l^m(\theta, \phi) \tag{9.38}$$

so that the quantities $\hbar m$ are eigenvalues of the z-component of the angular momentum. Thus, since for a given l there are $(2l + 1)$ values of m, any influence on an atom which completely removes the $(2l + 1)$-fold degeneracy lifts it into $(2l + 1)$ distinct energy levels; each one being labelled by m. Because this splitting can be achieved by means of a magnetic field, m is referred to as the magnetic quantum number.

From eqs. (9.30) to (9.32) and the basis functions $Y_l^m(\theta, \phi)$ we can readily establish that (cf. eq. (9.4)) $\hbar^2 l(l+1)$ is the eigenvalue of the operator $\mathscr{L}_x^2 + \mathscr{L}_y^2 + \mathscr{L}_z^2$. However, although we will not prove it here, it can be shown by using the properties of I_x, I_y, and I_z (and hence by group theory alone) that $l(l+1)$ is the eigenvalue of $(I_x^2 + I_y^2 + I_z^2)$.

Having now seen the relationship between the angular momentum operators and the infinitesimal rotation operators we will now establish that rotational invariance is synonymous with conservation of angular momentum. In order to do this we consider a system in a state $\psi(\mathbf{r}, t)$ which is a solution of the time-dependent Schrödinger equation

$$H\psi(\mathbf{r}, t) = i\hbar\frac{\partial}{\partial t}\psi(\mathbf{r}, t) \tag{9.39}$$

where H is the Hamiltonian. We now suppose that another operator A exists such that $A\psi$ is still a solution of eq. (9.39) in which case we can write

$$AH\psi = i\hbar A\frac{\partial\psi}{\partial t} \tag{9.40}$$

$$HA\psi = i\hbar\frac{\partial}{\partial t}(A\psi) \tag{9.41}$$

from which we obtain

$$\begin{aligned}\frac{\mathrm{d}A}{\mathrm{d}t} &= \frac{1}{i\hbar}(AH - HA) + \frac{\partial A}{\partial t} \\ &= \frac{1}{i\hbar}[A, H] + \frac{\partial A}{\partial t}\end{aligned} \tag{9.42}$$

where we have used the identity $\mathrm{d}/\mathrm{d}t\langle A\rangle = \langle \mathrm{d}A/\mathrm{d}t\rangle$; the brackets $\langle\ \rangle$ denoting expectation value. Thus any time-independent operator which commutes with the Hamiltonian must have time-independent eigenvalues.

It now follows that since the Hamiltonian is invariant under all rotations it must commute with $(\mathbf{u}\cdot\mathscr{L})$ showing that angular momentum is a constant of the motion of an electron in an atom.

9.5 Spin

It became apparent in the early days of atomic physics that it was necessary to introduce an extra source of angular momentum which was called electron spin. Classically this spin really has no meaning but it is often naively thought of as an intrinsic spin of the electron about its own axis. We can appreciate how spin came to be introduced into atomic physics by noting that, even before the invention of quantum mechanics, the fine structure of spectral lines (for example that of the sodium D line which is really two closely spaced lines) could not be accounted for solely with the aid of the integral quantum numbers n, l, and m. In fact it became necessary, empirically, to introduce *half-integral* quantum numbers by means of new labels s and m_s where s could only have one value equal to $\frac{1}{2}$ and m_s could only have two values, namely $\pm\frac{1}{2}$. It was s that was considered to be the spin angular momentum quantum number. This is in sharp contrast to the numbers l and m which can have arbitrarily large values. Hence as we make a transition from a quantum to a classical regime, i.e., $l \to \infty$ and $\hbar \to 0$ we can see that $l\hbar$ tends to a finite *classical* value. On the other hand as $\hbar \to 0$ the spin angular momentum clearly tends to zero showing that spin has *no* classical counterpart. Because of this we cannot therefore begin with a classical expression for the spin and invoke the correspondence principle to obtain the corresponding quantum mechanical operators. However, we overcome this difficulty by postulating that (by *analogy* with the orbital angular momentum operators)

$$s^2 u_{\pm} = \hbar^2(\tfrac{1}{2})(\tfrac{1}{2} + 1)u_{\pm} \tag{9.43}$$

$$s_z u_+ = \hbar\tfrac{1}{2}u_+ \tag{9.44}$$

$$s_z u_- = -\hbar\tfrac{1}{2}u_- \tag{9.45}$$

where $s^2 = s_x^2 + s_y^2 + s_z^2$ is the square of the total spin operator and s_z is the z-component. Clearly s_x, s_y, and s_z are entirely analogous to $\mathscr{L}_x$, $\mathscr{L}_y$, and $\mathscr{L}_z$ but now describe rotations in a two-dimensional spin-space spanned by the basis functions u_+ and u_- which have been termed *spinors*. If this statement seems rather strange all that we have to remember is that 'spin rotations' cannot be rotations in the normal sense or exist in the real space spanned by basis functions $Y_l^m(\theta, \phi)$ because if they were indeed real rotations then they would have a classical analogue.

Thus in order to introduce spin into a group theoretical framework it is necessary to postulate the existence of a two-dimensional spin space spanned by u_+ and u_- in which we can find a representation for R_3. This representation, which we will call $\mathbf{D}^s(R) \equiv \mathbf{D}^{1/2}(R)$, is obviously (cf. eq. (9.8)) two-dimensional and irreducible and is quite distinct from the one we could obtain from

$$\mathbf{D}^0(R) \oplus \mathbf{D}^0(R) = \begin{bmatrix} 1 & 0 \\ 0 & 1 \end{bmatrix} \tag{9.46}$$

which is the two-dimensional direct sum matrix formed from the one-dimensional representations $\mathbf{D}^l(R)$, $l = 0$. The representation $\mathbf{D}^{1/2}(\alpha)$ can now be found by following our earlier treatment using the axial group C_∞^z of which a typical element P_α is a rotation of α about the z axis. The representations of C_∞^z are one-dimensional and, by analogy with eq. (9.9), are $e^{i\alpha/2}$ and $e^{-i\alpha/2}$ so that

$$\mathbf{D}^{1/2}(\alpha) = \begin{bmatrix} e^{i\alpha/2} & 0 \\ 0 & e^{-i\alpha/2} \end{bmatrix} \tag{9.47}$$

which has a character (trace) equal to

$$\begin{aligned} \chi^{1/2}(\alpha) &= 2\cos(\alpha/2) \\ &= \frac{\sin(s + \frac{1}{2})\alpha}{\sin \alpha/2} \end{aligned} \tag{9.48a}$$

where $s = \frac{1}{2}$ (cf. eq. (9.10)). The introduction of spin has now led to $\mathbf{D}^{1/2}(R)$ whose dimension is *even* whereas previously we were restricted to $\mathbf{D}^{l}(R)$ whose dimension, namely $2l + 1$, could only be odd. Thus it seems reasonable that the extra degree of freedom introduced by the spin opens up the possibility of producing even-dimensional representations of the full rotation group. In general therefore it would appear that the representations of the full-rotation group are $\mathbf{D}^{j}(R)$ with dimension $(2j + 1)$ where j can be either integral or half-integral. We should observe here that the argument we have presented is at best only plausible, but it can be rigorously established by developing the correspondence between the operators of the rotation group and the so-called SU_2 matrices [two-dimensional (special) unitary matrices with determinant equal to $+1$]. In spite of our admission, we will not go beyond our present argument but will simply state that the general expression for the character of the representations of the full rotation group is

$$\chi^{j}(\alpha) = \frac{\sin(j + \frac{1}{2})\alpha}{\sin \alpha/2} \tag{9.48b}$$

It is interesting now to discover that representations with j equal to a half-integer are not ordinary representations. We can soon see this by examining (9.47) for a few special values of α, i.e.,

$$\begin{aligned} \mathbf{D}^{1/2}(0) &= \begin{bmatrix} 1 & 0 \\ 0 & 1 \end{bmatrix} \\ \mathbf{D}^{1/2}(2\pi) &= \begin{bmatrix} -1 & 0 \\ 0 & -1 \end{bmatrix} \\ \mathbf{D}^{1/2}(4\pi) &= \begin{bmatrix} 1 & 0 \\ 0 & 1 \end{bmatrix} \end{aligned} \tag{9.49}$$

Equation (9.49) now shows that a rotation of 2π which we would normally expect to bring us back to the identity $\mathbf{D}^{1/2}(0)$ actually produces a new result. Indeed we can only retain closure by *extending* the rotation to 4π.

Hence if we include spin in a description of a physical problem, such as the crystal field splitting of an atomic state with $j = \frac{3}{2}$, say, then we have to construct what is called the *double* point group. For example let us consider the point group C_2 containing the operations E and C_2 where $C_2^2 = E$. Then, from eq. (9.49) we can see that we now have a new operation β with representation (reducible)

$$\boldsymbol{\Gamma}(\beta) = \begin{bmatrix} -1 & 0 \\ 0 & -1 \end{bmatrix} \quad \text{such that} \quad \{\boldsymbol{\Gamma}(\beta)\}^2 = \begin{bmatrix} 1 & 0 \\ 0 & 1 \end{bmatrix} = \boldsymbol{\Gamma}(E)$$

Therefore in place of the group C_2 we now have the *double group* containing the elements $E = C_2^4$, C_2, $\beta = C_2^2$, $\beta C_2 = C_2^3$.

Now all the irreducible $\boldsymbol{\Gamma}(R)$ of this cyclic group are, since it is Abelian, one-dimensional. Also we know that $\{\boldsymbol{\Gamma}(C_2)\}^4 = 1$. Therefore $\boldsymbol{\Gamma}(C_2) = [e^{ik2\pi/4}] = [e^{ik\pi/2}]$, where k goes from 1 to 4. Also since the representations are one-dimensional they are equal to the characters, so we can therefore see that we have the 4×4 character table given in table 9.1. Actually the double group obtained from C_2 must, since it is a cyclic group of order 4, have exactly the same character table as C_4.

Table 9.1 The character table of the double group constructed from the point group C_2

	E	C_2	β	βC_2
General form of character	$e^{ik2\pi}$	$e^{ik\pi/2}$	$e^{ik\pi}$	$e^{ik3\pi/2}$
$k = 1$	1	i	-1	$-i$
$k = 2$	1	-1	1	-1
$k = 3$	1	$-i$	-1	i
$k = 4$	1	1	1	1

This can be readily checked by comparing table 9.1 with the character table of C_4 given in appendix C.

The production of double point groups from point groups containing improper rotations (e.g., rotation–reflection operations) is not so easy. However, we should recall that many such groups are direct product groups formed from proper rotation groups and the group S_2. The remaining groups containing improper rotations have a one-to-one correspondence with proper rotation groups. We can therefore always express a group in terms of proper rotations; hence we can use the techniques described above to construct the character tables of double groups.

9.6 Vector model of the atom

If we wish to consider the total angular momentum of an atom we must find a way of combining orbital and spin angular momenta. However, in the case of spectra arising from the behaviour of a single electron the problem reduces to finding a method of combining l and s. The classical vector model was developed as a solution to this problem. In this model, as the name implies, the spin and orbital angular momenta are imagined to be vectors in a purely classical sense. For example a multi-electron atom can be handled by vectorially combining angular momenta according to certain rules. However, much of the procedure is apparently rather *ad hoc* and it is our intention here to examine, in the light of our knowledge of group theory, the justification of this model.

Initially we will consider a single electron in the state for which $l = 1$, $s = \frac{1}{2}$ whose state functions must contain products of the basis functions for $\mathbf{D}^{1/2}(R)$ and $\mathbf{D}^1(R)$. Now in section 3.6 we showed that whenever we have a situation like this, the product functions became basis functions for the direct product representation which in this case is $\mathbf{D}^1(R) \otimes \mathbf{D}^{1/2}(R)$. This is a reducible representation which can be written as

$$\mathbf{D}^{1\otimes 1/2}(R) = \mathbf{D}^1(R) \otimes \mathbf{D}^{1/2}(R) = \sum_j a_j \mathbf{D}^j(R) \tag{9.50}$$

where a_j is the number of times $\mathbf{D}^j(R)$ occurs in the direct product representation. If we consider the characters we find that eq. (9.50) implies that

$$\begin{aligned}\chi^{1\otimes 1/2}(\alpha) = \chi^1(\alpha)\chi^{1/2}(\alpha) &= \sum_{m=-1}^{1} \mathrm{e}^{im\alpha} \sum_{m'=-1/2}^{1/2} \mathrm{e}^{im'\alpha} \\ &= (\mathrm{e}^{i\alpha} + 1 + \mathrm{e}^{-i\alpha})(\mathrm{e}^{i(\alpha/2)} + \mathrm{e}^{-i(\alpha/2)}) \\ &= \mathrm{e}^{i(3\alpha/2)} + 2\mathrm{e}^{i(\alpha/2)} + 2\mathrm{e}^{-i(\alpha/2)} + \mathrm{e}^{-i(3\alpha/2)} \\ &= \sum_j a_j \sum_{k=-j}^{j} \mathrm{e}^{ik\alpha}\end{aligned} \tag{9.51}$$

The maximum value of k in eq. (9.51) is clearly $\frac{3}{2}$ and in fact $a_{3/2} = 1$ so that we can now re-write the right-hand side of eq. (9.51) as

$$\begin{aligned}S &= a_{3/2} \sum_{k=-3/2}^{3/2} \mathrm{e}^{ik\alpha} + \sum_{j\neq 3/2} a_j \sum_{k=-j}^{j} \mathrm{e}^{ik\alpha} \\ &= (\mathrm{e}^{i(3\alpha/2)} + \mathrm{e}^{i(\alpha/2)} + \mathrm{e}^{-i(\alpha/2)} + \mathrm{e}^{-i(3\alpha/2)}) + \sum_{j\neq 3/2} a_j \sum_{k=-j}^{j} \mathrm{e}^{ik\alpha}\end{aligned} \tag{9.52}$$

Hence it is apparent that

$$\mathrm{e}^{i(\alpha/2)} + \mathrm{e}^{-i(\alpha/2)} = \sum_{j\neq 3/2} a_j \sum_{k=-j}^{j} \mathrm{e}^{ik\alpha} \tag{9.53}$$

which can only have the solution $j = \frac{1}{2}$, $a_{1/2} = 1$ implying that

$$\chi^{1 \otimes 1/2} = \chi^{3/2} + \chi^{1/2} \tag{9.54}$$

Generalizing this result we now state (without proof) that the general solution for the number of irreducible components of the direct product of two irreducible representations is

$$\begin{aligned} \mathbf{D}^{j_1 \otimes j_2}(R) &= \sum_{J=|j_1 - j_2|}^{j_1 + j_2} \mathbf{D}^J(R) \\ &= \mathbf{D}^{j_1 + j_2}(R) \oplus \mathbf{D}^{j_1 + j_2 - 1}(R) \ldots \mathbf{D}^{|j_1 - j_2|}(R) \end{aligned} \tag{9.55}$$

where j_1 and j_2 are two integral or half-integral quantum numbers. $\mathbf{D}^{j_1 \otimes j_2}(R)$ of eq. (9.55) is of course in block form which implies that the basis functions must be symmetrized (section 3.5). These symmetrized basis functions are suitable linear combinations of the products of the basis functions of the respective representations $\mathbf{D}^{j_1}(R)$ and $\mathbf{D}^{j_2}(R)$. The coefficients of these linear combinations are called Wigner coefficients and are in fact quite complicated. However, since we do not require them for our purposes we will not pursue them any further.

We return instead to eq. (9.55) which we will interpret in terms of a *vector model* in which the angular momenta are vectorially combined. Hence we consider j_1 and j_2 as the moduli of the vectors $\mathbf{j}_1$ and $\mathbf{j}_2$ labelling two angular momentum states which, when combined, form a resultant vector $\mathbf{J} = \mathbf{j}_1 + \mathbf{j}_2$. Obviously the maximum modulus of $\mathbf{J}$ occurs when $\mathbf{j}_1$ and $\mathbf{j}_2$ are parallel so that $|\mathbf{J}|_{\text{max}} = j_1 + j_2$ and the minimum value is $|\mathbf{J}|_{\text{min}} = j_1 - j_2$, i.e., when $\mathbf{j}_1$ and $\mathbf{j}_2$ are antiparallel and $j_1 > j_2$. Between these two limits $\mathbf{J}$ takes on a set of discrete values corresponding to certain allowed orientations of $\mathbf{j}_1$ to $\mathbf{j}_2$ given by eq. (9.55). The number of allowed values is $(j_1 + j_2 - |j_1 - j_2| + 1)$ which is equal to $(2j_2 + 1)$ if $j_1 > j_2$ and $(2j_1 + 1)$ if $j_2 > j_1$.

Up to now we have not really specified $\mathbf{j}_1$ and $\mathbf{j}_2$ but as an example we will consider a one-electron state, with $\mathbf{j}_1 = \boldsymbol{l}$ and $\mathbf{j}_2 = \mathbf{s}$, for which the possible electronic states as revealed by eq. (9.55) are labelled by $l + \frac{1}{2}$ and $l - \frac{1}{2}$. Now if $l = 0$ we can *only* have $J = \frac{1}{2}$, a fact which we can readily appreciate both from the group theoretical and the vector model point of view. These results also show that for $l > 0$ one-electron states consist of two energy levels, i.e., each state is a doublet. These types of levels give rise to the two sodium D-lines, arising from the transition from $l = 1$ to $l = 0$.

In a many-electron atom we have to deal with the behaviour of several electrons whose respective orbital angular momentum and spin quantum numbers are $l_1, l_2, l_3, \ldots$, and $s_1, s_2, s_3, \ldots$ A particular state of the atom is now described by the product of all the basis functions of $\mathbf{D}^{l_1}(R), \mathbf{D}^{l_2}(R) \ldots$ $\mathbf{D}^{s_1}(R), \mathbf{D}^{s_2}(R) \ldots$ thus producing the representation $\mathbf{D}^{l_1 \otimes l_2 \otimes \ldots s_1 \otimes s_2 \ldots}(R)$

which has irreducible components which can be found by successive application of eq. (9.55). For example let us consider

$$\begin{aligned}
\mathbf{D}^{1\otimes 2\otimes 1/2}(R) &= (\mathbf{D}^{3}(R) \oplus \mathbf{D}^{2}(R) \oplus \mathbf{D}^{1}(R)) \otimes \mathbf{D}^{1/2}(R) \\
&= \mathbf{D}^{3\otimes 1/2}(R) \oplus \mathbf{D}^{2\otimes 1/2}(R) \oplus \mathbf{D}^{1\otimes 1/2}(R) \\
&= (\mathbf{D}^{7/2}(R) \oplus \mathbf{D}^{5/2}(R)) \oplus (\mathbf{D}^{5/2}(R) \oplus \mathbf{D}^{3/2}(R)) \\
&\quad \oplus (\mathbf{D}^{3/2}(R) \oplus \mathbf{D}^{1/2}(R)) \\
&= \mathbf{D}^{7/2}(R) \oplus 2\mathbf{D}^{5/2}(R) \oplus 2\mathbf{D}^{3/2}(R) \oplus \mathbf{D}^{1/2}(R)
\end{aligned} \tag{9.56}$$

The order in which one combines the states is not really important from a group theoretical point of view. However, we could combine l_1, l_2, etc., separately to form the reducible representation $\mathbf{D}^L$ and then combine s_1, s_2, etc., to form $\mathbf{D}^S$ and then finally reduce $\mathbf{D}^{L\otimes S}$. This procedure amounts to first forming a vector $\mathbf{L} = \boldsymbol{l}_1 + \boldsymbol{l}_2 + \boldsymbol{l}_3 + \cdots$ which is called the total orbital angular momentum and then forming a vector $\mathbf{S} = \mathbf{s}_1 + \mathbf{s}_2 + \mathbf{s}_3 + \cdots$ which is the total spin angular momentum and finally combining $\mathbf{L}$ and $\mathbf{S}$ to form $\mathbf{J} = \mathbf{L} + \mathbf{S}$. This type of coupling is known as L–S or Russell–Saunders coupling. Alternatively we could first of all combine $\boldsymbol{l}_1 + \mathbf{s}_1$, $\boldsymbol{l}_2 + \mathbf{s}_2, \ldots$ and so on to form vectors $\mathbf{j}_1, \mathbf{j}_2, \ldots$ and then form the reducible representation $\mathbf{D}^{j_1\otimes j_2\otimes j_3\otimes\cdots}$. This latter procedure amounts, vectorially, to forming $\mathbf{J} = \mathbf{j}_1 + \mathbf{j}_2 + \cdots$ and is called j–j coupling. We emphasize that in either case from a group theoretical viewpoint we will obtain the *same* set of energy levels. However,

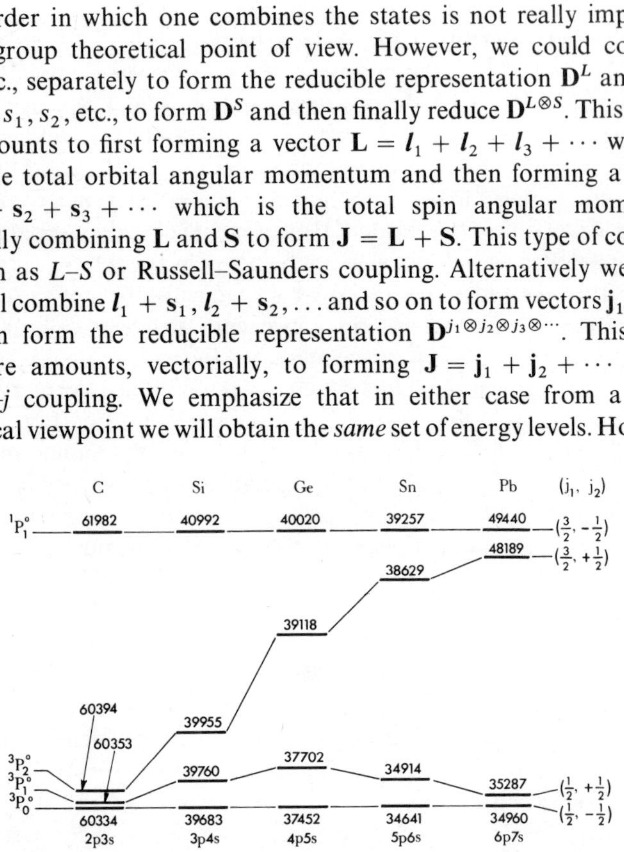

Fig. 9.1 Transition from L–S (Russell-Saunders) to j–j coupling as portrayed by the elements of the IVth column of the periodic table from carbon to lead. The energy level values, in wave-numbers, have been taken from Atomic Energy Levels published by the National Bureau of Standards United States Department of Commerce. There are two electrons involved labelled as np(n + 1)s where n is a principal quantum number. At the left-hand side there is L–S coupling which leads to the singlet state $^1P_1$ and the triplet states $^3P_2$, $^3P_1$, $^3P_0$. At the right-hand side there is j–j coupling which leads to states labelled by j_1 and j_2, i.e., the j-values of the two electrons.

a numerical computation would reveal that L–S and j–j coupling leads to sets of energy levels with *different* spacings.

Physically j–j coupling arises when the electron interacts strongly with its own orbital motion, i.e., the orbital motion produces a magnetic field at the electron which in turn experiences a strong interaction. This is known as spin–orbit coupling. On the other hand L–S coupling arises when spin–orbit coupling is negligible compared to the residual electrostatic energy which is 'left over' when we have constructed the central field $V(r)$. In an L–S coupled state the electron spin and the orbital motion are independent and **L** and **S** are approximately separate constants of the motion showing that L and S are nearly good quantum numbers. As we go from light to heavy atoms we progress from L–S to j–j coupling. This is physically reasonable if we recall that in a light atom the electrons have a strong *electrostatic* interaction. The transition from L–S to j–j coupling is shown in Fig. 9.1, as it occurs in practice.

9.7 The Pauli exclusion principle and the permutation group

The group $R_3 \otimes S_2$ does not exhaust all the operations which commute with the Hamiltonian of a *many-electron* atom. This is because electrons are indistinguishable from one another implying that it cannot make any difference to the Hamiltonian if we simply interchange or *permute* the positions of the electrons. The Hamiltonian then, is invariant under the group of operators which permute the coordinates of the electrons. This group of operators is called the permutation group and is designated as P_n, where n is the number of objects to be permuted (in our case the number of valence electrons in the atom). P_n, then, is the group of operators which permutes n objects in all possible (i.e., $n!$) ways, and therefore has $n!$ elements, i.e., one operator for each permutation.

As an example of a permutation operator let us consider the group P_4 whose operators we imagine to act on the sequence of numbers (1 2 3 4), labelling four objects in the order 1 to 4, to produce another sequence, say (3 1 2 4). We will write such an operator as $\begin{pmatrix} 1 & 2 & 3 & 4 \\ 3 & 1 & 2 & 4 \end{pmatrix}$ which has the interpretation that when $\begin{pmatrix} 1 & 2 & 3 & 4 \\ 3 & 1 & 2 & 4 \end{pmatrix}$ acts on the sequence (1 2 3 4) it produces the sequence (3 1 2 4). Note that $\begin{pmatrix} 1 & 2 & 3 & 4 \\ 3 & 1 & 2 & 4 \end{pmatrix}$ is *not* a matrix and (3 1 2 4) is not a vector.

The action of $\begin{pmatrix} 1 & 2 & 3 & 4 \\ 3 & 1 & 2 & 4 \end{pmatrix}$ is therefore as follows:

the third element of 1 2 3 4 goes into position one
the first element of 1 2 3 4 goes into position two

the second element of 1 2 3 4 goes into position three
the fourth element of 1 2 3 4 goes into position four.

We can now see that the action of $\begin{pmatrix} 1 & 2 & 3 & 4 \\ 3 & 1 & 2 & 4 \end{pmatrix}$ on some other sequence such as for example $(2 \quad 4 \quad 1 \quad 3)$ is

$$\begin{pmatrix} 1 & 2 & 3 & 4 \\ 3 & 1 & 2 & 4 \end{pmatrix}(2 \quad 4 \quad 1 \quad 3) = (1 \quad 2 \quad 4 \quad 3) \tag{9.57}$$

Now the sequence $(2 \quad 4 \quad 1 \quad 3)$ is actually the result of the operation $\begin{pmatrix} 1 & 2 & 3 & 4 \\ 2 & 4 & 1 & 3 \end{pmatrix}$ on $(1 \quad 2 \quad 3 \quad 4)$ so that we can re-write the eq. (9.57) as

$$\begin{pmatrix} 1 & 2 & 3 & 4 \\ 3 & 1 & 2 & 4 \end{pmatrix}\begin{pmatrix} 1 & 2 & 3 & 4 \\ 2 & 4 & 1 & 3 \end{pmatrix}(1 \quad 2 \quad 3 \quad 4) = (1 \quad 2 \quad 4 \quad 3) \tag{9.58}$$

Also since

$$(1 \quad 2 \quad 4 \quad 3) = \begin{pmatrix} 1 & 2 & 3 & 4 \\ 1 & 2 & 4 & 3 \end{pmatrix}(1 \quad 2 \quad 3 \quad 4) \tag{9.59}$$

we can conclude from eq. (9.58) that

$$\begin{pmatrix} 1 & 2 & 3 & 4 \\ 3 & 1 & 2 & 4 \end{pmatrix}\begin{pmatrix} 1 & 2 & 3 & 4 \\ 2 & 4 & 1 & 3 \end{pmatrix} = \begin{pmatrix} 1 & 2 & 3 & 4 \\ 1 & 2 & 4 & 3 \end{pmatrix} \tag{9.60}$$

Hence the product of two operators is equal to another operator from P_4.

In any permutation the minimum interchange we can effect involves two objects only. If we call the permutation operators which effect such changes the *simple operators*, then it is clear that a permutation which involves *more* than two objects can be constructed from successive applications of simple operators.

An operator is now said to be *odd* if it is a product of an odd number of simple operators. Similarly an operator is said to be *even* if it is a product of an even number of simple operators. Obviously an odd operator multiplying another odd operator will produce an even operator and an odd operator multiplying an even operator will produce an odd operator and so on. In fact in order to preserve the closure of the permutation group it must contain an equal number of odd and even operators. We see that this is so by postulating the converse, e.g., if P_3 contains two even operators and four odd operators then a product of an odd operator with all these elements of P_3 would now produce two odd operators and four even operators which is obviously not the same group. Hence our initial statement must be true.

It is now necessary to devise matrix representations of the permutation operators. One representation can of course, as always, be obtained by representing each element by 1 in which case we obtain the totally symmetric representation. However, as it happens, because of the fact that the group 'splits' into two equal sets of odd and even operators we can also devise an alternative one-dimensional representation by replacing the even operators by 1 and the odd operators by -1. The latter representation is called the antisymmetric representation.

Let us consider a two-electron system whose Hamiltonian is invariant under the group P_2. This group contains two elements, the identity R_{12}, an even operator, and an odd operator R_{21} which interchanges the two electrons, and can therefore only possess two irreducible representations namely the symmetric and the antisymmetric representations. A two-electron state $\psi(\mathbf{r}_1, \mathbf{r}_2)$ must now be a basis function for one of these irreducible representations. Thus, using R_{21} we have

$$R_{21}\psi(\mathbf{r}_1, \mathbf{r}_2) = \psi(\mathbf{r}_2, \mathbf{r}_1) = \psi(\mathbf{r}_1, \mathbf{r}_2)(\pm 1) \tag{9.61}$$

Hence,

$$\psi(\mathbf{r}_2, \mathbf{r}_1) = \pm\psi(\mathbf{r}_1, \mathbf{r}_2) \tag{9.62}$$

If $\psi(\mathbf{r}_1, \mathbf{r}_2)$ is a basis function for the symmetric representation we use the plus sign and if $\psi(\mathbf{r}_1, \mathbf{r}_2)$ is a basis function, for the antisymmetric representation we use the minus sign, i.e.,

$$S: \text{ SYMMETRIC} \qquad \psi_S(\mathbf{r}_2, \mathbf{r}_1) = \psi_S(\mathbf{r}_1, \mathbf{r}_2) \tag{9.63}$$

$$A: \text{ ANTISYMMETRIC} \qquad \psi_A(\mathbf{r}_2, \mathbf{r}_1) = -\psi_A(\mathbf{r}_1, \mathbf{r}_2) \tag{9.64}$$

We cannot at this stage say whether electrons have *S*- or *A*-functions but we can say that particles such as electrons must have one *or* the other, and that the symmetry is conserved. Indeed particles which have *A*-functions are called fermions and obey Fermi–Dirac statistics and those with *S*-functions are called bosons and obey Bose–Einstein statistics. Experimentally, electrons have been found to be fermions and hence their state functions must belong to the antisymmetric representation of the associated permutation group of the Hamiltonian.

We will now consider the construction of two-electron functions, i.e., $\psi(\mathbf{r}_1, \mathbf{r}_2)$ from the one-electron states $\psi^N(\mathbf{r}_1)$ and $\psi^{N'}(\mathbf{r}_2)$ where N and N' represent sets of quantum numbers, $N = nlms$ and $N' = n'l'm's'$. Obviously we can immediately construct an unsymmetrized basis function $\psi^N(\mathbf{r}_1)\psi^{N'}(\mathbf{r}_2)$ which we can then turn into an antisymmetric function by means of the projection operator O^A where (section 3.3)

$$O^A = \tfrac{1}{2}\{(+1)R_{12} + (-1)R_{21}\} \tag{9.65}$$

O^A will therefore act on $\psi^N(\mathbf{r}_1)\psi^{N'}(\mathbf{r}_2)$ to produce

$$\begin{aligned} O^A\psi^N(\mathbf{r}_1)\psi^{N'}(\mathbf{r}_2) &= \tfrac{1}{2}[\psi^N(\mathbf{r}_1)\psi^{N'}(\mathbf{r}_2) - \psi^N(\mathbf{r}_2)\psi^{N'}(\mathbf{r}_1)] \\ &= \psi^{N,N'}(\mathbf{r}_1, \mathbf{r}_2) \end{aligned} \tag{9.66}$$

where $\psi^{N,N'}(\mathbf{r}_1\mathbf{r}_2)$ is now an antisymmetric function. It is interesting now to note that we can re-write eq. (9.66) in the determinantal form

$$\psi^{N,N'}(\mathbf{r}_1\mathbf{r}_2) \propto \begin{vmatrix} \psi^{N}(\mathbf{r}_1) & \psi^{N}(\mathbf{r}_2) \\ \psi^{N'}(\mathbf{r}_1) & \psi^{N'}(\mathbf{r}_2) \end{vmatrix} \tag{9.67}$$

and in general we can write a many-electron function as

$$\psi^{N_1,N_2,N_3,\ldots,N_j}(\mathbf{r}_1,\mathbf{r}_2,\ldots,\mathbf{r}_j) \propto \begin{vmatrix} \psi^{N_1}(\mathbf{r}_1)\psi^{N_1}(\mathbf{r}_2)\cdots\psi^{N_1}(\mathbf{r}_j) \\ \vdots \qquad\qquad \vdots \\ \psi^{N_j}(\mathbf{r}_1)\psi^{N_j}(\mathbf{r}_2)\cdots\psi^{N_j}(\mathbf{r}_j) \end{vmatrix} \tag{9.68}$$

Equation (9.68) is often called the Slater determinant and this form of the wave-function is a basis for the antisymmetric representation of the associated permutation group. This can be seen by the fact that if two electrons are interchanged, two columns of the determinant are also interchanged, thus multiplying the determinant by -1. In addition we can see that, if two or more electrons possess identical sets of quantum numbers N_i then at least two columns of the determinant are equal; thus causing the many-electron function to vanish. The determinantal form of eq. (9.64) therefore incorporates the *Pauli exclusion principle* which states that no two electrons of a system can possess the same set of quantum numbers.

9.8 Selection rules

The interpretation of the spectra of atoms requires not only a knowledge of the energy levels but also a knowledge of the rules which allow an electron to go from one level to another. These rules put certain restrictions on the amounts by which quantum numbers can change during an electronic transition. In fact, given an energy level diagram we can use these rules to *select* the levels between which transitions may occur.

The probability of a transition from a state ϕ_i to a state ϕ_j is proportional to $|\langle\phi_i|H'|\phi_j\rangle|^2$ where H' is the *influence* which initiates the transition and we can determine whether $\langle\phi_i|H'|\phi_j\rangle$ is zero or not from a consideration of the orthogonality of basis functions. Thus if ϕ_i is orthogonal to $H'\phi_j$ the matrix element will clearly vanish. Now products of functions form basis functions for direct product representations (cf. section 3.6d) so that if ϕ_i belongs $\mathbf{\Gamma}^i(R)$ and $H'\phi_j$ to $\mathbf{\Gamma}^k(R)$ then the product $\phi_i H'\phi_j$ must belong to the direct product representation $\mathbf{\Gamma}^{i*}(R) \otimes \mathbf{\Gamma}^k(R)$. Also assuming that $\mathbf{\Gamma}^i(R)$ is irreducible and that $\mathbf{\Gamma}^k(R)$ is reducible we know that ϕ_i and $H'\phi_j$ are orthogonal unless $\mathbf{\Gamma}^k(R)$ contains $\mathbf{\Gamma}^i(R)$. This can be seen by realizing that if $\mathbf{\Gamma}^k(R)$ contains $\mathbf{\Gamma}^i(R)$, e.g., $\mathbf{\Gamma}^k(R) = \mathbf{\Gamma}^i(R) \oplus \mathbf{\Gamma}^l(R)$, where

$\mathbf{\Gamma}^l(R)$ is some other representation, then $H'\phi_j$ can be expressed as a linear combination of functions belonging to $\mathbf{\Gamma}^i(R)$ and $\mathbf{\Gamma}^l(R)$; thus if $\mathbf{\Gamma}^k(R)$ contains $\mathbf{\Gamma}^i(R)$ there is always a *part* of $H'\phi_j$ which belongs to $\mathbf{\Gamma}^i(R)$; hence the integral $\langle\phi_i|H'|\phi_j\rangle$ will *not* vanish. We can now formalize this statement by saying that if $\mathbf{\Gamma}^k(R)$ contains $\mathbf{\Gamma}^i(R)$ then $\mathbf{\Gamma}^{i*}(R) \otimes \mathbf{\Gamma}^k(R)$ must contain a term $\mathbf{\Gamma}^{i*}(R) \otimes \mathbf{\Gamma}^i(R)$ which in turn (as shown in section 3.11) must contain $\mathbf{\Gamma}^1(R)$, the totally symmetric representation. Hence a necessary condition for a transition to occur is that $\mathbf{\Gamma}^{i*}(R) \otimes \mathbf{\Gamma}^k(R)$ contains $\mathbf{\Gamma}^1(R)$. We use the word 'necessary' advisedly because the fact that $\mathbf{\Gamma}^{i*}(R) \otimes \mathbf{\Gamma}^k(R)$ contains $\mathbf{\Gamma}^1(R)$ is not a sufficient condition for the existence of a matrix element. For example if we consider two normalized basis functions ϕ_l^i and ϕ_m^k, where i and k label the irreducible representations and l and m label columns within these representations, then their scalar product is $\langle\phi_l^i|\phi_m^k\rangle = \delta_{ik}\delta_{lm}$ which is only non-zero provided $i = k$ *and* $l = m$. However the condition that $\mathbf{\Gamma}^{i*}(R) \otimes \mathbf{\Gamma}^k(R)$ contains $\mathbf{\Gamma}^1(R)$ is only enough to ensure that $i = k$, it does not ensure that $l = m$.

Electronic transitions, in an atom, can involve an excitation from a lower level via the action of an electromagnetic field. This, as shown in chapter 8, is equivalent, to first order, to a dipole interaction. We will therefore assume that all transitions either from a lower to an upper level or vice versa take place via a dipole transition. Obviously the dipole interaction is an approximation but we do not wish to dwell on this point except to say that other interactions exist.

The dipole interaction actually gives rise to a Hamiltonian which is proportional to the vector $\mathbf{r} = (x, y, z)$ so that the matrix element $\langle\phi_i|H'|\phi_j\rangle$ is proportional to $\langle\phi_i|\mathbf{r}|\phi_j\rangle$. Hence we have to consider the terms $\langle\phi_i|x|\phi_j\rangle$, $\langle\phi_i|y|\phi_j\rangle$ and $\langle\phi_i|z|\phi_j\rangle$ where the functions x, y, z form a basis for a three-dimensional representation of the rotation group, i.e. x, y, and z form a basis for $\mathbf{D}^1(R)$ where $\mathbf{D}^1(R)$ could be the standard rotation matrix of eq. (A.183) which takes (x, y, z) to (x', y', z'). Obviously since we have a product of three functions in the matrix element this product must be a basis function for the direct product representation $\mathbf{\Gamma}^{i*}(R) \otimes \mathbf{D}^1(R) \otimes \mathbf{\Gamma}^j(R)$. We have now replaced $\mathbf{\Gamma}^k(R)$ of our previous argument by $\mathbf{D}^1(R) \otimes \mathbf{\Gamma}^j(R)$, so that all we have to do now is to determine whether or not $\mathbf{\Gamma}^{i*}(R) \otimes \mathbf{D}^1(R) \otimes \mathbf{\Gamma}^j(R)$ contains $\mathbf{\Gamma}^1(R)$, in order to discover whether a transition from ϕ_i to ϕ_j is allowed.

Let us consider a transition from a level labelled by l_1 to a level labelled by l_2, where $\mathbf{\Gamma}^i(R) = \mathbf{D}^{l_1}(R)$ and $\mathbf{\Gamma}^k(R) = \mathbf{D}^1(R) \otimes \mathbf{D}^{l_2}(R)$. Now the totally symmetric representation is, in this case, $\mathbf{\Gamma}^1(R) = \mathbf{D}^0(R)$ and we have to find out whether or not $\mathbf{D}^{l_1*}(R) \otimes \mathbf{D}^1(R) \otimes \mathbf{D}^{l_2}(R)$ contains $\mathbf{D}^0(R)$.

We can do this using the result of eq. (9.55) from which we can see that

$$\mathbf{D}^{l_1}(R) \otimes \mathbf{D}^1(R) = \mathbf{D}^{l_1+1}(R) \oplus \mathbf{D}^{l_1}(R) \oplus \mathbf{D}^{l_1-1}(R) \qquad (9.69)$$

Hence for $\mathbf{D}^{l_1 *}(R) \otimes \mathbf{D}^1(R) \otimes \mathbf{D}^{l_2}(R)$ to contain $\mathbf{D}^0(R)$, $\mathbf{D}^{l_2}(R)$ must be $\mathbf{D}^{l_1+1}(R)$, $\mathbf{D}^{l_1}(R)$, or $\mathbf{D}^{l_1-1}(R)$ so that if the transition is to be allowed we require that

$$\Delta l = l_2 - l_1 = \pm 1, 0 \tag{9.70}$$

However, if $l_1 = 0$ then $\mathbf{D}^0(R) \otimes \mathbf{D}^1(R) = \mathbf{D}^1(R)$ so that $l_2 = 1$ showing that Δl must be unity.

The selection rules above have been obtained purely from a consideration of the rotation group. A further restriction therefore enters when we consider the role of the space-inversion group S_2. We can include the effects of space-inversion symmetry by noting that only transitions between states of different parity are allowed (cf. section 3.11). The parity of an atomic state is given by $(-1)^l$ so that for an allowed transition Δl must be an odd number. The combination of this result and eq. (9.70) gives the complete selection rule for these dipole transitions as

$$\Delta l = \pm 1 \tag{9.71}$$

If we consider products of both spin and orbital angular momentum eigenfunctions, they will form a basis for representations labelled by j where, for a one-electron system, $j = l \pm \frac{1}{2}$. We are thus led to a new selection rule which is obtained by determining whether or not the direct product $\mathbf{D}^{j_1 *}(R) \otimes \mathbf{D}^1(R) \otimes \mathbf{D}^{j_2}(R)$ contains $\mathbf{D}^0(R)$. This can be done by a similar argument to that employed for states labelled with l. The result is

$$\Delta j = 0, \pm 1 \tag{9.72}$$

Since j does not determine the parity $\Delta j = 0$ is also a possible transition.

If $j_1 = 0$ and $j_2 = 0$ then since $\mathbf{D}^{0*}(R) \otimes \mathbf{D}^1(R) \otimes \mathbf{D}^0(R) = \mathbf{D}^1(R)$, which does not contain $\mathbf{D}^0(R)$, the transition $j_1 = 0$ to $j_2 = 0$ is therefore forbidden.

In many-electron atoms we have to consider electrons with orbital angular momentum quantum numbers $l_1, l_2, \ldots$ and spin quantum numbers $s_1, s_2, \ldots$ so that we have to select a means of combining these elemental quantum numbers to form total quantum numbers J; or L and S, if we use Russell–Saunders coupling. In the latter coupling we combine the $l_1, l_2, \ldots$ to form L and then combine $s_1, s_2, \ldots$ to form S. We then form J from L and S. These combinations can be found using the vector model discussed earlier on. We should observe that it is implicit in the Russell–Saunders treatment that, to an approximation anyway, we can *regard* L, S, and J as good quantum numbers so we will find selection rules giving us ΔL, ΔS, and ΔJ. Clearly L and J label representations in the same way as do l and j so that by analogy to our previous discussion

$$\Delta L = 0, \pm 1 \tag{9.73}$$

$$\Delta J = 0, \pm 1 \tag{9.74}$$

where the transition $\Delta L = 0$ is now allowed because although we still require $\Delta l = \pm 1$ for the electron involved in an actual transition, due to the imposition of the parity selection rule, we can now have several levels labelled by $L = 1$, say, such that $\Delta l = \pm 1$ while $\Delta L = 0$.

If we only consider spin then we find that the representation of x, y, and z of the dipole interaction belongs to $\mathbf{D}^0(R)$ instead of $\mathbf{D}^1(R)$. This is because under transformations in spin-space only, x and y and z will, being functions in real space, remain invariant. Hence a transition from S_1 to S_2 is allowed provided $\mathbf{D}^{S_1}(R) \otimes \mathbf{D}^0(R) \otimes \mathbf{D}^{S_2}(R)$ contains $\mathbf{D}^0(R)$. This can only be true if

$$\Delta S = 0 \tag{9.75}$$

Strictly speaking L and S are not, from a group theoretical point of view, good quantum numbers. This is because physically the orbital and spin angular momentum are not separately conserved, although obviously the sum $(\mathbf{L} + \mathbf{S})$ is always conserved. A *good* quantum number should be interpreted as a label for conserved quantities.

If it becomes apparent that it is no longer viable to add the orbital angular momenta and the spin angular momenta separately, then we should combine l_1 with s_1, l_2 with s_2, and so on and then consider the coupling of $j_1, j_2, \ldots$ to form $\mathbf{J}$. This is known as *j–j* coupling and is applicable to heavy atoms in which there is strong spin–orbit coupling. For this type of coupling we therefore lose the selection rules $\Delta L = 0, \pm 1$ and $\Delta S = 0$ (which were approximate anyway) but retain the selection rule

$$\begin{aligned} &\Delta J = 0, \pm 1 \\ &J = 0 \to J = 0 \quad \text{forbidden} \end{aligned} \tag{9.76}$$

In the case of an atom immersed in an external magnetic field (Zeeman effect) all the degenerate levels for a given J separate into $(2J + 1)$ distinct levels labelled by M_J. This is because we have reduced the symmetry group from $R_3 \otimes S_2$ to C_∞ which has only one-dimensional representations labelled by M_J. Electronic transitions between the separated M_J states can take place, by dipole interactions, according to the rules $\Delta M_J = 0, \pm 1$. This rule can be established as follows.

The irreducible representations $\mathbf{\Gamma}^{M_J}(\alpha)$ which are labelled by M_J are $e^{iM_J\alpha}$ where α is the angle of rotation about the z-axis. Also x and y form a basis for the two-dimensional reduced representation

$$\begin{bmatrix} e^{i\alpha} & 0 \\ 0 & e^{-i\alpha} \end{bmatrix} = \begin{bmatrix} \mathbf{\Gamma}^1(\alpha) & 0 \\ 0 & \mathbf{\Gamma}^{-1}(\alpha) \end{bmatrix} \tag{9.77}$$

Therefore x and y belong to $\mathbf{\Gamma}^1(\alpha) \oplus \mathbf{\Gamma}^{-1}(\alpha)$. On the other hand z is invariant under rotations about the z axis and must therefore belong to

$\Gamma^0(\alpha)$. Thus for a transition between M_{J_1} and M_{J_2} we require

$$e^{-iM_{J1}} \otimes e^{\pm i\alpha} \otimes e^{iM_{J2}} = e^{i[M_{J2}-M_{J1}\pm 1]\alpha} = e^0 \tag{9.78}$$

$$e^{-iM_{J1}} \otimes e^0 \otimes e^{iM_{J2}} = e^{i[M_{J2}-M_{J1}]} = e^0 \tag{9.79}$$

(N.B. since all representations are one-dimensional, direct products are also one-dimensional and are therefore *irreducible*.)

The selection rules are therefore

$$\begin{aligned} \Delta M_J &= M_{J_2} - M_{J_1} = \pm 1 \\ \Delta M_J &= M_{J_2} - M_{J_1} = 0 \end{aligned} \tag{9.80}$$

where the first selection rule corresponds to the emission or absorption of radiation polarized with electric field vectors perpendicular to the magnetic field and the second selection rule refers to electric field vectors parallel to the magnetic field.

9.9 Problems

9.1 Consider an atom in a crystal possessing the point group symmetry T_d. Determine the number of levels into which the atomic D state will split.

9.2 Examine the matrix $\mathbf{h}$ where

$$\mathbf{h} = x\begin{bmatrix} 0 & -i \\ i & 0 \end{bmatrix} + y\begin{bmatrix} 0 & 1 \\ 1 & 0 \end{bmatrix} + z\begin{bmatrix} 1 & 0 \\ 0 & -1 \end{bmatrix} \equiv \mathbf{r}\,.\,\boldsymbol{\sigma}$$

and show that

$$\mathbf{h}' = \mathbf{u}^{-1}\mathbf{h}\mathbf{u} \equiv \mathbf{r}'\,.\,\boldsymbol{\sigma}$$

where

$$\mathbf{u} = \begin{bmatrix} a & -b \\ b^* & a^* \end{bmatrix}$$

such that $aa^* + bb^* = 1$. Hence show that $\mathbf{r}' = \mathbf{R}\mathbf{r}$ where $\mathbf{R}$ represents a real rotation belonging to the group R_3. (The three components of the matrix $\boldsymbol{\sigma}$ are the Pauli spin matrices; and the matrix $\mathbf{u}$ is known as a two-dimensional special unitary matrix—often termed an SU_2 matrix).

9.3 Show that the character table of the double group constructed from C_3 is the same as that of the group C_6.

9.4 By using the methods of the first part of section 9.6 show that

$$\mathbf{D}^{1\otimes 1}(R) = \mathbf{D}^2(R) \oplus D^1(R) \oplus \mathbf{D}^0(R)$$

9.5 Write down the six operators of the permutation group P_3. Determine which of these are simple operators and hence find the odd and even operators. Determine the multiplication table of this group and hence find the class structure. Finally show that the group has only two one-dimensional representations namely the symmetric and antisymmetric representations.

9.6 A two-electron $(2s3s)$ state can be described in terms of the wave-functions

$$\Phi_{2s}(r_1)Y_0^0(\theta_1, \phi_1)u^+(\zeta_1), \quad \Phi_{2s}(r_1)Y_0^0(\theta_1, \phi_1)u^-(\zeta_1)$$
$$\Phi_{3s}(r_2)Y_0^0(\theta_2, \phi_2)u^+(\zeta_2), \quad \Phi_{3s}(r_2)Y_0^0(\theta_2, \phi_2)u^-(\zeta_2)$$

Find the four possible product functions which will describe the state and by using the projection operators of the group P_2 determine the symmetrized wave-functions. To which irreducible representations of the group R_3 will these functions belong?

9.7 How are the results of problem 6 modified if the 3*s* electron is replaced by a 2*s* electron (i.e., if the two electrons become *equivalent*)?

Appendix A

A.1 Definition of a matrix

A matrix is an array of elements set out in rectangular form as for example:

$$\text{(i) } [1], \quad \text{(ii) } \begin{bmatrix} 0 \\ 1 \end{bmatrix}, \quad \text{(iii) } \begin{bmatrix} 1 & -2 & 3 \\ -4 & 0 & -6 \end{bmatrix}, \quad \text{(iv) } \begin{bmatrix} x & x+1 \\ x^2 - x & x^2 - 1 \end{bmatrix},$$

$$\text{(v) } \begin{bmatrix} a & b & c \\ d & e & f \\ g & h & i \end{bmatrix} \tag{A.1}$$

The elements may be (a) real or complex numbers, (b) polynomials, or (c) other matrices.

An arbitrary matrix **A** can be written as:

$$\mathbf{A} = [A_{ij}] = \begin{bmatrix} A_{11} & A_{12} & \ldots & A_{1n} \\ A_{21} & A_{22} & \ldots & A_{2n} \\ A_{m1} & A_{m2} & \ldots & A_{mn} \end{bmatrix} \tag{A.2}$$

where **A** is known as an $(m \times n)$ matrix and possesses m rows and n columns. The element A_{ij} which is sometimes used to typify the matrix is in the ith row and the jth column.

If $m = n$ the matrix is said to be square and the value of n is the order or dimension of the matrix. Also *if* m *(or* n*)* $= 1$ the matrix has only one row or column respectively and is known as a row or column *vector*, which we shall discuss later. The elements A_{ii} (A_{11}, A_{22}, etc.) are known as the elements on the *leading diagonal* and the sum of these elements is known as the *trace* of the matrix which we shall write as

$$\text{trace } \mathbf{A} = \sum_i A_{ii} \tag{A.3}$$

The trace is important in the theory of group representations (chapter 3).

A.2 Arithmetic operations

The standard arithmetic operations namely addition, subtraction, multiplication, and division are also applicable to matrices. We shall only consider the first three for the moment, deferring the question of division until after the introduction of determinants in section A.3.

(a) Addition and Subtraction

The sum of the two matrices is obtained by adding corresponding elements; thus

$$\mathbf{C} = \mathbf{A} \pm \mathbf{B} \tag{A.4}$$

is the matrix whose elements are,

$$C_{ij} = A_{ij} \pm B_{ij} \tag{A.5}$$

For example,

$$\begin{bmatrix} 1 & 2 \\ 5 & 4 \end{bmatrix} + \begin{bmatrix} -1 & 0 \\ 2 & 3 \end{bmatrix} = \begin{bmatrix} 0 & 2 \\ 7 & 7 \end{bmatrix} \tag{A.6}$$

It should be obvious that the operation of addition is

Commutative, i.e., $\mathbf{A} + \mathbf{B} = \mathbf{B} + \mathbf{A}$ (A.7)

and

Associative, i.e., $(\mathbf{A} + \mathbf{B}) + \mathbf{C} = \mathbf{A} + (\mathbf{B} + \mathbf{C})$

$$= \mathbf{A} + \mathbf{B} + \mathbf{C} \tag{A.8}$$

(b) Multiplication

We can define three types of multiplication namely scalar, matrix, and direct product multiplication.

Scalar multiplication. If k is a number or function other than a matrix then the equation

$$\mathbf{B} = k\mathbf{A} \tag{A.9}$$

means that a typical element of B is

$$B_{ij} = kA_{ij} \tag{A.10}$$

For example

$$3\begin{bmatrix} 1 & 2 \\ 5 & 4 \end{bmatrix} = \begin{bmatrix} 3 & 6 \\ 15 & 12 \end{bmatrix} \tag{A.11}$$

Matrix multiplication. Given two matrices **A** and **B** we may multiply them together to form a product which we write simply as **AB**. The multiplication involves a special procedure in which each *row* of **A** is multiplied element by element with each *column* of **B** to form a *single* element of **C** consisting of the sum of the product. Thus each element of

$$\mathbf{C} = \mathbf{AB} \tag{A.12}$$

is obtained from those of **A** and **B** by the equation

$$C_{ik} = \sum_j A_{ij}B_{jk} \tag{A.13}$$

This is best illustrated by way of the example

$$\mathbf{A} = \begin{bmatrix} 1 & 2 \\ 3 & 0 \\ 2 & -1 \end{bmatrix}, \quad \mathbf{B} = \begin{bmatrix} 1 & 2 \\ 5 & 4 \end{bmatrix}, \quad \mathbf{C} = \begin{bmatrix} 11 & 10 \\ 3 & 6 \\ -3 & 0 \end{bmatrix}. \tag{A.14}$$

The element C_{32} in eq. (A.14) is

$$\begin{aligned} C_{32} &= \sum_j A_{3j}B_{j2} \\ &= A_{31}B_{12} + A_{32}B_{22} \\ &= (2)(2) + (-1)(4) \\ &= 0 \end{aligned} \tag{A.15}$$

where the third *row* of **A** has been multiplied element by element with the second *column* of **B** to form the element in the third *row* and second *column* of **C**.

In order that a matrix multiplication **AB** may be defined we must be able to form the sum over the suffix j which appears as a column symbol in A_{ij} and as a row symbol in B_{jk}. This means that *the number of columns of* **A** *must equal the number of rows of* **B**. Two matrices that satisfy this condition are said to be *conformable*. We should, however, point out that although matrices **A** and **B** may be conformable for the product **AB** they are not necessarily conformable for the product **BA** (i.e., the product **BA** may not exist). Square matrices, if they are of the same order, are conformable in either direction, i.e., both **AB** and **BA** exist though in general the two products need *not* be the same. For example

$$\begin{bmatrix} 1 & 2 \\ 3 & 4 \end{bmatrix} \begin{bmatrix} 2 & 4 \\ -1 & -3 \end{bmatrix} = \begin{bmatrix} 0 & -2 \\ 2 & 0 \end{bmatrix} \tag{A.16}$$

whereas

$$\begin{bmatrix} 2 & 4 \\ -1 & -3 \end{bmatrix} \begin{bmatrix} 1 & 2 \\ 3 & 4 \end{bmatrix} = \begin{bmatrix} 14 & 20 \\ -10 & -14 \end{bmatrix} \tag{A.17}$$

Multiple products (**AB**) **C** may also be formed provided all the matrices are conformable and it can easily be shown that (**AB**) **C** = **A** (**BC**). Thus a multiple product can be written in general as **A B C** . . . without the need to resort to brackets; hence if

$$\mathbf{D} = (\mathbf{AB})\,\mathbf{C} \tag{A.18}$$

then

$$\begin{aligned} D_{il} &= \sum_k \left(\sum_j A_{ij}B_{jk} \right) C_{kl} \\ &= \sum_k \sum_j A_{ij}B_{jk}C_{kl} \\ &= \sum_j A_{ij} \sum_k B_{jk}C_{kl} \end{aligned} \tag{A.19}$$

so that

$$\mathbf{D} = \mathbf{A}\,(\mathbf{BC}) \tag{A.20}$$

Similarly we may also show that

$$\mathbf{A}(\mathbf{B} + \mathbf{C}) = \mathbf{AB} + \mathbf{AC} \tag{A.21}$$

hence the operation of matrix multiplication is

Associative, i.e., $(\mathbf{AB})\mathbf{C} = \mathbf{A}(\mathbf{BC}) = \mathbf{ABC}$ (A.22)

Distributive, i.e., $\mathbf{A}(\mathbf{B} + \mathbf{C}) = \mathbf{AB} + \mathbf{AC}$ (A.23)

However, as previously stated, matrix multiplication is *not* in general *commutative*, that is usually

$$\mathbf{AB} \neq \mathbf{BA} \tag{A.24}$$

If we do find two matrices **A** and **B** such that **AB** = **BA** we then say that they *commute*. For example if

$$\mathbf{A} = \begin{bmatrix} 1 & 2 \\ 5 & 4 \end{bmatrix}, \qquad \mathbf{B} = \begin{bmatrix} 1 & 4 \\ 10 & 7 \end{bmatrix} \tag{A.25}$$

then

$$\mathbf{AB} = \mathbf{BA} = \begin{bmatrix} 21 & 18 \\ 45 & 48 \end{bmatrix} \tag{A.26}$$

Direct Product Multiplication. The *direct product* may be formed from any number of matrices, whether or not they are conformable. The direct product of two matrices **A** and **B** will be written as

$$\mathbf{D} = \mathbf{A} \otimes \mathbf{B} \tag{A.27}$$

where each element of **D** is formed by replacing each element A_{ij} of **A** by the matrix $A_{ij}\mathbf{B}$.

Thus, if

$$\mathbf{A} = \begin{bmatrix} 1 & 2 \\ 3 & 0 \\ 2 & -1 \end{bmatrix}, \qquad \mathbf{B} = \begin{bmatrix} a & b \\ c & d \end{bmatrix} \tag{A.28}$$

the direct product of **A** and **B** is

$$\mathbf{D} = \begin{bmatrix} \mathbf{B} & 2\mathbf{B} \\ 3\mathbf{B} & \mathbf{0} \\ 2\mathbf{B} & -\mathbf{B} \end{bmatrix} = \begin{bmatrix} a & b & 2a & 2b \\ c & d & 2c & 2d \\ 3a & 3b & 0 & 0 \\ 3c & 3d & 0 & 0 \\ 2a & 2b & -a & -b \\ 2c & 2d & -c & -d \end{bmatrix} \tag{A.29}$$

A further example of a direct product in the case when **AB** does not exist, can be constructed by using

$$\mathbf{A} = \begin{bmatrix} A_{11} & A_{12} \\ A_{21} & A_{22} \end{bmatrix}, \qquad \mathbf{B} = \begin{bmatrix} B_{11} & B_{12} \\ B_{21} & B_{22} \\ B_{31} & B_{32} \end{bmatrix} \tag{A.30}$$

which gives

$$\mathbf{D} = \mathbf{A} \otimes \mathbf{B} = \begin{bmatrix} A_{11}B_{11} & A_{11}B_{12} & A_{12}B_{11} & A_{12}B_{12} \\ A_{11}B_{21} & A_{11}B_{22} & A_{12}B_{21} & A_{12}B_{22} \\ A_{11}B_{31} & A_{11}B_{32} & A_{12}B_{31} & A_{12}B_{32} \\ A_{21}B_{11} & A_{21}B_{12} & A_{22}B_{11} & A_{22}B_{12} \\ A_{21}B_{21} & A_{21}B_{22} & A_{22}B_{21} & A_{22}B_{22} \\ A_{21}B_{31} & A_{21}B_{32} & A_{22}B_{31} & A_{22}B_{32} \end{bmatrix} \tag{A.31}$$

The elements of **D** are conveniently specified by a double suffix notation such that

$$D_{ii'jj'} = A_{ij}B_{i'j'} \tag{A.32}$$

and **D** may be written as

$$\begin{bmatrix} D_{1111} & D_{1112} & D_{1121} & D_{1122} \\ D_{1211} & D_{1212} & D_{1221} & D_{1222} \\ D_{1311} & D_{1312} & D_{1321} & D_{1322} \\ D_{2111} & D_{2112} & D_{2121} & D_{2122} \\ D_{2211} & D_{2212} & D_{2221} & D_{2222} \\ D_{2311} & D_{2312} & D_{2321} & D_{2322} \end{bmatrix} \tag{A.33}$$

where we should now observe that each row and column is specified by the same *pair* of numbers.
It is obvious from the definition that, in general

$$\mathbf{A} \otimes \mathbf{B} \neq \mathbf{B} \otimes \mathbf{A} \tag{A.34}$$

Also continued products $\mathbf{A} \otimes \mathbf{B} \otimes \mathbf{C} \otimes \cdots$ can be formed. Matrix and direct product multiplication may be mixed provided we adhere to the rule of conformability. A particular relation, that arises in group theory, is

$$(\mathbf{A}\,\mathbf{C}) \otimes (\mathbf{B}\,\mathbf{D}) = (\mathbf{A} \otimes \mathbf{B})(\mathbf{C} \otimes \mathbf{D}) \tag{A.35}$$

where obviously **C** must be conformable to **A**, and **B** must be conformable to **D**. If the elements of **A**, **B**, **C**, and **D** are respectively A_{ij}, $B_{i'j'}$, C_{jk}, and $D_{j'k'}$ we can see the validity of eq. (A.35) from the fact that the element in the ii' row and kk' column of $(\mathbf{AC}) \otimes (\mathbf{BD})$ is

$$\begin{aligned} (\mathbf{AC}) \otimes (\mathbf{BD})_{ii'kk'} &= \sum_j A_{ij}C_{jk} \sum_{j'} B_{i'j'}D_{j'k'} \\ & \sum_j \sum_{j'} A_{ij}B_{i'j'}C_{jk}D_{j'k'} \\ &= \sum_{jj'} (\mathbf{A} \otimes \mathbf{B})_{ii'jj'}(\mathbf{C} \otimes \mathbf{D})_{jj'kk'} \end{aligned} \tag{A.36}$$

which is a typical element in the ii' row and kk' column of $(\mathbf{A} \otimes \mathbf{B})(\mathbf{C} \otimes \mathbf{D})$.

A.3 Determinant of a matrix

The elements of a *square* matrix can be combined to produce an algebraic form known as its *determinant* which we shall write as $|\mathbf{A}|$.

We do not consider it worthwhile to examine here the theory of determinants from a general standpoint, we shall, instead, discuss their principal features with the aid of a few examples. The determinant of a (1×1) matrix **A** is defined as

$$|\mathbf{A}| = A_{11} \tag{A.37}$$

The determinant of a (2×2) matrix **A** is written as

$$|\mathbf{A}| = \begin{vmatrix} A_{11} & A_{12} \\ A_{21} & A_{22} \end{vmatrix} \tag{A.38}$$

and is defined to be

$$|\mathbf{A}| = (A_{11}A_{22} - A_{21}A_{12}) \tag{A.39}$$

For example the matrix **A**

$$\mathbf{A} = \begin{bmatrix} 1 & 2 \\ 5 & 4 \end{bmatrix}$$

has a determinant

$$|\mathbf{A}| = -6 \tag{A.40}$$

Let us now consider the 3 × 3 matrix:

$$\mathbf{A} = \begin{bmatrix} A_{11} & A_{12} & A_{13} \\ \vdots & & \\ A_{21} \cdots & A_{22} \cdots & A_{23} \\ \vdots & & \\ A_{31} & A_{32} & A_{33} \end{bmatrix} \tag{A.41}$$

If we now choose, for example, the element A_{21} and imagine that we eliminate the row and column that includes it, (as indicated by the dotted lines), then the four elements which remain can be considered to form a 2 × 2 matrix, which is a sub-matrix of the original one. The determinant of this sub-matrix is called the *minor* m_{21} of the element A_{21} and from this we can form what is known as the *cofactor* α_{21} by multiplying m_{21} by either +1 or −1 depending on the position of the element that is eliminated. Specifically, for a general element A_{ij}, the cofactor is

$$\alpha_{ij} = (-1)^{i+j} m_{ij} \tag{A.42}$$

The determinant of the 3 × 3 matrix in terms of the cofactors is

$$|\mathbf{A}| = \sum_{i=1}^{3} A_{ij}\alpha_{ij} = \sum_{j=1}^{3} A_{ij}\alpha_{ij} \tag{A.43}$$

For example, if $j = 2$ then

$$|\mathbf{A}| = \sum_{i=1}^{3} A_{i2}\alpha_{i2} = A_{12}\alpha_{12} + A_{22}\alpha_{22} + A_{32}\alpha_{32} \tag{A.44}$$

As a numerical example, consider

$$\mathbf{A} = \begin{bmatrix} 1 & 2 & 3 \\ -3 & 2 & 1 \\ 0 & 1 & 2 \end{bmatrix} \tag{A.45}$$

For which

$$A_{12} = 2, \qquad A_{22} = 2, \qquad A_{32} = 1 \tag{A.46}$$

and

$$\alpha_{12} = -\begin{vmatrix} -3 & 1 \\ 0 & 2 \end{vmatrix} = 6, \qquad \alpha_{22} = +\begin{vmatrix} 1 & 3 \\ 0 & 2 \end{vmatrix} = 2, \qquad \alpha_{32} = -\begin{vmatrix} 1 & 3 \\ -3 & 1 \end{vmatrix} = -10 \tag{A.47}$$

Hence

$$|\mathbf{A}| = 2 \times 6 + 2 \times 2 + 1 \times -10 = 6 \tag{A.48}$$

The correct sign to be attached to the minors may also be found by placing + or − according to the following arrangements.

$$\begin{array}{ccc} 2\times 2 & 3\times 3 & 4\times 4 \\ \begin{vmatrix} + & - \\ - & + \end{vmatrix} & \begin{vmatrix} + & - & + \\ - & + & - \\ + & - & + \end{vmatrix} & \begin{vmatrix} + & - & + & - \\ - & + & - & + \\ + & - & + & - \\ - & + & - & + \end{vmatrix} \end{array} \tag{A.49}$$

where the extension to higher orders is obvious.

The evaluation of the determinant in the above example has been done by an expansion involving the use of the second column. It may also be evaluated using any of the other rows or columns, depending on the value of i or j chosen in eq. (A.43). The determinants of matrices of arbitrary order n can be evaluated by the use of eq. (A.43) in its general form, i.e.,

$$|\mathbf{A}| = \sum_{i=1}^{n} A_{ij}\alpha_{ij} = \sum_{j=1}^{n} A_{ij}\alpha_{ij} \tag{A.50}$$

where for the matrix of order n the cofactors are of order $n-1$, and may themselves be evaluated using eq. (A.43) (summed from 1 to $n-1$) in terms of cofactors of order $n-2$, and so on.

In practice this procedure may be tedious, however there are certain methods of reducing determinants to simpler forms which alleviate the amount of work involved in their calculation.

These methods make use of the following properties, (which we shall quote here without proof).

1. A determinant is unchanged in value if a constant multiple of one row or column is added to another row or column respectively.
2. If a row or column of a determinant has a common factor k then this may be taken outside, for example:

$$\begin{vmatrix} 1 & 2 \\ 3k & 4k \end{vmatrix} = k\begin{vmatrix} 1 & 2 \\ 3 & 4 \end{vmatrix} \tag{A.51}$$

3. A determinant is zero if two rows or columns are equal. This could be used to show that if each element of *one* row or column is multiplied by the cofactors of the corresponding element in *another* row or column then the sum of these is zero. That is

$$\sum_{i} A_{ij}\alpha_{ik} = 0 \tag{A.52}$$

This result follows from the fact that taking the cofactors of another row is equivalent to replacing the former row by the latter, so that two rows are equal and the determinant is zero.

A particular property of determinants that we shall have occasion to use, but whose proof is not simple, is that if two matrices **A** and **B** are multiplied together then

$$|\mathbf{AB}| = |\mathbf{A}|\,|\mathbf{B}| \tag{A.53}$$

A.4 Special matrices

Certain varieties of *square* matrix occur frequently and have special properties; they are referred to by particular names and are detailed below.

(a) Null matrix

All the elements of this matrix which we write as **0** are zero. It is possible to have two non-zero matrices whose product is zero.

(b) Diagonal matrix

In this matrix all the elements are zero except those on the leading diagonal, that is

$$A_{ij} = 0 \qquad (i \neq j) \tag{A.54}$$

For example

$$\mathbf{A} = \begin{bmatrix} a & 0 & 0 \\ 0 & b & 0 \\ 0 & 0 & c \end{bmatrix} \tag{A.55}$$

is diagonal and is sometimes written as

$$\mathbf{A} = \text{diag}\,(a, b, c) \tag{A.56}$$

(c) Unit matrix

This is a diagonal matrix in which the elements on the leading diagonal are unity, that is

$$A_{ij} = \delta_{ij} \tag{A.57}$$

where δ_{ij} is the Kronecker delta defined as

$$\begin{aligned} \delta_{ij} &= 0 \qquad i \neq j, \\ \delta_{ij} &= 1 \qquad i = j \end{aligned} \tag{A.58}$$

If we are not using the suffix notation the unit matrix is usually given the symbol **I** (for historical reasons, in the theory of group representations, the symbol **E** is used). As may be readily checked, the unit matrix plays the same role in matrix multiplication as unity does in arithmetic, thus

$$\mathbf{AI} = \mathbf{IA} = \mathbf{A} \tag{A.59}$$

A constant multiple $k\mathbf{I}$ of the unit matrix is called a *constant matrix*.

(d) Direct sum matrix

This matrix is similar to the diagonal matrix but the elements on the leading diagonal are themselves matrices. For example the matrix **A**, i.e.,

$$\mathbf{A} = \begin{bmatrix} a & b & 0 & 0 & 0 & 0 \\ c & d & 0 & 0 & 0 & 0 \\ 0 & 0 & e & 0 & 0 & 0 \\ 0 & 0 & 0 & f & g & h \\ 0 & 0 & 0 & i & j & k \\ 0 & 0 & 0 & l & m & n \end{bmatrix} = \begin{bmatrix} \mathbf{A}_1 & \mathbf{0} & \mathbf{0} \\ \mathbf{0} & \mathbf{A}_2 & \mathbf{0} \\ \mathbf{0} & \mathbf{0} & \mathbf{A}_3 \end{bmatrix} \tag{A.60}$$

is a direct sum matrix and may be written as diag $(\mathbf{A}_1\ \mathbf{A}_2\ \mathbf{A}_3)$ or as

$$\mathbf{A} = \mathbf{A}_1 \oplus \mathbf{A}_2 \oplus \mathbf{A}_3 \tag{A.61}$$

where

$$\mathbf{A}_1 = \begin{bmatrix} a & b \\ c & d \end{bmatrix}, \qquad \mathbf{A}_2 = [e], \quad \mathbf{A}_3 = \begin{bmatrix} f & g & h \\ i & j & k \\ l & m & n \end{bmatrix}$$

It can be shown that

$$|\text{diag}\,(\mathbf{A}_1\ \mathbf{A}_2 \cdots \mathbf{A}_n)| = |\mathbf{A}_1|\,|\mathbf{A}_2| \cdots |\mathbf{A}_n| \tag{A.62}$$

(e) Transposed matrix

The transpose of a matrix is formed by interchanging its rows and columns and is denoted by $\tilde{\mathbf{A}}$. According to the definition if an element of $\mathbf{A}$ is A_{ij} then an element of the transpose $\tilde{\mathbf{A}}$ is A_{ji}. Thus, for example

$$\mathbf{A} = \begin{bmatrix} 1 & 2 & 3 \\ 4 & 5 & 6 \\ 7 & 8 & 9 \end{bmatrix}, \qquad \tilde{\mathbf{A}} = \begin{bmatrix} 1 & 4 & 7 \\ 2 & 5 & 8 \\ 3 & 6 & 9 \end{bmatrix} \tag{A.63}$$

Transpose of a Product. If $\mathbf{C} = \mathbf{AB}$ then a typical element of $\mathbf{C}$ is $C_{ik} = \sum_j A_{ij}B_{jk}$ and a typical element of the transpose of C is

$$\begin{aligned} C_{ki} &= \sum_j A_{kj}B_{ji} \\ &= \sum_j \tilde{A}_{jk}\tilde{B}_{ij} \\ &= \sum_j \tilde{B}_{ij}\tilde{A}_{jk} \end{aligned} \tag{A.64}$$

which is an element of $\tilde{\mathbf{B}}\tilde{\mathbf{A}}$; hence

$$\widetilde{\mathbf{AB}} = \tilde{\mathbf{B}}\tilde{\mathbf{A}} \tag{A.65}$$

That is *the transpose of a product is the product of the transposes in the reverse order.*

(f) Symmetric and antisymmetric matrices

A symmetric matrix is one for which $\tilde{\mathbf{A}} = \mathbf{A}$.
An antisymmetric matrix is one for which $\tilde{\mathbf{A}} = -\mathbf{A}$.

(g) Hermitian matrix

If we denote the complex conjugate of an element A_{ij} by A_{ij}^* and the corresponding matrices by $\mathbf{A}$ and $\mathbf{A}^*$ then a Hermitian matrix is one for which

$$\tilde{\mathbf{A}}^* = \mathbf{A} \tag{A.66}$$

In order to avoid unnecessary duplication of superscripts, it is customary to denote the complex conjugate of the transposed matrix by $\mathbf{A}^\dagger$. (It is referred to in most physics texts as the *adjoint* matrix.) An example of a Hermitian matrix is

$$\begin{bmatrix} a & c + id \\ c - id & b \end{bmatrix} \tag{A.67}$$

where $i = \sqrt{-1}$. It is evident that the diagonal elements of a Hermitian matrix must be real, because from the definition $A_{ii} = A_{ii}^*$. The symmetric matrix of case (f) is a special case of a Hermitian matrix in which *all* the elements are real.

An antisymmetric Hermitian matrix is one for which

$$\mathbf{A}^\dagger = -\mathbf{A} \tag{A.68}$$

(h) Adjugate matrix

This is a matrix constructed from a given matrix **A** by replacing each element by its cofactor and then *transposing* the result. (Mathematical texts often refer to the adjugate matrix as the adjoint matrix.) Thus if an element of **A** is A_{ij} then an element of the adjugate is α_{ji}.

As an example consider

$$\mathbf{A} = \begin{bmatrix} 1 & 2 & 3 \\ -3 & 2 & 1 \\ 0 & 1 & 2 \end{bmatrix} \tag{A.69}$$

for which the matrix of the cofactors is

$$\begin{bmatrix} 3 & 6 & -3 \\ -1 & 2 & -1 \\ -4 & -10 & 8 \end{bmatrix} \tag{A.70}$$

and the adjugate matrix is

$$\textbf{adj}\,\mathbf{A} = \begin{bmatrix} 3 & -1 & -4 \\ 6 & 2 & -10 \\ -3 & -1 & 8 \end{bmatrix} \tag{A.71}$$

(i) Inverse of a matrix

This is a matrix which we write as $\mathbf{A}^{-1}$ such that for a given matrix **A** we have

$$\mathbf{A}\mathbf{A}^{-1} = \mathbf{A}^{-1}\mathbf{A} = \mathbf{I} \tag{A.72}$$

The method of constructing the inverse of a matrix is given in the following section.

(j) Unitary matrix

This is a matrix usually written as **U** for which the adjoint matrix is equal to the inverse, that is

$$\mathbf{U}^\dagger = \mathbf{U}^{-1} \tag{A.73}$$

(k) Orthogonal matrix

This is a particular case of a unitary matrix in which all the elements are real, in which case

$$\tilde{\mathbf{U}} = \mathbf{U}^{-1} \tag{A.74}$$

A.5 Inverse of a matrix

In this section we shall only consider square matrices and assert that if matrices **A** and **B** exist such that

$$\mathbf{AB} = \mathbf{BA} = \mathbf{I} \tag{A.75}$$

then **B** is the inverse of **A**, and is written as $\mathbf{A}^{-1}$. We will now show that one method of finding the inverse is to make use of the adjugate matrix. Thus from eqs. (A.69) and (A.71) we find that

$$\mathbf{A}(\mathbf{adj\ A}) = \begin{bmatrix} 1 & 2 & 3 \\ -3 & 2 & 1 \\ 0 & 1 & 2 \end{bmatrix} \begin{bmatrix} 3 & -1 & -4 \\ 6 & 2 & -10 \\ -3 & -1 & 8 \end{bmatrix} = \begin{bmatrix} 6 & 0 & 0 \\ 0 & 6 & 0 \\ 0 & 0 & 6 \end{bmatrix} = 6 \begin{bmatrix} 1 & 0 & 0 \\ 0 & 1 & 0 \\ 0 & 0 & 1 \end{bmatrix}$$
$$= |\mathbf{A}|\mathbf{I} \tag{A.76}$$

Hence

$$\frac{1}{|\mathbf{A}|}\mathbf{A}\ \mathbf{adj\ A} = \mathbf{I} \tag{A.77}$$

so that

$$\mathbf{A}^{-1} = \mathbf{adj\ A}/|\mathbf{A}| \tag{A.78}$$

The general proof for a 3 × 3 matrix is as follows. If

$$\mathbf{A} = \begin{bmatrix} A_{11} & A_{12} & A_{13} \\ A_{21} & A_{22} & A_{23} \\ A_{31} & A_{32} & A_{33} \end{bmatrix} \tag{A.79}$$

then

$$\mathbf{adj\ A} = \begin{bmatrix} \alpha_{11} & \alpha_{21} & \alpha_{31} \\ \alpha_{12} & \alpha_{22} & \alpha_{32} \\ \alpha_{13} & \alpha_{23} & \alpha_{33} \end{bmatrix} \tag{A.80}$$

and

$$\mathbf{A}(\mathbf{adj\ A}) = \begin{vmatrix} \sum_j A_{1j}\alpha_{1j} & \sum_j A_{1j}\alpha_{2j} & \sum_j A_{1j}\alpha_{3j} \\ \sum_j A_{2j}\alpha_{1j} & \sum_j A_{2j}\alpha_{2j} & \sum_j A_{2j}\alpha_{3j} \\ \sum_j A_{3j}\alpha_{1j} & \sum_j A_{3j}\alpha_{2j} & \sum_j A_{3j}\alpha_{3j} \end{vmatrix} = \begin{vmatrix} |\mathbf{A}| & 0 & 0 \\ 0 & |\mathbf{A}| & 0 \\ 0 & 0 & |\mathbf{A}| \end{vmatrix} \tag{A.81}$$

which follows from eqs. (A.50) and (A.52). Similarly we can find

$$(\mathbf{adj\ A})\mathbf{A} = |\mathbf{A}|\mathbf{I}$$

The extension to higher order matrices is obvious, and eq. (A.78) is quite general. Also apart from giving a formula for the inverse, eq. (A.78) is particularly revealing because it demonstrates that $\mathbf{A}^{-1}$ only exists if $|\mathbf{A}| \neq 0$. A matrix whose inverse does not exist, i.e., whose determinant is zero is said to be *singular*; conversely a matrix which possesses an inverse has a non-zero determinant and is said to be *non-singular*.

The inverse of a product **AB** is $\mathbf{B}^{-1}\mathbf{A}^{-1}$ a result which we can readily see from

$$\begin{aligned}(\mathbf{B}^{-1}\mathbf{A}^{-1})(\mathbf{AB}) &= \mathbf{B}^{-1}\mathbf{A}^{-1}\mathbf{AB} \\ &= \mathbf{B}^{-1}\mathbf{IB} \\ &= \mathbf{B}^{-1}\mathbf{B} \\ &= \mathbf{I} \end{aligned} \tag{A.82}$$

In general, for a multiple product *the inverse of a product is the product of the inverses in the reverse order.*

A.6 Vectors

Before the theory of matrices can be discussed any further it is necessary to consider the quantities known as *vectors*, which may be written as single row $(1 \times n)$ or single column $(n \times 1)$ matrices. They will often be referred to as row or column vectors.

If a vector has only three real elements then each element can be considered to be a component of a familiar three-dimensional Cartesian vector. However, in general, it is convenient to define a general vector which may have any number of elements which may themselves be not only real or complex numbers, but also functions or other vectors.

We shall consider first those vectors whose elements are real or complex numbers and denote them by one of the following forms:

(a) An arbitrary vector will be indicated by the use of black type, viz. **a** or **x**,

(b) The notation introduced by Dirac will be used in which a column vector is represented as

$$|x\rangle = \begin{bmatrix} x_1 \\ x_2 \\ \vdots \\ x_n \end{bmatrix} \tag{A.83}$$

where x_i are called the components. Related to $|x\rangle$ is a row vector $\langle x|$ known as its *dual vector* where

$$\langle x| = [\overset{*}{x}_1 \overset{*}{x}_2 \ldots \overset{*}{x}_n] \tag{A.84}$$

(note that the complex conjugate is not indicated in the notation but is taken to be understood).

These forms are known as the ket and bra vectors respectively (from a division of the word bra(c)ket).

A.7 Basis vectors

We shall now consider in more detail what is meant by the components x_i of the vector **x**. A simple way of doing this is by means of the example provided by the two-dimensional space formed from the familiar displacement vectors in a plane.

Let $\mathbf{e}_1$ and $\mathbf{e}_2$ be two vectors of unit length lying along two perpendicular Cartesian axes; as shown in Fig. A.1. Any other vector **x** in the plane of $\mathbf{e}_1$ and $\mathbf{e}_2$ can be expressed as a sum of suitable multiples of $\mathbf{e}_1$ and $\mathbf{e}_2$, i.e.,

$$\mathbf{x} = x_1\mathbf{e}_1 + x_2\mathbf{e}_2 \tag{A.85}$$

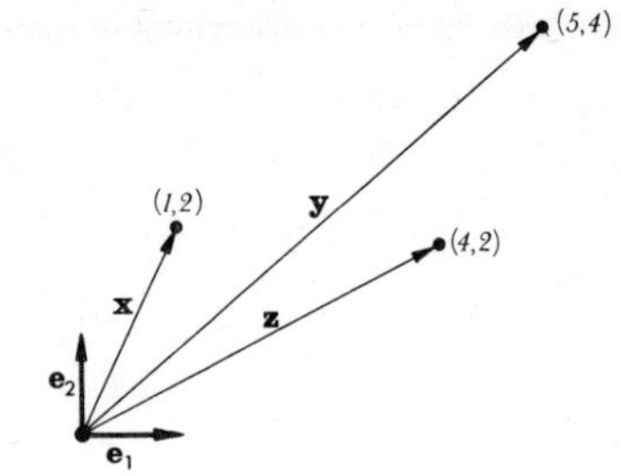

Fig. A.1 *Vectors x, y, and z in the basis* $\mathbf{e}_1, \mathbf{e}_2$.

The vectors $\mathbf{e}_1$ and $\mathbf{e}_2$ are said to be *basis vectors* which *span* the two-dimensional space of the plane, whilst the numbers x_1 and x_2 are the *components* of $\mathbf{x}$ *in terms of this basis*. Provided the basis is specified, the numbers x_1 and x_2 uniquely define the vector $\mathbf{x}$ which can then be written in the row or column forms of eqs. (A.83) and (A.84).

Let us now consider two vectors $\mathbf{x}$ and $\mathbf{y}$ (cf. Fig. A.1) given by

$$\mathbf{x} = \mathbf{e}_1 + 2\mathbf{e}_2 \tag{A.86}$$

$$\mathbf{y} = 5\mathbf{e}_1 + 4\mathbf{e}_2 \tag{A.87}$$

which can be written as

$$|x\rangle = \begin{bmatrix} 1 \\ 2 \end{bmatrix}, \qquad |y\rangle = \begin{bmatrix} 5 \\ 4 \end{bmatrix} \tag{A.88}$$

Consider now a further vector $\mathbf{z}$ such as

$$\begin{aligned} \mathbf{z} &= -\mathbf{x} + \mathbf{y} \\ &= 4\mathbf{e}_1 + 2\mathbf{e}_2 \end{aligned} \tag{A.89}$$

where $\mathbf{z}$, when referred to the basis $\mathbf{e}_1$ and $\mathbf{e}_2$, is

$$|z\rangle = \begin{bmatrix} 4 \\ 2 \end{bmatrix} \tag{A.90}$$

However, since we could equally well choose the vectors $\mathbf{x}$ and $\mathbf{y}$ as basis vectors we can also write

$$|z\rangle = \begin{bmatrix} -1 \\ 1 \end{bmatrix} \tag{A.91}$$

Obviously eq. (A.91) is an equally valid representation of $\mathbf{z}$ but of course the basis vectors are no longer of unit length, nor are they orthogonal.

The above example illustrates the fact that the representation of a vector by the form shown in eqs. (A.83) and (A.84) depends on the choice of basis and in general a vector $\mathbf{x}$ of dimension n with components x_i can be expressed in terms of n basis vectors $\mathbf{e}_i$ as

$$\begin{aligned} \mathbf{x} &= x_1\mathbf{e}_1 + x_2\mathbf{e}_2 + \cdots + x_n\mathbf{e}_n \\ &= \sum_{i=1}^{n} x_i\mathbf{e}_i \end{aligned} \tag{A.92}$$

In the Dirac notation eq. (A.92) is

$$|x\rangle = \sum_i x_i|e_i\rangle \tag{A.93}$$

and the dual vector is

$$\langle x| = \sum_i \langle e_i|x_i^* \tag{A.94}$$

A.8 Linear dependence of vectors

We have already seen how vectors can be expressed as a combination of other vectors, cf. eq. (A.89). For example

$$2\begin{bmatrix}1\\2\\3\end{bmatrix} + 3\begin{bmatrix}4\\5\\6\end{bmatrix} - 7\begin{bmatrix}3\\2\\1\end{bmatrix} = \begin{bmatrix}-7\\5\\17\end{bmatrix} \tag{A.95}$$

We could also use the same set of vectors with different coefficients and obtain a null vector. For instance

$$7\begin{bmatrix}1\\2\\3\end{bmatrix} - 4\begin{bmatrix}4\\5\\6\end{bmatrix} + 3\begin{bmatrix}3\\2\\1\end{bmatrix} = 7|x_1\rangle - 4|x_2\rangle + 3|x_3\rangle = \begin{bmatrix}0\\0\\0\end{bmatrix} \tag{A.96}$$

If, as in this case, any one of the vectors $|x_1\rangle, |x_2\rangle, |x_3\rangle$ can be expressed in terms of the other two then the set of vectors is said to be *linearly dependent*. In general a set of vectors $|x_i\rangle$ is linearly dependent if coefficients a_i can be found such that

$$\sum_i a_i|x_i\rangle = 0 \tag{A.97}$$

where we exclude the trivial case in which *all* the coefficients are zero. Conversely if a set a_i cannot be found which satisfies eq. (A.97) then the set is said to be *linearly* independent.

Number of linearly dependent vectors

Suppose we have r vectors each of dimension n. Then we can establish the following results

(a) $r \leqslant n$ The vectors may or may not be linearly dependent. For example if $r = 2$, $n = 3$ the vectors [1 2 3] and [4 5 6] are linearly independent whereas the vectors [1 2 3] and [2 4 6] are linearly dependent.

(b) $r > n$ The vectors are of necessity linearly dependent. We can show this by setting the vectors out as an $(n \times r)$ matrix and performing successive operations to reduce at least one of the columns to zero. For example let us consider the vectors $\mathbf{x}_1, \mathbf{x}_2, \ldots, \mathbf{x}_n, \ldots, \mathbf{x}_r$, i.e.,

$$\begin{array}{rcccccccc} & |x_1\rangle & |x_2\rangle & \cdots & |x_n\rangle & \cdots & |x_r\rangle \\ \text{components of vector} \left\{\begin{array}{c} \\ \\ \\ \\ \end{array}\right. & \begin{array}{c}\alpha_1\\ \alpha_2\\ \vdots\\ \alpha_n\end{array} & \begin{array}{c}\beta_1\\ \beta_2\\ \vdots\\ \beta_n\end{array} & \begin{array}{c}\cdots\\ \cdots\\ \\ \cdots\end{array} & \begin{array}{c}\gamma_1\\ \gamma_2\\ \vdots\\ \gamma_n\end{array} & \begin{array}{c}\cdots\\ \cdots\\ \\ \cdots\end{array} & \begin{array}{c}\delta_1\\ \delta_2\\ \vdots\\ \delta_n\end{array} \end{array} \tag{A.98}$$

Suitable multiples of $|x_i\rangle$ may now be subtracted from each other so as to form new vectors $|x'_j\rangle$ with every element in the first row equal to zero, i.e.,

$$\begin{matrix} |x_1\rangle & |x'_2\rangle & \cdots & |x'_n\rangle & \cdots & |x'_r\rangle \\ \alpha_1 & 0 & \cdots & 0 & \cdots & 0 \\ \alpha_2 & \beta'_2 & \cdots & \gamma'_2 & \cdots & \delta'_1 \\ \vdots & \vdots & & \vdots & & \vdots \\ \alpha_n & \beta'_n & & \gamma'_n & & \delta'_n \end{matrix} \tag{A.99}$$

where $\beta'_2 = \beta_2 - (\beta_2/\alpha_1)\alpha_2$ and $\gamma'_2 = \gamma_2 - (\beta_1/\alpha_1)\gamma_1$, etc. This leaves one vector of dimension n, and $r - 1$ vectors with one zero component and the process can now be repeated n times with each vector in turn to produce at least $r - n$ null vectors. These null vectors are linear combinations of the vectors $\mathbf{x}_1$ to $\mathbf{x}_n$. Thus $\mathbf{x}_1$ to $\mathbf{x}_n$ are linearly dependent. The remaining n vectors may, or may not, be (cf. (a)) linearly dependent. We conclude therefore that *there are at most n linearly independent vectors of dimension n.*

A.9 Vector spaces

From the above discussion we can now see that *a set of r linearly independent vectors will form a basis for a space of dimension r* and the vectors are said to span this space. In addition any set of r linearly independent vectors formed by linear combination of the original set may also form a basis. For example the vectors $\langle x_1| = [1\ 2\ 3]$ and $\langle x_2| = [4\ 5\ 6]$ which are each of dimension $n = 3$ span a space of dimension $r = 2$; in Cartesian coordinates these two vectors define a plane in a three-dimensional space. Also any vector in this plane can be expressed as a linear combination of $\langle x_1|$ and $\langle x_2|$ so that any linear combination may also serve as a basis.

A.10

(a) Scalar product

In elementary vector analysis, using a three-dimensional space we can geometrically define orthogonal vectors, i.e., those that are perpendicular to each other. This statement can be formalized by means of the scalar product $\mathbf{x}\,.\,\mathbf{y}$ of two vectors; defined as

$$\mathbf{x}\,.\,\mathbf{y} = xy\cos\theta \tag{A.100}$$

where x is the length of $\mathbf{x}$, y is the length of $\mathbf{y}$, and θ is the angle between the vectors; two vectors $\mathbf{x}$ and $\mathbf{y}$ are then said to be orthogonal if their scalar product is zero.

In a general vector space of arbitrary dimension and basis, the idea of orthogonal vectors presents more difficulty because the scalar product itself can only be completely defined in terms of the basis.

The scalar product $\langle e_i|e_j\rangle$ of two basis vectors $\mathbf{e}_i$ and $\mathbf{e}_j$ is usually written as

$$\mathbf{e}_i\,.\,\mathbf{e}_j = \langle e_i|e_j\rangle = g_{ij} \tag{A.101}$$

where g_{ij} is an element of the so-called metric matrix. Hence the scalar product of two arbitrary vectors **x** and **y** in terms of eqs. (A.93) and (A.94), is

$$\begin{aligned}\langle x|y\rangle &= \sum_i \sum_j \langle e_i|x_i^* y_j|e_j\rangle \\ &= \sum_i \sum_j x_i^* y_j \langle e_i|e_j\rangle \\ &= \sum_i \sum_j x_i^* y_j g_{ij}\end{aligned} \tag{A.102}$$

The vectors $|x\rangle$ and $|y\rangle$ are then said to be orthogonal if their scalar product is zero.

(b) Orthonormal basis

The expression for the scalar product takes a particularly simple form if the basis is orthonormal, that is if:
Two different basis vectors $\mathbf{e}_i$ and $\mathbf{e}_j$ have zero scalar product, i.e.,

$$\langle e_i|e_j\rangle = 0 \tag{A.103}$$

Each basis vector has a scalar product with itself equal to unity, i.e.,

$$\langle e_i|e_i\rangle = 1 \tag{A.104}$$

These two conditions are summarized as

$$\langle e_i|e_j\rangle = \delta_{ij} \tag{A.105}$$

where δ_{ij} is zero if $i \neq j$ and unity if $i = j$. In this case the scalar product of two vectors **x** and **y** is simply

$$\langle x|y\rangle = \sum_i x_i^* y_i \tag{A.106}$$

Note that the sum $\sum_i x_i^* y_i$ can be written as

$$[x_1^* x_2^* \cdots x_n^*]\begin{bmatrix} y_1 \\ y_2 \\ \vdots \\ y_n \end{bmatrix} \tag{A.107}$$

which is the justification for writing the scalar product as $\langle x|y\rangle$.

(c) Reciprocal basis

If the basis vectors **e** are not orthonormal then one can introduce a second basis $\bar{\mathbf{e}}$ which is defined so that

$$\bar{\mathbf{e}}_j \cdot \mathbf{e}_i = \langle \bar{e}_j|e_i\rangle = \delta_{ij} \tag{A.108}$$

The set $\bar{\mathbf{e}}$ is called the *reciprocal basis*; notice that an orthonormal basis is self reciprocal; for example if we have an orthonormal set $\mathbf{e}_1$ and $\mathbf{e}_2$ in a two-dimensional space, then $\mathbf{e}_1 \cdot \mathbf{e}_2 = 0$ and $\bar{\mathbf{e}}_2 = \mathbf{e}_1$. In terms of the reciprocal basis a vector **x** will have new components $\bar{x}_i$ and in analogy with eqs. (A.92), (A.93), and (A.94) we can write

$$\mathbf{x} = \sum_i \bar{x}_i \bar{\mathbf{e}}_i \tag{A.109}$$

$$|x\rangle = \sum_i \bar{x}_i |\bar{e}_i\rangle \tag{A.110}$$

$$\langle x| = \sum_i \langle \bar{e}_i|\bar{x}_i^* \tag{A.111}$$

The scalar product of two vectors $|x\rangle$ and $|y\rangle$ can then be written in terms of a basis and the reciprocal of this basis as

$$\begin{aligned}\langle x|y\rangle &= \sum_i \sum_j \langle \bar{e}_i|\bar{x}_i^* y_j|e_j\rangle \\ &= \sum_i \sum_j \bar{x}_i^* y_j \langle \bar{e}_i|e_j\rangle \\ &= \sum_i \sum_j \bar{x}_i^* y_j \delta_{ij} \\ &= \sum_i \bar{x}_i^* y_i\end{aligned} \tag{A.112}$$

(d) Length of a vector

A vector $|x\rangle$ has a length L where

$$L^2 = \langle x|x\rangle \tag{A.113}$$

Note that, even when the components of a vector are complex, its length is real, and greater than or equal to zero. For example if

$$\langle x| = [i \quad (1 + i)] \tag{A.114}$$

then

$$\langle x|x\rangle = [i \quad (1 + i)]\begin{bmatrix} -i \\ 1 - i\end{bmatrix} = 3 \tag{A.115}$$

Hence the length L is $\sqrt{3}$.

A vector is said to be a *unit* vector if it has a length of unity; and any vector may be *normalized* to unit length by dividing each component by L. Therefore

$$\langle y| = \begin{bmatrix} \dfrac{i}{\sqrt{3}} & \dfrac{1 + i}{\sqrt{3}}\end{bmatrix} \tag{A.116}$$

is a unit vector.

(e) Construction of an orthonormal basis

From a given set of basis vectors $\mathbf{x}_i$ we can always form another, mutually orthogonal, set $\mathbf{y}_i$ that span the same space. This is done by means of the Gram–Schmidt procedure in which $\mathbf{y}_i$ is constructed according to the formula.

$$\mathbf{y}_i = \mathbf{x}_i - \sum_{j=1}^{i-1} \frac{(\mathbf{y}_j \,.\, \mathbf{x}_i)}{(\mathbf{y}_j \cdot \mathbf{y}_j)}\mathbf{y}_j \tag{A.117}$$

It can be seen from equation (A.117) that
Each vector $\mathbf{y}_i$ is a linear combination of the vectors $\mathbf{x}_i$ and hence the set of vectors $\mathbf{y}_i$ span the same space.
The vectors $\mathbf{y}_i$ are mutually orthogonal because

$$\mathbf{y}_j \cdot \mathbf{y}_i = 0 \tag{A.118}$$

From the *orthogonal* basis $\mathbf{y}_i$ we can then form an *orthonormal* basis $\mathbf{z}_i$ by dividing each vector by its length, i.e.,

$$\mathbf{z}_i = \mathbf{y}_i/(\mathbf{y}_i \cdot \mathbf{y}_i)^{1/2} \tag{A.119}$$

As an example let us consider the vectors

$$\begin{aligned} \mathbf{x}_1 &= [1 \quad -1 \quad 0] \\ \mathbf{x}_2 &= [2 \quad -1 \quad -2] \\ \mathbf{x}_3 &= [1 \quad -1 \quad -2] \end{aligned} \tag{A.120}$$

where the components of each are expressed in a Cartesian basis. If we choose $\mathbf{y}_1 = \mathbf{x}_1$ then from eq. (A.117), we obtain

$$\begin{aligned} \mathbf{y}_2 &= [2 \quad -1 \quad 2] - \tfrac{3}{2}[1 \quad -1 \quad 0] \\ &= [\tfrac{1}{2} \quad \tfrac{1}{2} \quad -2] \\ &= [1 \quad 1 \quad -4] \end{aligned} \tag{A.121}$$

The last step follows because any multiple of the components defines a vector in the same space. Applying eq. (A.117) again we find that

$$\begin{aligned} \mathbf{y}_3 &= [1 \quad -1 \quad -2] - \tfrac{2}{2}[1 \quad -1 \quad 0] - \tfrac{8}{18}[1 \quad 1 \quad -4] \\ &= [-\tfrac{4}{9} \quad -\tfrac{4}{9} \quad -\tfrac{2}{9}] \\ &= [2 \quad 2 \quad 1] \end{aligned} \tag{A.122}$$

The vectors

$$\begin{aligned} \mathbf{y}_1 &= [1 \quad -1 \quad 0] \\ \mathbf{y}_2 &= [1 \quad 1 \quad -4] \\ \mathbf{y}_3 &= [2 \quad 2 \quad 1] \end{aligned} \tag{A.123}$$

now span the same space as $\mathbf{x}_1\,\mathbf{x}_2\,\mathbf{x}_3$ and are mutually orthogonal. The corresponding orthonormal vectors are

$$\begin{aligned} \mathbf{z}_1 &= \left[\frac{1}{\sqrt{2}} \quad \frac{1}{\sqrt{2}} \quad 0\right] \\ \mathbf{z}_2 &= \left[\frac{1}{3\sqrt{2}} \quad \frac{1}{3\sqrt{2}} \quad -\frac{4}{3\sqrt{2}}\right] \\ \mathbf{z}_3 &= [\tfrac{2}{3} \quad \tfrac{2}{3} \quad \tfrac{1}{3}] \end{aligned} \tag{A.124}$$

A.11 Change of components in a given basis

We have stated (eq. (A.92)) that the components x_i of a vector $\mathbf{x}$ depend on the basis $\mathbf{e}_i$. In order to calculate the effect of a change of basis on the components of $\mathbf{x}$ let us suppose that a new basis $\mathbf{e}'$ is formed from the old one $\mathbf{e}$ according to the transformation

$$[\mathbf{e}'_1\mathbf{e}'_2 \cdots \mathbf{e}'_n] = [\mathbf{e}_1\mathbf{e}_2 \cdots \mathbf{e}_n] \begin{bmatrix} T_{11} & T_{12} & \cdots & T_{1n} \\ T_{21} & T_{22} & \cdots & T_{2n} \\ \vdots & \vdots & & \vdots \\ T_{n1} & T_{n2} & \cdots & T_{nn} \end{bmatrix} \tag{A.125}$$

which can be symbolized as

$$\widetilde{(\mathbf{e}')} = \widetilde{(\mathbf{e})}\mathbf{T} \tag{A.126}$$

or

$$(\mathbf{e}') = \mathbf{T}^{\dagger}(\mathbf{e}) \tag{A.127}$$

Then we have in terms of the new basis

$$\mathbf{x} = x'_1\mathbf{e}'_1 + x'_2\mathbf{e}'_2 + \cdots + x'_n\mathbf{e}'_n \tag{A.128}$$

which can be formally written as

$$\begin{aligned} \mathbf{x} &= [\mathbf{e}'_1\mathbf{e}'_2 \cdots \mathbf{e}'_n]\begin{bmatrix} x'_1 \\ x'_2 \\ \vdots \\ x'_n \end{bmatrix} \\ &= (\widetilde{\mathbf{e}})\mathbf{T}|x'\rangle \end{aligned} \tag{A.129}$$

However, in the original basis

$$\mathbf{x} = (\widetilde{\mathbf{e}})|x\rangle \tag{A.130}$$

and hence we can see that the components change as

$$|x\rangle = \mathbf{T}|x'\rangle \tag{A.131}$$

or

$$|x'\rangle = \mathbf{T}^{-1}|x\rangle \tag{A.132}$$

and

$$\langle x'| = \langle x|(\mathbf{T}^{-1})^{\dagger} \tag{A.133}$$

As an example consider the vectors of section A.7 again. Then, (writing $\mathbf{e}'_1$ for $\mathbf{x}$ and $\mathbf{e}'_2$ for $\mathbf{y}$) we have

$$\begin{aligned} \mathbf{e}'_1 &= \mathbf{e}_1 + 2\mathbf{e}_2 \\ \mathbf{e}'_2 &= 5\mathbf{e}_1 + 4\mathbf{e}_2 \end{aligned} \tag{A.134}$$

or

$$[\mathbf{e}'_1 \quad \mathbf{e}'_2] = [\mathbf{e}_1 \quad \mathbf{e}_2]\begin{bmatrix} 1 & 5 \\ 2 & 4 \end{bmatrix} \tag{A.135}$$

i.e.,

$$(\widetilde{\mathbf{e}'}) = (\widetilde{\mathbf{e}})\mathbf{T} \tag{A.136}$$

Thus

$$\mathbf{T} = \begin{bmatrix} 1 & 5 \\ 2 & 4 \end{bmatrix} \quad \text{and} \quad (\mathbf{T}^{-1})^{\dagger} = \tfrac{1}{6}\begin{bmatrix} -4 & 2 \\ 5 & -1 \end{bmatrix} \tag{A.137}$$

Now the components of $\mathbf{z}$ (cf. eq. (A.89)) in the original basis $\mathbf{e}$ are $z_1 = 4$, $z_2 = 2$. Hence by eq. (A.133)

$$\begin{aligned} [z'_1 \quad z'_2] &= [4 \quad 2]\tfrac{1}{6}\begin{bmatrix} -4 & 2 \\ 5 & -1 \end{bmatrix} \\ &= [1 - \quad 1] \end{aligned} \tag{A.138}$$

in agreement with eq. (A.91).

A.12 Change of components in the reciprocal basis

The components in the reciprocal basis $\bar{\mathbf{e}}$ behave in a different way from those in the original basis $\mathbf{e}$. Thus since eq. (A.108) must still be true in the new basis $\mathbf{e}'$ we can say that

$$\mathbf{e}'_i \, . \, \bar{\mathbf{e}}'_j = \delta_{ij} = \mathbf{e}_i \, . \, \bar{\mathbf{e}}_j \tag{A.139}$$

Hence, from eq. (A.126)

$$\widetilde{(\mathbf{e})}\mathbf{T}(\bar{\mathbf{e}}') = \widetilde{(\mathbf{e})}(\bar{\mathbf{e}}) \tag{A.140}$$

therefore

$$\mathbf{T}(\bar{\mathbf{e}}') = (\bar{\mathbf{e}}) \tag{A.141}$$

or

$$(\bar{\mathbf{e}}') = \mathbf{T}^{-1}(\bar{\mathbf{e}}) \tag{A.142}$$

and

$$\widetilde{(\bar{\mathbf{e}}')} = \widetilde{(\bar{\mathbf{e}})}(\mathbf{T}^{-1})^{\dagger} \tag{A.143}$$

Now, if we compare eq. (A.142) with eq. (A.132) we can see that the *reciprocal basis* $\bar{\mathbf{e}}_i$ changes in the same way as the *components* x_i in the *basis* $\mathbf{e}_i$; and it should therefore be obvious that the converse is true, i.e., the components x_i in the reciprocal basis $\bar{\mathbf{e}}_i$ will vary in the same way as the basis $\mathbf{e}$, viz.:

$$\langle \bar{x}'| = \langle \bar{x}|\mathbf{T} \tag{A.144}$$

or

$$|\bar{x}'\rangle = \mathbf{T}^{\dagger}|\bar{x}\rangle \tag{A.145}$$

If we adopt a double suffix notation then the above equations can be rewritten as

$$\mathbf{e}'_j = \sum_i \mathbf{e}_i T_{ij} \tag{A.146}$$

$$x'_j = \sum_i (T^{-1})_{ji} x_i \tag{A.147}$$

$$\bar{x}'_j = \sum_i \bar{x}_i T_{ij} \tag{A.148}$$

and we may draw attention to the fact that the components x_i vary *inversely* as the bases $\mathbf{e}_i$ whilst the components $\bar{x}_i$ vary in the *same way*; these are therefore, respectively, referred to as the contravariant and covariant components of the vector $\mathbf{x}$. We shall refer to them again in section A.21.

A.13 Change of a matrix

Suppose two vectors $|x\rangle$ and $|y\rangle$ are related to each other via a matrix $\mathbf{A}$ so that

$$|y\rangle = \mathbf{A}|x\rangle \tag{A.149}$$

Then under a change of basis, by eq. (A.132)

$$|y'\rangle = \mathbf{T}^{-1}|y\rangle \tag{A.150}$$

and

$$|x'\rangle = \mathbf{T}^{-1}|x\rangle \tag{A.151}$$

On substituting in (A.149) we therefore find that

$$\mathbf{T}|y'\rangle = \mathbf{AT}|x'\rangle \tag{A.152}$$

so that

$$|y'\rangle = \mathbf{T}^{-1}\mathbf{AT}|x'\rangle \tag{A.153}$$
$$= \mathbf{A}'|x\rangle$$

Thus

$$\mathbf{A}' = \mathbf{T}^{-1}\mathbf{AT} \tag{A.154}$$

represents the transformation of **A** under a change of basis. A transformation of matrices of the above type is known as a *similarity transformation* and the matrices **A**′ and **A** of eq. (A.154) are said to be *similar*.

A.14 Change of the metric matrix

We defined earlier (eq. (A.101)) the quantities

$$g_{ij} = \mathbf{e}_i \cdot \mathbf{e}_j = \langle e_i|e_j\rangle \tag{A.155}$$

which are the elements of the metric matrix **g** which can, in two dimensions for example, be written as

$$\mathbf{g} = \begin{bmatrix} g_{11} & g_{12} \\ g_{21} & g_{22} \end{bmatrix}$$
$$= \begin{bmatrix} \mathbf{e}_1 \cdot \mathbf{e}_1 & \mathbf{e}_1 \cdot \mathbf{e}_2 \\ \mathbf{e}_2 \cdot \mathbf{e}_1 & \mathbf{e}_2 \cdot \mathbf{e}_2 \end{bmatrix} \tag{A.156}$$

which can be written as

$$\mathbf{g} = \begin{bmatrix} \mathbf{e}_1 \\ \mathbf{e}_2 \end{bmatrix} [\mathbf{e}_1 \quad \mathbf{e}_2] \tag{A.157}$$

Under a change of basis we obtain a new metric matrix **g**′ where

$$\mathbf{g}' = \begin{bmatrix} \mathbf{e}'_1 \\ \mathbf{e}'_2 \end{bmatrix} [\mathbf{e}'_1 \mathbf{e}'_2] \tag{A.158}$$

which, using eqs. (A.126) and (A.127) becomes

$$\mathbf{g}' = \mathbf{T}^\dagger \begin{bmatrix} \mathbf{e}_1 \\ \mathbf{e}_2 \end{bmatrix} [\mathbf{e}_1 \quad \mathbf{e}_2]\mathbf{T}$$
$$= \mathbf{T}^\dagger \mathbf{g} \mathbf{T} \tag{A.159}$$

This transform is known as a *congruence transformation*, and we should note the manner in which it differs from the similarity transformation of the previous section. If the original basis is orthonormal then

$$\mathbf{g} = \mathbf{I} \tag{A.160}$$

Hence, if we also require the new basis to be orthonormal, we must have

$$\mathbf{g}' = \mathbf{I} \tag{A.161}$$

and, from eq. (A.159)

$$\mathbf{T}^\dagger \mathbf{T} = \mathbf{I} \tag{A.162}$$

A matrix that satisfies the eq. (A.162) is said to be a *unitary matrix* (see section A.4j).
If the elements of **T** are all real then $\mathbf{T}^\dagger = \tilde{\mathbf{T}} = \mathbf{T}^{-1}$ in which case, the matrix is said to be *orthogonal*. Also when **T** is unitary, i.e., $\mathbf{T}^\dagger = \mathbf{T}^{-1}$, there is no difference between the similarity and congruence transformation.

A.15 Functions as bases

It is useful in the theory of groups to generalize the concept of a vector to include vectors whose bases are continuous functions. A vector $\langle\phi|$ is then a linear combination of a set of basis vectors $\langle\phi_i(x_j)|$.

The definition of the scalar product must now be altered, in the usual manner, by making the transition from summation to integration, so that given two vectors $\langle\phi|$ and $\langle\psi|$, the scalar product is defined as

$$\langle\phi|\psi\rangle = \int \phi^*(x_j)\psi(x_j)\,\mathrm{d}x_j \tag{A.163}$$

between appropriate limits.

The most well known orthonormal functions are the sine and cosine functions for which

$$\frac{1}{\pi}\int_{-\pi}^{+\pi} \sin mx \sin nx\,\mathrm{d}x = \delta_{mn} \tag{A.164}$$

$$\frac{1}{\pi}\int_{-\pi}^{+\pi} \cos mx \cos nx\,\mathrm{d}x = \delta_{mn} \tag{A.165}$$

$$\frac{1}{\pi}\int_{-\pi}^{+\pi} \sin mx \cos nx\,\mathrm{d}x = 0 \text{ for all } m \text{ and } n \tag{A.166}$$

and the well known technique of Fourier analysis allows us to express any function in the range $-\pi$ to π in terms of a linear combination of the basis functions $\sin mx$ and $\cos mx$. For example, within the range $-\pi$ to π the function x^2 can be expressed as

$$x^2 = \frac{\pi^2}{3} - 4(\cos x - \tfrac{1}{4}\cos 2x + \tfrac{1}{9}\cos 3x - \cdots) \tag{A.167}$$

which means that we can write x^2 as the vector

$$\langle x^2| = \left[\frac{\pi^2}{3}, -4, 1, -\frac{4}{9}, \ldots\right] \tag{A.168}$$

in terms of the basis function $\cos nx$.

A.16 Rotation and reflection matrices

There are three changes of basis in a three-dimensional Cartesian space that are of particular importance in group theory. These are *rotation* about an axis, *reflection* in a plane, and *inversion* through a point. We shall now obtain the matrices representing these operations by considering the effect of a change of basis on the components x, y, z of a vector **r**.

(a) Rotation

We shall consider first a rotation about the coordinate axis Oz. Thus let a point P in the $x - y$ plane be a distance r from the origin O such that $P\hat{O}x = \phi$ and, by reference to Fig. A.2

$$x = r\cos\phi \qquad y = r\sin\phi \tag{A.169}$$

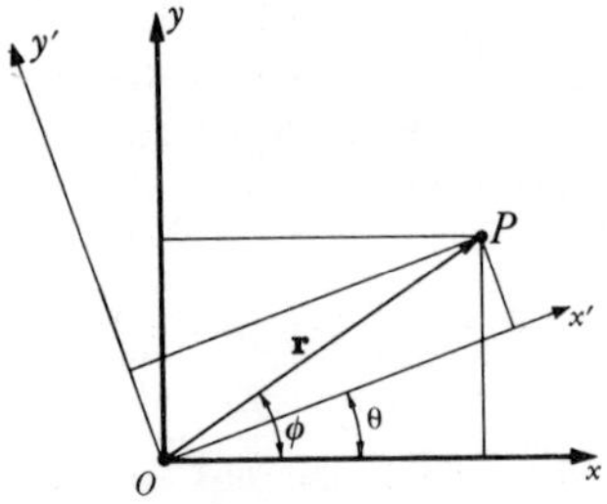

Fig. A.2 *Vector* **r** *referred to the axes* (x, y) *and in the rotated axes* (x', y').

Hence if the axes are rotated through an angle θ then the new coordinates of P will be

$$\begin{aligned} x' &= r\cos(\phi - \theta) = r\cos\phi\cos\theta + r\sin\phi\sin\theta \\ y' &= r\sin(\phi - \theta) = r\sin\phi\cos\theta - r\cos\phi\sin\theta \end{aligned} \tag{A.170}$$

Therefore

$$\begin{aligned} x' &= x\cos\theta + y\sin\theta \\ y' &= -x\sin\theta + y\cos\theta \end{aligned} \tag{A.171}$$

which can be written as

$$\begin{bmatrix} x' \\ y' \end{bmatrix} = \begin{bmatrix} \cos\theta & \sin\theta \\ -\sin\theta & \cos\theta \end{bmatrix} \begin{bmatrix} x \\ y \end{bmatrix} \tag{A.172}$$

and if we include the fact that in this example the value of z is unchanged we obtain

$$\begin{bmatrix} x' \\ y' \\ z' \end{bmatrix} = \begin{bmatrix} \cos\theta & \sin\theta & 0 \\ -\sin\theta & \cos\theta & 0 \\ 0 & 0 & 1 \end{bmatrix} \begin{bmatrix} x \\ y \\ z \end{bmatrix} \tag{A.173}$$

or

$$|x'\rangle = \mathbf{R}^{-1}|x\rangle \tag{A.174}$$

where $\mathbf{R}$ is the matrix representing a rotation about the z axis (cf. eq. (A.132)). It can be seen that $\mathbf{R}$ *is unitary*, with *trace* equal to $1 + 2\cos\theta$ and determinant equal to $+1$. The matrices representing rotations about the Ox or Oy axes can be obtained in a similar manner. However the problem of finding the matrix description of a rotation about an arbitrary axis is more difficult.

We can solve the latter problem in the following way. Let the axis OQ in Fig. A.3 have direction cosines l, m, n. Then if Q has spherical polar coordinates (r, θ, ϕ)

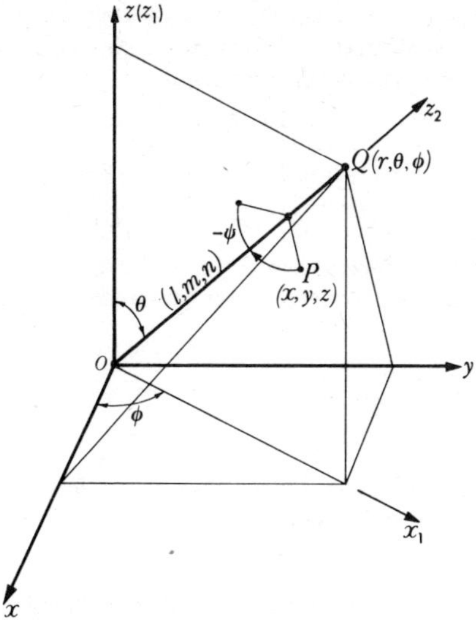

Fig. A.3 *General rotation about an arbitrary axis OQ with direction cosines (l, m, n).*

we can easily show that

$$\begin{aligned} l &= \sin\theta\cos\phi \\ m &= \sin\theta\sin\phi \\ n &= \cos\theta \end{aligned} \tag{A.175}$$

Let P be a point with coordinates x, y, z. Then we are seeking the matrix $\mathbf{R}^{-1}(\psi, lmn)$ such that

$$\begin{bmatrix} x' \\ y' \\ z' \end{bmatrix} = \mathbf{R}^{-1}(\psi, lmn)\begin{bmatrix} x \\ y \\ z \end{bmatrix} \tag{A.176}$$

where x', y', z' are the new coordinates of P after the axes $Oxyz$ have been rotated through a positive angle ψ (i.e., in the direction of a right-handed screw) about the axis OQ to form new axes $Ox'\,y'\,z'$. We may note that this is equivalent to rotating the vector OP through a negative angle $-\psi$ about OQ and it is this situation that is shown in Fig. A.3.

The matrix $\mathbf{R}^{-1}(\psi, lmn)$ can be found by breaking down the required rotation into the following sequence:

(1) Rotate the axes through a positive angle ϕ about Oz so as to form new axes $Ox_1y_1z_1$ where x_1 is as shown in Fig. A.3 and $z_1 \equiv z$.

The matrix $\mathbf{R}^{-1}(\phi, Oz)$ describing this operation is

$$\mathbf{R}^{-1}(\phi, Oz) = \begin{bmatrix} \cos\phi & \sin\phi & 0 \\ -\sin\phi & \cos\phi & 0 \\ 0 & 0 & 1 \end{bmatrix} \tag{A.177}$$

(2) Rotate the axes $Ox_1y_1z_1$, about the Oy_1 axis, through a positive angle θ, so as to form new axes $Ox_2y_2z_2$ such that Oz_2 coincides with OQ as shown in the figure. The matrix $\mathbf{R}^{-1}(\theta, Oy_1)$ describing this operation is

$$\mathbf{R}^{-1}(\theta, Oy_1) = \begin{bmatrix} \cos\theta & 0 & -\sin\theta \\ 0 & 1 & 0 \\ \sin\theta & 0 & \cos\theta \end{bmatrix} \tag{A.178}$$

(3) The axis OQ is now the principal axis Oz_2 and the rotation of the axes through a positive angle ψ about Oz_2 is described by the matrix $\mathbf{R}^{-1}(\psi, Oz_2)$ where

$$\mathbf{R}^{-1}(\psi, Oz_2) = \begin{bmatrix} \cos\psi & \sin\psi & 0 \\ -\sin\psi & \cos\psi & 0 \\ 0 & 0 & 1 \end{bmatrix} \tag{A.179}$$

In order to find the coordinates of P when referred to the rotated axes $Ox'y'z'$ obtained from $Oxyz$, we must undo the effect of operations (1) and (2). Thus we must continue with the sequence as follows:

(4) Restore the axes $Ox_1y_1z_1$ by performing the inverse operation $\mathbf{R}(\theta, Oy_1)$ where

$$\mathbf{R}(\theta, Oy_1) = \begin{bmatrix} \cos\theta & 0 & \sin\theta \\ 0 & 1 & 0 \\ -\sin\theta & 0 & \cos\theta \end{bmatrix} \tag{A.180}$$

(5) Recover the original axes $Oxyz$ by performing the inverse operation

$$\mathbf{R}(\phi, Oz) = \begin{bmatrix} \cos\phi & -\sin\phi & 0 \\ \sin\phi & \cos\phi & 0 \\ 0 & 0 & 1 \end{bmatrix} \tag{A.181}$$

The matrix $\mathbf{R}^{-1}(\psi, lmn)$ that describes the effect of a rotation of axes through angle ψ about an axis with direction cosines (l, m, n) is

$$\mathbf{R}^{-1}(\psi, lmn) = \mathbf{R}(\phi, Oz)\mathbf{R}(\theta, Oy_1)\mathbf{R}^{-1}(\psi, Oz_2)\mathbf{R}^{-1}(\theta, Oy_1)\mathbf{R}^{-1}(\phi, Oz) \tag{A.182}$$

and by multiplying together the matrices of eqs. (A.177) to (A.181) in the order given by eq. (A.182) we find that

$$\begin{bmatrix} \cos\psi + l^2(1-\cos\psi) & lm(1-\cos\psi) + n\sin\psi & ln(1-\cos\psi) - m\sin\psi \\ lm(1-\cos\psi) - n\sin\psi & \cos\psi + m^2(1-\cos\psi) & mn(1-\cos\psi) + l\sin\psi \\ ln(1-\cos\psi) + m\sin\psi & mn(1-\cos\psi) - l\sin\psi & \cos\psi + n^2(1-\cos\psi) \end{bmatrix} = \mathbf{R}^{-1}(\psi, lmn) \tag{A.183}$$

(b) Reflection

Suppose a point P in the $x - y$ plane is reflected to P' as shown in Fig. A.4 so that

$$\begin{aligned} x &= r\cos\phi \\ y &= r\sin\phi \end{aligned} \tag{A.184}$$

The coordinates of P' are

$$\begin{aligned} x' &= r\cos(2\theta - \phi) = r\cos\phi\cos 2\theta + r\sin\phi\sin 2\theta \\ y' &= r\sin(2\theta - \phi) = r\cos\phi\sin 2\theta - r\sin\phi\cos 2\theta \end{aligned} \tag{A.185}$$

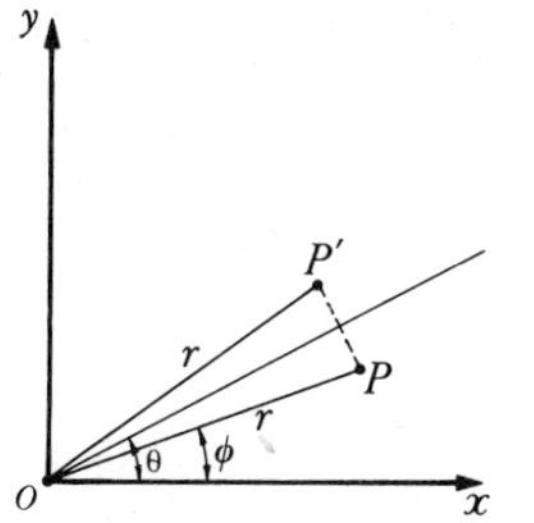

Fig. A.4 *Reflection of the point P to P′ in a plane perpendicular to the x − y plane.*

therefore

$$\begin{bmatrix} x' \\ y' \end{bmatrix} = \begin{bmatrix} \cos 2\theta & \sin 2\theta \\ \sin 2\theta & -\cos 2\theta \end{bmatrix} \begin{bmatrix} x \\ y \end{bmatrix} \tag{A.186}$$

and in three dimensions by including the fact that z is unchanged we obtain

$$\begin{bmatrix} x' \\ y' \\ z' \end{bmatrix} = \begin{bmatrix} \cos 2\theta & \sin 2\theta & 0 \\ \sin 2\theta & -\cos 2\theta & 0 \\ 0 & 0 & 1 \end{bmatrix} \begin{bmatrix} x \\ y \\ z \end{bmatrix} \tag{A.187}$$

i.e.,

$$|x'\rangle = \mathbf{S}^{-1}|x\rangle \tag{A.188}$$

where $\mathbf{S}$ is called the reflection matrix. Observe that $\mathbf{S}$ is *unitary* with a *trace* equal to $+1$ and a *determinant* equal to -1.

A particular case of the reflection matrix is that for reflection in planes containing two of the coordinate axes. For example, reflection in the Oyz plane means putting $\theta = 90°$; this is usually indicated by the symbol σ_v which is represented by the matrix

$$\mathbf{\Gamma}(\sigma_v) = \mathbf{S} = \begin{bmatrix} 1 & 0 & 0 \\ 0 & -1 & 0 \\ 0 & 0 & 1 \end{bmatrix} \tag{A.189}$$

The z axis is conventionally regarded as vertical and is indicated by the suffix v.

Reflection σ_h in the *horizontal* plane Oxy is represented by the matrix

$$\mathbf{\Gamma}(\sigma_h) = \begin{bmatrix} 1 & 0 & 0 \\ 0 & 1 & 0 \\ 0 & 0 & -1 \end{bmatrix} \tag{A.190}$$

(c) Inversion

If the coordinate axes are reversed so that a right-handed set becomes a left-handed set then all the coordinates of a point P are reversed. This is represented by the inversion matrix $\mathbf{\Gamma}(i)$

$$\mathbf{\Gamma}(i) = \begin{bmatrix} -1 & 0 & 0 \\ 0 & -1 & 0 \\ 0 & 0 & -1 \end{bmatrix} \tag{A.191}$$

This matrix is *unitary*, possesses a *trace* equal to -3 and a *determinant* equal to -1.

A.17 Eigenvalues and eigenvectors

When a matrix operates on a vector it produces a new vector and in certain cases the new vector is simply a multiple of the original one, as for example

$$\begin{bmatrix} 1 & 2 \\ 5 & 4 \end{bmatrix}\begin{bmatrix} 2 \\ 5 \end{bmatrix} = \begin{bmatrix} 12 \\ 30 \end{bmatrix} = 6\begin{bmatrix} 2 \\ 5 \end{bmatrix} \tag{A.192}$$

If this occurs the vector is said to be an *invariant* vector or *eigenvector* of the matrix and the corresponding multiplier is called the *eigenvalue*. It is said to be invariant because it remains in the same space when operated on by the matrix.

Any constant multiple of an eigenvector is obviously also an eigenvector and there are in general, different row eigenvectors and column eigenvectors of a matrix; however, the eigenvalues are unique. The general method of finding the eigenvalues and the eigenvectors is as follows; suppose (for the case of a column vector) that

$$\mathbf{A}|x\rangle = \lambda|x\rangle \tag{A.193}$$

then as λ can be written as $\lambda\mathbf{I}$

$$[\mathbf{A} - \lambda\mathbf{I}]|x = 0 \tag{A.194}$$

This is a set of simultaneous homogeneous equations in the components of $|x\rangle$, i.e.,

$$\begin{bmatrix} a_{11} - \lambda & a_{12} & \cdots & a_{1n} \\ a_{21} & a_{22} - \lambda & & a_{2n} \\ \vdots & \vdots & & \vdots \\ a_{n1} & a_{n2} & \cdots & a_{nn} - \lambda \end{bmatrix}\begin{bmatrix} x_1 \\ x_2 \\ \vdots \\ x_n \end{bmatrix} = 0 \tag{A.195}$$

so that for a unique solution we must have

$$|\mathbf{A} - \lambda\mathbf{I}| = 0 \tag{A.196}$$

The solutions of this determinantal equation, known as the *secular equation* give the eigenvalues λ. These values of λ when substituted back into eq. (A.194) will give the ratios of the components of the vectors corresponding to each root. In this way we can find both the eigenvalues and the eigenvectors. As an example of the above discussion let us consider the matrix

$$\mathbf{A} = \begin{bmatrix} 1 & 2 \\ 5 & 4 \end{bmatrix} \tag{A.197}$$

which gives rise to the secular equation

$$\begin{vmatrix} 1 - \lambda & 2 \\ 5 & 4 - \lambda \end{vmatrix} = 0 \tag{A.198}$$

which produces the eigenvalues

$$\lambda = 6, \qquad \lambda = -1 \tag{A.199}$$

The eigenvector associated with $\lambda = 6$ can be obtained from the matrix equation

$$[\mathbf{A} - \lambda\mathbf{I}]|x\rangle = \begin{bmatrix} -5 & 2 \\ 5 & -2 \end{bmatrix}\begin{bmatrix} x_1 \\ x_2 \end{bmatrix} = \begin{bmatrix} 0 \\ 0 \end{bmatrix} \tag{A.200}$$

which shows that

$$-5x_1 + 2x_2 = 0 \tag{A.201}$$

so that a suitable solution (in integers) is

$$x_1 = 2, \qquad x_2 = 5 \tag{A.202}$$

Similarly for $\lambda = -1$ it is readily shown that

$$x_1 = 1, \qquad x_2 = -1 \tag{A.203}$$

A.18 Diagonalization of matrices

An important procedure in matrix methods is the reduction of a matrix to diagonal form (where this is possible). We can gain an insight into this procedure by considering further the example already discussed. Let us form a matrix **R** of the column eigenvectors, i.e.,

$$\mathbf{R} = \begin{bmatrix} 2 & \vdots & 1 \\ 5 & \vdots & -1 \end{bmatrix} \tag{A.204}$$

so that

$$\mathbf{R}^{-1} = \tfrac{1}{7}\begin{bmatrix} 1 & 1 \\ 5 & -2 \end{bmatrix} \tag{A.205}$$

we then find that

$$\mathbf{R}^{-1}\mathbf{A}\mathbf{R} = \tfrac{1}{7}\begin{bmatrix} 1 & 1 \\ 5 & -2 \end{bmatrix}\begin{bmatrix} 1 & 2 \\ 5 & 4 \end{bmatrix}\begin{bmatrix} 2 & 1 \\ 5 & -1 \end{bmatrix} \tag{A.206}$$

or

$$\mathbf{R}^{-1}\mathbf{A}\mathbf{R} = \begin{bmatrix} 6 & 0 \\ 0 & -1 \end{bmatrix} = \mathbf{D} \tag{A.207}$$

where $\mathbf{D} = \text{diag}\,(\lambda_1 \lambda_2)$ and λ_1 and λ_2 are the eigenvalues of **A**. Thus the matrix **A** has been converted to the diagonal form **D** with the eigenvalues set out along the diagonal. Reference to eq. (A.155) shows that this has been done by a similarity transform on the matrix **A**; hence **D** is similar to **A**. We should note that the diagonalization is only possible if $\mathbf{R}^{-1}$ exists, i.e., if **R** is non-singular.

A.19 Some theorems on eigenvalues and eigenvectors

(a) Hermitian matrices

(1) *The eigenvalues of a Hermitian matrix are real.* Let $|x\rangle$ be an eigenvector with eigenvalue λ of a Hermitian matrix **H** so that

$$\mathbf{H}|x\rangle = \lambda|x\rangle \tag{A.208}$$

Then by taking the complex conjugate of the transpose of each side we have

$$\langle x|\mathbf{H}^\dagger = \lambda^*\langle x| \tag{A.209}$$

But $\mathbf{H}^\dagger = \mathbf{H}$ by definition (section A.4g), therefore

$$\langle x|\mathbf{H}|x\rangle = \lambda^*\langle x|x\rangle \tag{A.210}$$

and from eq. (A.208) we have

$$\langle x|\mathbf{H}|x\rangle = \lambda\langle x|x\rangle \tag{A.211}$$

Therefore by comparing eqs. (A.210) and (A.211) we can see that

$$\lambda = \lambda^* \tag{A.212}$$

because $\langle x|x\rangle \geqslant 0$. In the case $\langle x|x\rangle = 0$ $\langle x| = 0$ and $\lambda = 0$. Hence λ *is always real.* This theorem is important in quantum theory, where the fact that the operators of the theory are Hermitian, means that the observables of the theory (the eigenvalues or energies of the system) are real.

(2) *The eigenvectors corresponding to distinct roots of a Hermitian matrix are mutually orthogonal.* With the same notation as before we can write eq. (A.208) for a particular eigenvector $|x_i\rangle$ with eigenvalue λ_i as

$$\mathbf{H}|x_i\rangle = \lambda_i|x_i\rangle \tag{A.213}$$

Therefore

$$\langle x_j|\mathbf{H}|x_i\rangle = \lambda_i\langle x_j|x_i\rangle \tag{A.214}$$

but for a distinct root $\lambda_i \neq \lambda_j$, and for its eigenvector $|x_j\rangle$ we have

$$\mathbf{H}|x_j\rangle = \lambda_j|x_j\rangle \tag{A.215}$$

Therefore by forming the complex conjugate of the transpose of eq. (A.215) we find that

$$\langle x_j|\mathbf{H} = \lambda_j\langle x_j| \tag{A.216}$$

Because λ_j is real and $\mathbf{H}^\dagger = \mathbf{H}$. Hence on multiplying eq. (A.216) by $|x_i\rangle$ we obtain

$$\langle x_j|\mathbf{H}|x_i\rangle = \lambda_j\langle x_j|x_i\rangle \tag{A.217}$$

so that on comparing eqs. (A.214) and (A.217) we find that

$$\langle x_j|x_i\rangle = 0 \tag{A.218}$$

because $\lambda_i \neq \lambda_j$. Hence the vectors are orthogonal. It can be shown however even in the case of repeated roots, that the eigenvectors of a Hermitian matrix are linearly independent and can always be chosen to be orthogonal (using the Gram–Schmidt process, section A.10e). A consequence of this is that the matrix **R** (cf. eq. (A.207)) can be constructed from orthogonal vectors. Therefore the matrix which diagonalizes a Hermitian matrix can always be chosen to be a unitary matrix **T**. Hence

$$\mathbf{T}^\dagger\mathbf{H}\mathbf{T} = \mathbf{T}^{-1}\mathbf{H}\mathbf{T} = \text{diag}(\lambda_i) \tag{A.219}$$

Note that if **H** has only real elements then **H** is symmetric (section A.4f) and **T** will be orthogonal, i.e., $\mathbf{T}^\dagger = \tilde{\mathbf{T}} = \mathbf{T}^{-1}$.

(b) Similar matrices

(1) *The eigenvalues of similar matrices are the same.* If a matrix **B** is similar to a matrix **A** then (eq. (A.154))

$$\mathbf{B} = \mathbf{R}^{-1}\mathbf{A}\mathbf{R} \tag{A.220}$$

The eigenvalues of **B** can then be obtained from

$$|\mathbf{B} - \lambda\mathbf{I}| = 0 \tag{A.221}$$

hence

$$|\mathbf{R}^{-1}\mathbf{A}\mathbf{R} - \lambda\mathbf{R}^{-1}\mathbf{I}\mathbf{R}| = 0 \tag{A.222}$$

Thus

$$|\mathbf{R}^{-1}|\,|\mathbf{A} - \lambda\mathbf{I}|\,|\mathbf{R}| = 0 \tag{A.223}$$

Therefore

$$|\mathbf{A} - \lambda\mathbf{I}| = 0 \tag{A.224}$$

because

$$|\mathbf{R}^{-1}|\,|\mathbf{R}| = |\mathbf{R}^{-1}\mathbf{R}| = |\mathbf{I}| = 1 \tag{A.225}$$

Hence the roots of the secular equation for **B** are the same as those for **A**.

(2) *The sum of the eigenvalues of a matrix is equal to the trace.* Consider the roots of the secular equation in the form

$$f(\lambda) = |\lambda\mathbf{I} - \mathbf{A}|$$

$$= \begin{vmatrix} \lambda - A_{11} & -A_{22} & \cdots & -A_{1n} \\ -A_{21} & \lambda - A_{22} & \cdots & -A_{2n} \\ \vdots & & & \\ -A_{n1} & -A_{n2} & \cdots & \lambda - A_{nn} \end{vmatrix} = 0 \tag{A.226}$$

Now $f(\lambda)$ is a polynomial in λ, and by expanding the determinant about the first column we can see that it must take the form

$$f(\lambda) = \lambda^n - A_{11}\lambda^{n-1} + \cdots \tag{A.227}$$

whilst by expanding about the second column we obtain

$$f(\lambda) = \lambda^n - A_{22}\lambda^{n-1} + \cdots \tag{A.228}$$

Thus we must have

$$f(\lambda) = \lambda^n - \Big(\sum_i A_{ii}\Big)\lambda^{n-1} + \cdots \tag{A.229}$$

However, if the roots are $\lambda_1, \lambda_2, \ldots, \lambda_n$ then

$$f(\lambda) = (\lambda - \lambda_1)(\lambda - \lambda_2)\ldots(\lambda - \lambda_n) \tag{A.230}$$

so that

$$f(\lambda) = \lambda^n - \Big(\sum_i \lambda_i\Big)\lambda^{n-1} + \cdots \tag{A.231}$$

and hence on comparing eqs. (A.229) and (A.231) we find that

$$\sum_i \lambda_i = \sum_i A_{ii} = \text{trace}\ \mathbf{A} \tag{A.232}$$

(3) *Similar matrices have the same trace.* If **B** is similar to **A** then by (a) their roots are the same, i.e., $\lambda_i^B = \lambda_i^A$. Therefore

$$\text{trace}\ \mathbf{B} = \sum_i \lambda_i^B = \sum_i \lambda_i^A = \text{trace}\ \mathbf{A} \tag{A.233}$$

Hence we can see that the trace of a matrix is a quantity that remains invariant under a similarity transform.

A.20 Simultaneous diagonalization of two matrices

It can easily be proved, that two matrices can be simultaneously diagonalized by the same *similarity* transformation if, and only if, they commute; however, an important case in practice is that two *symmetric* matrices can be simultaneously reduced by a *congruence* transform (eq. (A.159)), such that one becomes a diagonal matrix and the other becomes a unit matrix, provided one of them is what is known as positive definite, i.e., its eigenvalues are all greater than zero. Thus, suppose **K** and **M** are two real symmetric matrices (i.e., Hermitian with real elements) then, as we showed in section A.19(a), an orthogonal matrix **P** exists (eq. (A.219)) such that

$$\tilde{\mathbf{P}}\mathbf{M}\mathbf{P} = \operatorname{diag}(\mu_m) \tag{A.234}$$

where μ_m are the eigenvalues of **M** and are all greater than zero. Now let

$$\mathbf{Q} = \operatorname{diag}(1/\mu_m^{1/2}) = \tilde{\mathbf{Q}} \tag{A.235}$$

then

$$\tilde{\mathbf{Q}}\tilde{\mathbf{P}}\mathbf{M}\mathbf{P}\mathbf{Q} = \mathbf{I} \tag{A.236}$$

showing that **M** can be reduced to a unit matrix. Also if we transform **K** by the matrices **Q** and **P** we obtain

$$\mathbf{K}' = \tilde{\mathbf{Q}}\tilde{\mathbf{P}}\mathbf{K}\mathbf{P}\mathbf{Q} \tag{A.237}$$

which is itself real symmetric and hence can be diagonalized by another orthogonal matrix **R**, i.e.,

$$\tilde{\mathbf{R}}\mathbf{K}'\mathbf{R} = \operatorname{diag}(\lambda_k) \tag{A.238}$$

where λ_k are the eigenvalues of **K**. Therefore

$$\tilde{\mathbf{R}}\tilde{\mathbf{Q}}\tilde{\mathbf{P}}\mathbf{K}\mathbf{P}\mathbf{Q}\mathbf{R} = \operatorname{diag}(\lambda_k) \tag{A.239}$$

and also because $\tilde{\mathbf{R}} = \mathbf{R}^{-1}$ we can write eq. (A.236) as

$$\tilde{\mathbf{R}}\tilde{\mathbf{Q}}\tilde{\mathbf{P}}\mathbf{M}\mathbf{P}\mathbf{Q}\mathbf{R} = \mathbf{I} \tag{A.240}$$

Therefore the matrix

$$\mathbf{S} = \mathbf{P}\mathbf{Q}\mathbf{R} \tag{A.241}$$

simultaneously reduces the symmetric matrices **K** and **M**, i.e.,

$$\tilde{\mathbf{S}}\mathbf{K}\mathbf{S} = \operatorname{diag}(\lambda_k) \tag{A.242}$$

$$\tilde{\mathbf{S}}\mathbf{M}\mathbf{S} = \mathbf{I} \tag{A.243}$$

As an example let us consider two matrices that occur in chapter 7 (eqs. (7.32) and (7.33)). These can be written in block form (with 2 × 2 block matrices) as:

$$\mathbf{K} = \begin{bmatrix} \mathbf{K}_{11} & & \\ & \mathbf{0} & \\ & & \mathbf{0} \end{bmatrix} \tag{A.244}$$

and

$$\mathbf{M} = \begin{bmatrix} \mathbf{M}_{11} & & \\ & \mathbf{M}_{11} & \\ & & \mathbf{M}_{11} \end{bmatrix} \tag{A.245}$$

where

$$\mathbf{K}_{11} = \begin{bmatrix} k & -k \\ -k & k \end{bmatrix} \tag{A.246}$$

and

$$\mathbf{M}_{11} = \begin{bmatrix} m_1 & 0 \\ 0 & m_2 \end{bmatrix} \tag{A.247}$$

It is therefore only necessary to reduce $\mathbf{M}_{11}$ and $\mathbf{K}_{11}$, which we can do as follows.

The matrix $\mathbf{M}_{11}$ is already diagonal, hence the matrix $\mathbf{P}$ of eq. (A.234) is simply a unit matrix, and it follows that the matrix $\mathbf{Q}$ is

$$\mathbf{Q} = \begin{bmatrix} 1/m_1^{1/2} & 0 \\ 0 & 1/m_2^{1/2} \end{bmatrix} \tag{A.248}$$

and

$$\tilde{\mathbf{Q}}\mathbf{M}_{11}\mathbf{Q} = \mathbf{I} \tag{A.249}$$

From eq. (A.237) we then obtain

$$\begin{aligned} \mathbf{K}' &= \tilde{\mathbf{Q}}\tilde{\mathbf{P}}\mathbf{K}_{11}\mathbf{P}\mathbf{Q} \\ &= k\begin{bmatrix} 1/m_1 & -1/(m_1 m_2)^{1/2} \\ -1/(m_1 m_2)^{1/2} & 1/m_2 \end{bmatrix} \end{aligned} \tag{A.250}$$

We can then find that $\mathbf{K}'$ has roots

$$\lambda_1 = k\left(\frac{1}{m_1} + \frac{1}{m_2}\right) \quad \text{and} \quad \lambda_2 = 0 \tag{A.251}$$

with corresponding normalized eigenvectors

$$\frac{1}{(m_1 + m_2)^{1/2}}\begin{bmatrix} m_2^{1/2} \\ -m_1^{1/2} \end{bmatrix} \quad \text{and} \quad \frac{1}{(m_1 + m_2)^{1/2}}\begin{bmatrix} m_1^{1/2} \\ m_2^{1/2} \end{bmatrix} \tag{A.252}$$

respectively.

Hence $\mathbf{K}'$ can be brought to diagonal form by the orthogonal matrix $\mathbf{R}$ where

$$\mathbf{R} = \frac{1}{(m_1 + m_2)^{1/2}}\begin{bmatrix} m_2^{1/2} & m_1^{1/2} \\ -m_1^{1/2} & m_2^{1/2} \end{bmatrix} \tag{A.253}$$

that is

$$\tilde{\mathbf{R}}\mathbf{K}'\mathbf{R} = \operatorname{diag}(\lambda_1 \lambda_2) \tag{A.254}$$

We can now form the matrix $\mathbf{S}$ of eq. (A.241), thus

$$\begin{aligned} \mathbf{S} &= \mathbf{PQR} \\ &= \frac{1}{(m_1 + m_2)^{1/2}}\begin{bmatrix} (m_2/m_1)^{1/2} & 1 \\ -(m_1/m_2)^{1/2} & 1 \end{bmatrix} \end{aligned} \tag{A.255}$$

and it is then easy to check that

$$\tilde{\mathbf{S}}\mathbf{M}_{11}\mathbf{S} = \mathbf{I} \tag{A.256}$$

$$\tilde{\mathbf{S}}\mathbf{K}_{11}\mathbf{S} = \operatorname{diag}(\lambda_1 \lambda_2) \tag{A.257}$$

The remaining blocks of $\mathbf{K}$ are null matrices so we need only convert the remaining blocks of $\mathbf{M}$ to unit matrices by the matrix $\mathbf{Q}$ (eq. (A.248)), hence the matrix that simultaneously reduces $\mathbf{K}$ (eq. (A.244)) to diagonal form and $\mathbf{M}$ (eq. (A.245)) to a unit matrix is the block matrix

$$\boldsymbol{\alpha} = \begin{bmatrix} \mathbf{S} & & \\ & \mathbf{Q} & \\ & & \mathbf{Q} \end{bmatrix} \tag{A.258}$$

A.21 Tensors

The discussion of the changes of the components of a vector under a change of basis which is given in sections A.11 and A.12 leads us naturally into a consideration of the quantities known as tensors.

We have seen (eqs. (A.146), (A.147), and (A.148)) how a vector can have two types of component that transform inversely (contravariant) or in the same way (covariant) as the basis does. An interesting geometrical distinction between these components is shown in Fig. A.5 where for non-orthogonal axes we see that the usual components

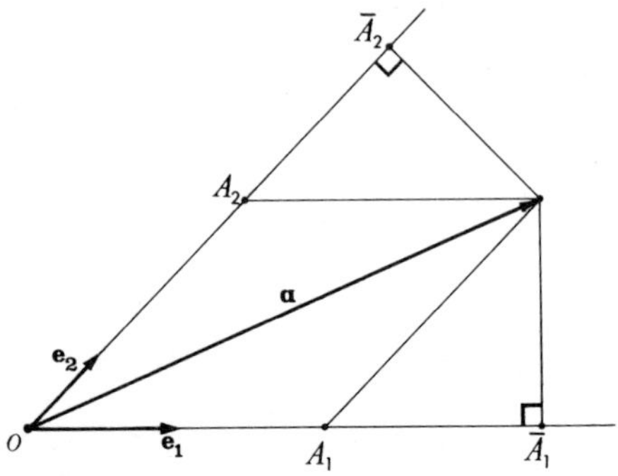

Fig. A.5 *Geometrical distinction between covariant (OA_1, OA_2) and contravariant ($O\bar{A}_1$, $O\bar{A}_2$) components of a vector* **a**.

OA_1 and OA_2 of the vector **a** are different from the perpendicular projections $O\bar{A}_1$ and $O\bar{A}_2$. If the usual components are taken as the contravariant ones then it can be shown that the projections on the axes are proportional to the covariant ones. In orthonormal axes, of course, the distinction between covariance and contravariance vanishes. The transformation of bases and vector components have already been written down in eqs. (A.146), (A.147), and (A.148) but we shall repeat them here *using the Einstein convention that summations take place over repeated indices.*

$$\mathbf{e}'_j = \mathbf{e}_i T_{ij} \quad \text{(A.259)}$$

$$x'_j = x_i T^{-1}_{ji} \quad \text{(A.260)}$$

$$\bar{x}'_j = \bar{x}_i T_{ij} \quad \text{(A.261)}$$

We should point out that the conventional notation is rather different from ours and is

$$\begin{aligned} &\mathbf{e}^j \text{ where we have } \mathbf{e}_j \\ &\mathbf{e}_j \text{ where we have } \bar{\mathbf{e}}_j \\ &x^j \text{ where we have } x_j \\ &x_j \text{ where we have } \bar{x}_j \end{aligned} \quad \text{(A.262)}$$

in which case eqs. (A.259), (A.260), and (A.261) would then read

$$\begin{aligned} \mathbf{e}'^j &= \mathbf{e}^i T^j_i \\ x'^j &= x^i (T^{-1})^j_i \\ x'_j &= x_i T^i_j \end{aligned} \quad \text{(A.263)}$$

However, for reasons that will appear immediately below, we have not found it necessary to introduce the suffix and superfix notation. In what follows we shall be restricting ourselves to the study of Cartesian tensors and shall only be considering:

(1) Orthornormal bases, which means that
  (a) there is no distinction between the covariant and contravariant components,
  (b) the metric matrix is a unit matrix.

(2) Real unitary (i.e., orthogonal) transforms of bases since these preserve the metric.

If the basis transformation matrix **T** is orthogonal then

$$T^{-1}_{ji} = T^{\dagger}_{ji} = \tilde{T}_{ji} = T_{ij} \tag{A.264}$$

because **T** is real, and the eqs. (A.259), (A.260), and (A.261) then reduce to

$$\mathbf{e}'_j = \mathbf{e}_i T_{ij} \tag{A.265}$$

$$x'_j = x_i T_{ij} \tag{A.266}$$

where we assume that $\mathbf{e}_j$ and x_j are written as single *row* matrices. Equivalently, in terms of single *column* matrices, the equations can be written as

$$\mathbf{e}'_j = \tilde{T}_{ji}\mathbf{e}_i \tag{A.267}$$

$$x'_j = \tilde{T}_{ji}x_i \tag{A.268}$$

We should note that in chapter 5, we use

$$a_{ji} = \tilde{T}_{ji} \tag{A.269}$$

We see that under a change of basis a quantity x_i, with *one* suffix is transformed by a quantity T_{ij} which possesses *two* suffixes.

Before we introduce any generalizations let us consider one more example. Equation (A.154) shows how a matrix **A** transforms to **A**′ under a basis transformation, i.e.,

$$\mathbf{A}' = \mathbf{T}^{-1}\mathbf{A}\mathbf{T} \tag{A.270}$$

If we write out one element A_{ij} of the matrix **A** using the Einstein convention, then we have

$$A'_{ij} = T^{-1}_{ik} A_{kl} T_{lj} \tag{A.271}$$

showing that

$$A'_{ij} = A_{kl} T_{ki} T_{lj} \tag{A.272}$$

Equation (A.272) now demonstrates that, under a change of basis, a two-suffix quantity A_{kl} is transformed via two transformation matrices. In general we shall call a quantity with *n* suffices a *tensor* of rank *n provided* that under a transformation of basis via a *matrix* **T** the tensor $A_{abc\ldots n}$ transforms as

$$A'_{a'b'c'\ldots n'} = A_{abc\ldots n} T_{aa'} T_{bb'} T_{cc'} \ldots T_{nn'} \tag{A.273}$$

A tensor of rank $n = 0$ has no suffix and is called a scalar, and a tensor of rank $n = 1$ is called a vector. In an *r*-dimensional space tensors of rank *n* have r^n components.

A.22

(a) Products of tensors

If two tensors of ranks *m* and *n* respectively are multiplied together then one obtains another tensor of rank $m + n$ known as the *outer product*. For example with tensors

A and B of rank 3 and 2 respectively, we can form

$$C_{ijklm} = A_{ijk}B_{lm} \tag{A.274}$$

where the fact that C is a tensor follows immediately from the transformation process; thus:

$$\begin{aligned} C'_{i'j'k'l'm'} &= A'_{i'j'k'}B'_{l'm'} \\ &= A_{ijk}T_{ii'}T_{jj'}T_{kk'}B_{lm}T_{ll'}T_{mm'} \\ &= A_{ijk}B_{lm}T_{ii'}T_{jj'}T_{kk'}T_{ll'}T_{mm'} \\ &= C_{ijklm}T_{ii'}T_{jj'}T_{kk'}T_{ll'}T_{mm'} \end{aligned} \tag{A.275}$$

which shows that C transforms as a tensor of rank 5.

(b) Contraction of tensors

If two indices are made equal then we must sum over this index, hence the resulting sum no longer depends on it and the rank of the resulting tensor is reduced by two; this procedure is known as contraction. As an example, contraction of the tensor C_{ijklm} over the indices i and l yields C_{ijkim}.

In a three-dimensional space this is written as

$$C_{1jk1m} + C_{2jk2m} + C_{3jk3m} = D_{jkm} \tag{A.276}$$

In a similar manner the outer product of eq. (A.275) can be contracted. Thus if i is set equal to l we obtain

$$\begin{aligned} C_{ijkim} &= A_{ijk}B_{im} \\ &= D_{jkm} \end{aligned} \tag{A.277}$$

The contracted product of eq. (A.277) is known as the *inner product*.

A.23 Tensor equations

We have seen (eq. (A.277)) how, by means of contraction, a relationship connecting two tensors of the *same rank* can be obtained. All physical laws can be written in this form, and we shall now show that this is so because the form of these laws is invariant under a change of basis, so that the form of the law is independent of the choice of coordinate system. For example, using eq. (A.277) we have

$$D_{jkm} = A_{ijk}B_{im} \tag{A.278}$$

If we now make a change of basis we find, in the new basis, that

$$A'_{i'j'k'} = A_{ijk}T_{ii'}T_{jj'}T_{kk'} \tag{A.279}$$

and

$$B'_{i'm'} = B_{im}T_{ii'}T_{mm'} \tag{A.280}$$

so that

$$\begin{aligned} A'_{i'j'k'}B'_{i'm'} &= A_{ijk}B_{im}T_{ii'}T_{jj'}T_{kk'}T_{ii'}T_{mm'} \\ &= D_{ijk}T_{jj'}T_{kk'}T_{mm'} \end{aligned} \tag{A.281}$$

because, as we shall show below,

$$T_{ii'}T_{ii'} = 1 \tag{A.282}$$

Hence

$$A'_{i'j'k'}B'_{i'm'} = D'_{j'k'm'} \tag{A.283}$$

which is of the same form as eq. (A.278).

In the above example we used the fact that $T_{ii'}T_{ii'} = 1$ which can be derived from the matrix representation of T (eq. (A.125)) and the fact that $\mathbf{T}$ is orthogonal. For instance in three dimensions we know that

$$\mathbf{T}\tilde{\mathbf{T}} = \mathbf{I} \tag{A.284}$$

i.e.,

$$\begin{bmatrix} T_{11} & T_{12} & T_{13} \\ T_{21} & T_{22} & T_{23} \\ T_{31} & T_{32} & T_{33} \end{bmatrix} \begin{bmatrix} T_{11} & T_{21} & T_{31} \\ T_{12} & T_{22} & T_{32} \\ T_{13} & T_{23} & T_{33} \end{bmatrix} = \begin{bmatrix} 1 & 0 & 0 \\ 0 & 1 & 0 \\ 0 & 0 & 1 \end{bmatrix} \tag{A.285}$$

so that the diagonal elements satisfy an equation such as (A.282).

A.24 The symmetry of tensors

For tensors of rank 2 and higher we can perform the operation of interchanging two indices. If this is done it will be found that some tensors either remain of the same sign or else change in sign; in the former case they are said to be symmetric, and in the latter case antisymmetric, with respect to an interchange of indices.

Thus if $A_{ijk} = +A_{jik}$ then A is symmetric in i and j whilst if $A_{ijk} = -A_{jik}$, A is said to be antisymmetric in i and j. Some of the commonly occurring basis transformations that are applicable in physics are the rotations, reflections, and inversions, and by reference to eqs. (A.173), (A.187), and (A.191) one can see that, written as tensors, these have some symmetric and some antisymmetric components which, if not immediately obvious, can be found by forming the sums:

$$A_{ij}^{+} = \tfrac{1}{2}(A_{ij} + A_{ji}) \quad \text{symmetric part} \tag{A.286}$$

$$A_{ij}^{-} = \tfrac{1}{2}(A_{ij} - A_{ji}) \quad \text{antisymmetric part} \tag{A.287}$$

Any tensor A (of rank 2) can therefore be represented as a sum of its symmetric and antisymmetric parts. Tensors of higher order have more complicated properties, because they may be symmetric with respect to certain indices and antisymmetric with respect to others; nevertheless any tensor can still be written as a sum of its symmetric and antisymmetric parts. One of the most important properties of tensors is that their symmetry properties are invariant under a change of basis. Thus suppose for example that

$$A_{ijk} = -A_{jik} \tag{A.288}$$

Then under a change of basis

$$\begin{aligned} A'_{i'j'k'} &= A_{ijk}T_{ii'}T_{jj'}T_{kk'} \\ &= -A_{jik}T_{ii'}T_{jj'}T_{kk'} \\ &= -A_{jik}T_{jj'}T_{ii'}T_{kk'} \end{aligned} \tag{A.289}$$

(where in the last line we have simply rewritten the T matrices in the same order as the suffices jik).

Hence

$$A'_{i'j'k'} = -A'_{j'i'k'} \tag{A.290}$$

so that in the new basis the tensor A_{ijk} has the same symmetry. As particular examples of symmetric and antisymmetric tensors we shall consider the relevant parts of the tensor A of rank two that can be constructed from two tensors a_i and b_i of rank one, in a three-dimensional space. Thus

$$\mathbf{A} = |a\rangle\langle b| = \begin{bmatrix} a_1 \\ a_2 \\ a_3 \end{bmatrix} [b_1 b_2 b_3] \tag{A.291}$$

$$\mathbf{A} = \begin{bmatrix} a_1b_1 & a_1b_2 & a_1b_3 \\ a_2b_1 & a_2b_2 & a_2b_3 \\ a_3b_1 & a_3b_2 & a_3b_3 \end{bmatrix} \tag{A.292}$$

This is a matrix representation of a second rank tensor, whose symmetric and anti-symmetric parts can be found from eqs. (A.286) and (A.287) to be

$$\mathbf{A}^+ = \begin{bmatrix} a_1b_1 & \frac{1}{2}(a_1b_2 + a_2b_1) & \frac{1}{2}(a_1b_3 + a_3b_1) \\ \frac{1}{2}(a_1b_2 + a_2b_1) & a_2b_2 & \frac{1}{2}(a_2b_3 + a_3b_2) \\ \frac{1}{2}(a_1b_3 + a_3b_1) & \frac{1}{2}(a_2b_3 + a_3b_2) & a_3n_3 \end{bmatrix} \tag{A.293}$$

and

$$\mathbf{A}^- = \begin{bmatrix} 0 & \frac{1}{2}(a_1b_2 - a_2b_1) & -\frac{1}{2}(a_3b_1 - a_1b_3) \\ -\frac{1}{2}(a_1b_2 - a_2b_1) & 0 & \frac{1}{2}(a_2b_3 - a_3b_2) \\ \frac{1}{2}(a_3b_1 - a_1b_3) & -\frac{1}{2}(a_2b_3 - a_3b_2) & 0 \end{bmatrix} \tag{A.294}$$

We may notice that the trace of $\mathbf{A}^+$ is the *scalar product* of the vectors **a** and **b**; whilst the off-diagonal parts of $2\mathbf{A}^-$ consist of three distinct elements (those above, or those below the diagonal) which can be taken to be the components of the vector usually called the *vector product* $\mathbf{a} \times \mathbf{b}$.

Now we know that if one of the vectors **a** or **b** is reversed in direction, as for example by reflection in a (vertical) mirror parallel to **a** and $\mathbf{a} \times \mathbf{b}$, then their vector product changes direction, which agrees with the behaviour of $\mathbf{A}^-$; hence under a *vertical* reflection σ_v the vector product changes in sign, whereas an ordinary vector does not. This is illustrated in Fig. A.6. The reverse is true under a horizontal reflection. We can see then that vectors can be divided into two types, which behave in an opposite fashion under reflection, and are known as *polar* and *axial* vectors. The polar vectors are the usual vectors representing translation, or velocity and force, whilst the *axial* vectors represent rotation, and are as shown above, strictly speaking antisymmetric tensors of rank 2. The transformation under reflections σ_v and σ_h of polar and axial vectors is illustrated in Fig. A.7.

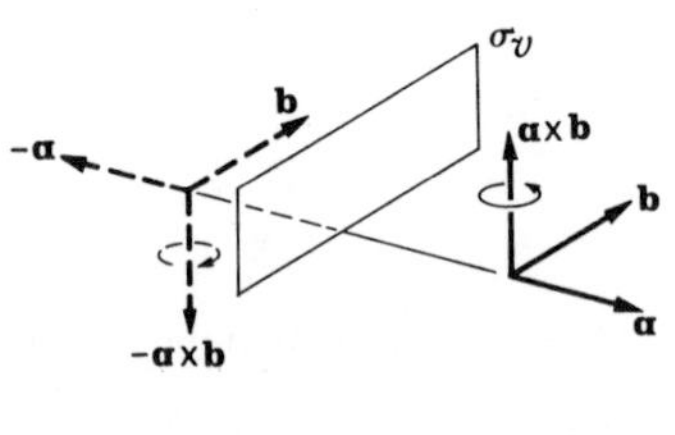

Fig. A.6 *Reversal of the sign of a vector product under a reflection.*

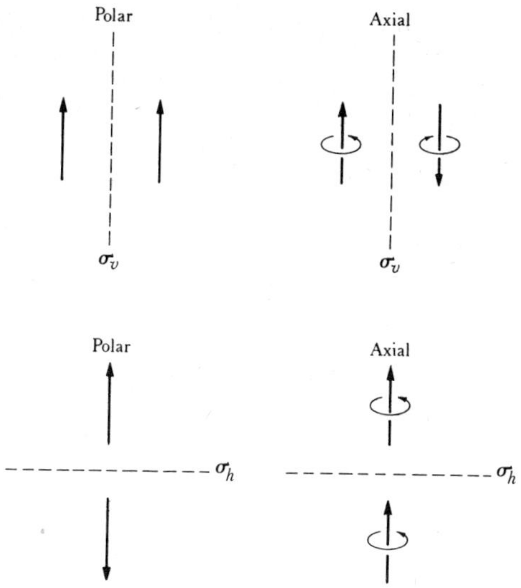

Fig. A.7 *Comparison of the behaviour of polar and axial vectors under reflections σ_v and σ_h.*

A.25 A note on notation

In the multiple suffix notation written above it is immaterial in which order the elements $A, B, C, \ldots$, etc., are written. However, for the two-suffix cases which can be written in matrix form it is essential to preserve the correct order of the matrices in forming the multiplication so that the sums over a repeated suffix agree with the eq. (A.13).

A.26 Problems

A.1 Show that $(\mathbf{A}^\dagger)^{-1} = (\mathbf{A}^{-1})^\dagger$

A.2 If $\mathbf{A}$ is an $(n \times n)$ matrix show that

(a) $|\mathbf{A}^*| = |\mathbf{A}|^*$

(b) $|\tilde{\mathbf{A}}| = |\mathbf{A}|$

(c) $|\mathbf{A}^{-1}| = |\mathbf{A}|^{-1}$

(d) $|k\mathbf{A}| = k^n|\mathbf{A}|$

A.3 If an eigenvalue of $\mathbf{A}$ is λ, find the eigenvalues of (a) $\mathbf{A}^\dagger$, (b) $\mathbf{A}^{-1}$, (c) adj $\mathbf{A}$, (d) $k\mathbf{A}$

A.4 Given

$$\mathbf{A} = \begin{bmatrix} 1 & 1 & -2 \\ 2 & 0 & -1 \\ 3 & -1 & 0 \end{bmatrix}, \qquad \mathbf{B} = \begin{bmatrix} 1 & 2 & 1 \\ 3 & 6 & 3 \\ 2 & 4 & 2 \end{bmatrix}$$

Show that

(a) $\mathbf{BA} \neq \mathbf{AB} = \mathbf{0}$

(b) $|\mathbf{A}| = |\mathbf{B}| = 0$

(c) $\mathbf{A}$ and $\mathbf{B}$ both have an eigenvalue $\lambda = 0$

A.5 If two matrices $\mathbf{A}$ and $\mathbf{B}$ commute (i.e., $\mathbf{AB} = \mathbf{BA}$) then prove that

(a) They have same eigenvectors

(b) They can be simultaneously diagonalized by the same similarity transform.
(c) The eigenvalue λ_{AB} of **AB** is given by

$$\lambda_{AB} = \lambda_A \lambda_B$$

where λ_A, λ_B are the eigenvalues of **A** and **B** respectively.

A.6 Test the results of problem A.5 by using
(a) the matrices of eqs. (A.25) and (A.26)
(b) the matrices

$$\mathbf{A} = \begin{bmatrix} -2 & 2 \\ 5 & 1 \end{bmatrix}, \quad \mathbf{B} = \begin{bmatrix} 1 & 2 \\ 5 & 4 \end{bmatrix}$$

A.7 (a) Show that if λ_A, λ_B, $|x_A\rangle$ and $|x_B\rangle$ are the eigenvalues and eigenvectors of **A** and **B**, then the eigenvalues and eigenvectors of $\mathbf{A} \otimes \mathbf{B}$ are $\lambda_A \lambda_B$ and $|x_A\rangle \otimes |x_B\rangle$.
(b) Show further that (i) trace $\mathbf{A} \otimes \mathbf{B}$ = trace **A** trace **B**
(ii) $(\mathbf{A} \otimes \mathbf{B}^{-1} = \mathbf{A}^{-1} \otimes \mathbf{B}^{-1}$, provided the inverses exist.

A.8 (a) Find the eigenvalues and the row and column eigenvectors of the following matrices

$$\mathbf{A} = \begin{bmatrix} 5 & -1 & -1 \\ -1 & 5 & -1 \\ -1 & -1 & 5 \end{bmatrix}, \quad \mathbf{D} = \begin{bmatrix} 1 & 0 & 2 \\ 0 & 1 & 1 \\ 2 & 0 & 1 \end{bmatrix},$$

$$\mathbf{E} = \begin{bmatrix} 1 & 1+i \\ 1-i & 0 \end{bmatrix}, \quad \mathbf{G} = \begin{bmatrix} 2i & 1 \\ 1 & 0 \end{bmatrix},$$

$$\mathbf{J} = \frac{1}{2}\begin{bmatrix} -4 & 2 & 2 \\ -3 & 3 & 1 \\ -3 & 1 & 3 \end{bmatrix}, \quad \mathbf{K} = \frac{1}{2\sqrt{2}}\begin{bmatrix} 1 & 2 & -\sqrt{3} \\ -1 & 2 & \sqrt{3} \\ \sqrt{6} & 0 & \sqrt{2} \end{bmatrix},$$

$$\mathbf{L} = \begin{bmatrix} 1 & 2 \\ 5 & 4 \end{bmatrix}$$

(b) Notice that (i) **A** is symmetric, **E** is Hermitian, **K** is unitary and hence can be diagonalized. Find a diagonalizing matrix and check that the trace of a matrix is equal to the sum of its eigenvalues.
(ii) **G** is symmetric but not Hermitian. Check that it cannot be diagonalized.
(iii) Check that **D** cannot be diagonalized.

A.9 Let the functions $\phi_n = x^n$ be used as a basis. Use the Gram–Schmidt procedure (eq. (A.117)) to produce an orthonormal basis in the range $-1 \leqslant x \leqslant 1$. (The resulting functions are the Legendre polynomials.)

A.10 Consider the vector **z** of eqs. (A.90) and (A.91) expressed in terms of the bases $\mathbf{e}_1, \mathbf{e}_2$ and $\mathbf{e}'_1, \mathbf{e}'_2$ of eqs. (A.85) and (A.135).
(a) Find the reciprocal basis $\bar{\mathbf{e}}'_1, \bar{\mathbf{e}}'_2$.
(b) Find the components of **z** in the reciprocal basis.
(c) Check that (a) and (b) agree with eqs. (A.127), (A.132), and (A.145).
(d) Find the metric matrix **g** (eq. (A.156)) in the bases $\mathbf{e}_1, \mathbf{e}_2; \mathbf{e}'_1, \mathbf{e}'_2; \bar{\mathbf{e}}'_1, \bar{\mathbf{e}}'_2$.
(e) Check that the scalar product $\langle z|z\rangle$ given by eq. (A.102) is the same in any basis.
(f) Check that the scalar product $\langle z|z\rangle$ is the same as given by eq. (A.112).

A.11 The so-called Pauli spin matrices are

$$\boldsymbol{\sigma}_1 = \begin{bmatrix} 0 & 1 \\ 1 & 0 \end{bmatrix}, \quad \boldsymbol{\sigma}_2 = \begin{bmatrix} 0 & -i \\ i & 0 \end{bmatrix}, \quad \boldsymbol{\sigma}_3 = \begin{bmatrix} 1 & 0 \\ 0 & -1 \end{bmatrix}$$

where $i = \sqrt{-1}$.

(a) Show that they all have the same eigenvalues and find their eigenvectors.

(b) Obtain the following results

(i) $\boldsymbol{\sigma}_1\boldsymbol{\sigma}_2 - \boldsymbol{\sigma}_2\boldsymbol{\sigma}_1 = 2i\boldsymbol{\sigma}_3$

(ii) $\Sigma_r \boldsymbol{\sigma}_r\boldsymbol{\sigma}_r^* = \mathbf{I}$

(iii) $\Sigma_r \boldsymbol{\sigma}_r\boldsymbol{\sigma}_r^\dagger = 3\mathbf{I}$

A.12 Show that the eigenvalues of a unitary matrix are of the form $e^{i\theta}$.

A.13 (a) Prove that the eigenrows of a *symmetric* matrix are the transposes of the eigencolumns belonging to the same eigenvalue.

(b) Prove that the eigenrows $\langle r_i|$ belonging to an eigenvalue λ_i of a matrix $\mathbf{A}$ are orthogonal to the eigencolumns $|c_j\rangle$ belonging to a *different* eigenvalue.

(c) A normalized projection operator $\mathbf{P}^j$ can be defined as

$$\mathbf{P}^j = \frac{|c_j\rangle\langle r_j|}{\langle c_j|r_j\rangle}$$

Show that

(i) If an arbitrary column vector is given by $|c\rangle = \sum_i a_i|c_i\rangle$ (i.e., a linear combination of the eigencolumns of $\mathbf{A}$) then the action of $\mathbf{P}^j$ on $|c\rangle$ is to project out that part belonging to λ_j

(ii) $\mathbf{P}^j\mathbf{P}^j = \mathbf{P}^j$

(iii) $\sum_j \mathbf{P}^j = \mathbf{I}$

(iv) $\sum_j \lambda_j\mathbf{P}^j = \mathbf{A}$

A.14 Test the conclusions of problem A.13 by using the matrices **A** and **L** of problem A.8. What can be said about matrix **D**?

A.15 Use the methods of section A.20 to reduce the following pairs of matrices to diagonal form, i.e., $\mathbf{K} \to \mathrm{diag}\,(\lambda_i)$, $\mathbf{M} \to \mathbf{I}$

(a) $\mathbf{K} = \begin{bmatrix} 5 & 3 \\ 3 & 5 \end{bmatrix}, \quad \mathbf{M} = \begin{bmatrix} 3 & 1 \\ 1 & 3 \end{bmatrix}$

(b) $\mathbf{K} = \begin{bmatrix} 2 & -1 & -1 \\ -1 & 1 & 0 \\ -1 & 0 & 1 \end{bmatrix}, \quad \mathbf{M} = \begin{bmatrix} 2 & 0 & 0 \\ 0 & 1 & 0 \\ 0 & 0 & 1 \end{bmatrix}$

(c) $\mathbf{K} = \frac{1}{2}\begin{bmatrix} 9 & -3\sqrt{3} & 0 \\ -3\sqrt{3} & 12 & 3 \\ 0 & 3 & 9 \end{bmatrix}, \quad \mathbf{M} = \frac{1}{3}\begin{bmatrix} 6 & -\sqrt{3} & \sqrt{3} \\ -\sqrt{3} & 8 & 1 \\ \sqrt{3} & 1 & 8 \end{bmatrix}$

A.16 Prove that the contravariant components of a vector in a non-orthogonal basis are proportional to the projections of the vector onto the bases (see Fig. A.5).

A.17 Prove that $T_{ii'}T_{ij'} = \delta_{i'j'}$.

A.18 Express the similarity transform $\mathbf{B} = \mathbf{R}^{-1}\mathbf{A}\mathbf{R}$ in tensor notation and hence show that

$$\text{trace } \mathbf{B} = \text{trace } \mathbf{A}$$

Appendix B Comparison of point group notations

$C_1 \equiv 1$	$C_2 \equiv 2$	$C_3 \equiv 3$	$C_4 \equiv 4$	$C_6 \equiv 6$
$C_{1h} \equiv m$	$C_{2h} \equiv 2/m$	$C_{3h} \equiv \bar{6}$	$C_{4h} \equiv 4/m$	$C_{6h} \equiv 6/m$
	$C_{2v} \equiv 2mm$	$C_{3v} \equiv 3m$	$C_{4v} \equiv 4mm$	$C_{6v} \equiv 6mm$
	$S_2 \equiv \bar{1}$		$S_4 \equiv \bar{4}$	$S_6 \equiv \bar{3}$
	$D_2 \equiv 222$	$D_3 \equiv 32$	$D_4 \equiv 422$	$D_6 \equiv 622$
	$D_{2h} \equiv mmm$	$D_{3h} \equiv \bar{6}m2$	$D_{4h} \equiv 4/mmm$	$D_{6h} \equiv 6/mmm$
	$D_{2d} \equiv \bar{4}2m$	$D_{3d} \equiv \bar{3}m$		
$T \equiv 23$	$T_d \equiv \bar{4}3m$	$T_h \equiv m3$	$O \equiv 432$	$O_h \equiv m3m$

The Schoenflies symbol is given on the left and the international notation on the right. In the international notation n denotes an n-fold proper rotation axis; $\bar{n}$ denotes an n-fold rotation–inversion axis, i.e., a rotation of $2\pi/n$ followed by inversion; m denotes a vertical mirror plane and $/m$ denotes a horizontal mirror plane. For example $\bar{4}2m$ indicates a point group containing a four-fold rotation–inversion axis, two-fold rotation axes, and vertical mirror planes.

Appendix C Character tables for the point groups

The character tables are given in the form illustrated in table 2.6 and are discussed in section 2.6. The labelling of the irreducible representations follows the notation of section 2.7. The chief symbols being A and B denoting one-dimensional representations, E denoting a two-dimensional representation, and T denoting a three-dimensional representation. g and u denote parity of the representation: g is even parity and u is odd parity.

Certain groups, containing three-fold and six-fold axes, contain non-integral characters. In these cases the character is one of the nth roots of unity. In this respect the symbol ε denotes $\exp(2\pi i/3)$ for a three-fold axis and $\exp(2\pi i/6)$ for a six-fold axis and ε^* is its complex conjugate.

Also in these groups one finds that a number of pairs of one-dimensional representation are coupled together and given a label E. This procedure is related to the fact that in the cyclic groups although *all* the elements belong to *different* classes there is actually no physical distinction between forward and backward rotations because of invariance with respect to time inversion. This feature creates a degeneracy amongst pairs of irreducible representations and they are therefore coupled together. At the left of the tables are found the functions x, y, and z and the rotation vectors $\mathbf{R}_x$, $\mathbf{R}_y$, and $\mathbf{R}_z$. These are assigned to their appropriate irreducible representations.

Table C.1 The groups C_n

C_1	E
A	1

C_2	E	C_2		
A	1	1	z	R_z
B	1	-1	x, y	R_x, R_y

C_3	E	C_3	C_3^2		
A	1	1	1	z	R_z
E	$\begin{Bmatrix}1\\1\end{Bmatrix}$	$\begin{matrix}\varepsilon\\\varepsilon^*\end{matrix}$	$\begin{matrix}\varepsilon^*\\\varepsilon\end{matrix}$	x, y	R_x, R_y

$\varepsilon = \exp(2\pi i/3)$

Table C.1 (cont.)

C_4	E	C_4	C_4^3	C_2		
A	1	1	1	1	z	R_z
B	1	-1	-1	1		
E	1	i	$-i$	-1	x, y	R_x, R_y
	1	$-i$	i	-1		

C_6	E	C_6	C_6^5	C_3	C_3^2	C_2		
A	1	1	1	1	1	1	z	R_z
B	1	-1	-1	1	1	-1		
E_1	1	ε	ε^*	$-\varepsilon^*$	$-\varepsilon$	-1	x, y	R_x, R_y
	1	ε^*	ε	$-\varepsilon$	$-\varepsilon^*$	-1		
E_2	1	$-\varepsilon$	$-\varepsilon^*$	$-\varepsilon^*$	$-\varepsilon$	1		
	1	$-\varepsilon^*$	$-\varepsilon$	$-\varepsilon$	$-\varepsilon^*$	1		

$\varepsilon = \exp(2\pi i/6)$

Table C.2 The groups C_{nv}

C_{2v}	E	C_2	σ_v'	σ_v''		
A_1	1	1	1	1	z	
A_2	1	1	-1	-1		R_z
B_1	1	-1	1	-1	x	R_y
B_2	1	-1	-1	1	y	R_x

C_{3v}	E	$2C_3$	$3\sigma_v$		
A_1	1	1	1	z	
A_2	1	1	-1		R_z
E	2	-1	0	(x, y)	R_x, R_y

C_{4v}	E	$2C_4$	C_2	$2\sigma_v$	$2\sigma_d$		
A_1	1	1	1	1	1	z	
A_2	1	1	1	-1	-1		R_z
B_1	1	-1	1	1	-1		
B_2	1	-1	1	-1	1		
E	2	0	-2	0	0	x, y	R_x, R_y

C_{6v}	E	$2C_6$	$2C_3$	C_2	$3\sigma_v$	$3\sigma_d$		
A_1	1	1	1	1	1	1	z	
A_2	1	1	1	1	-1	-1		R_z
B_1	1	-1	1	-1	1	-1		
B	1	-1	1	-1	-1	1		
E_1	2	1	-1	-2	0	0	x, y	R_x, R_y
E_2	2	-1	-1	2	0	0		

Table C.3 The groups C_{nh}

C_{1h}	E	σ_h		
A'	1	1	x, y	R_z
A''	1	-1	z	R_x, R_y

C_{2h}	E	C_2	i	σ_h		
A_g	1	1	1	1		R_z
B_g	1	-1	1	-1		R_x, R_y
A_u	1	1	-1	-1	z	
B_u	1	-1	-1	1	x, y	

$$C_{2h} = C_2 \otimes S_2$$

C_{3h}	E	C_3	C_3^2	σ_h	S_3	S_3^5		
A'	1	1	1	1	1	1		R_z
A''	1	1	1	-1	-1	-1	z	
E'	1	ε	ε^*	1	ε	ε^*	x, y	
	1	ε^*	ε	1	ε^*	ε		
E''	1	ε	ε^*	-1	$-\varepsilon$	$-\varepsilon^*$		R_x, R_y
	1	ε^*	ε	-1	$-\varepsilon^*$	$-\varepsilon$		

$$C_{3h} = C_3 \otimes C_{1h} \qquad \varepsilon = \exp(2\pi i/3)$$

C_{4h}	E	C_4	C_4^3	C_2	i	S_4^3	S_4	σ_h		
A_g	1	1	1	1	1	1	1	1		R_z
B_g	1	-1	-1	1	1	-1	-1	1		
A_u	1	1	1	1	-1	-1	-1	-1	z	
B_u	1	-1	-1	1	-1	1	1	-1		
E_g	1	i	$-i$	-1	1	i	$-i$	-1		R_x, R_y
	1	$-i$	i	-1	1	$-i$	i	-1		
E_u	1	i	$-i$	-1	-1	$-i$	i	1	x, y	
	1	$-i$	i	-1	-1	i	$-i$	1		

$$C_{4h} = C_4 \otimes S_2$$

C_{6h}	E	C_6	C_6^5	C_3	C_3^2	C_2	i	S_6^5	S_6	S_3^2	S_3	σ_h		
A_g	1	1	1	1	1	1	1	1	1	1	1	1		R_z
B_g	1	-1	-1	1	1	-1	1	-1	-1	1	1	-1		
A_u	1	1	1	1	1	1	-1	-1	-1	-1	-1	-1	z	
B_u	1	-1	-1	1	1	-1	-1	1	1	-1	-1	1		
E_{1g}	1	ε	ε^*	$-\varepsilon^*$	$-\varepsilon$	-1	1	ε	ε^*	$-\varepsilon^*$	$-\varepsilon$	-1		R_x, R_y
	1	ε^*	ε	$-\varepsilon$	$-\varepsilon^*$	-1	1	ε^*	ε	$-\varepsilon$	$-\varepsilon^*$	-1		
E_{2g}	1	$-\varepsilon$	$-\varepsilon^*$	$-\varepsilon^*$	$-\varepsilon$	1	1	$-\varepsilon$	$-\varepsilon^*$	$-\varepsilon^*$	$-\varepsilon$	1		
	1	$-\varepsilon^*$	$-\varepsilon$	$-\varepsilon$	$-\varepsilon^*$	1	1	$-\varepsilon^*$	$-\varepsilon$	$-\varepsilon$	$-\varepsilon^*$	1		
E_{1u}	1	ε	ε^*	$-\varepsilon^*$	$-\varepsilon$	-1	-1	$-\varepsilon$	$-\varepsilon^*$	ε^*	ε	1	x, y	
	1	ε^*	ε	$-\varepsilon$	$-\varepsilon^*$	-1	-1	$-\varepsilon^*$	$-\varepsilon$	ε	ε^*	1		
E_{2u}	1	$-\varepsilon$	$-\varepsilon^*$	$-\varepsilon^*$	$-\varepsilon$	1	-1	ε	ε^*	ε^*	ε	-1		
	1	$-\varepsilon^*$	$-\varepsilon$	$-\varepsilon$	$-\varepsilon^*$	1	-1	ε^*	ε	ε	ε^*	-1		

$$C_{6h} = C_6 \otimes S_2 \qquad \varepsilon = \exp(2\pi i/6)$$

Table C.4 The groups S_n

S_2	E	i		
A_g	1	1		R_x, R_y, R_z
A_u	1	-1	x, y, z	

S_4	E	S_4	S_4^3	C_2		
A	1	1	1	1		R_z
B	1	-1	-1	1	z	
E	$\left\{\begin{matrix}1 \\ 1\end{matrix}\right.$	$\begin{matrix}i \\ -i\end{matrix}$	$\begin{matrix}-i \\ i\end{matrix}$	$\left.\begin{matrix}-1 \\ -1\end{matrix}\right\}$	x, y	R_x, R_y

S_6	E	C_3	C_3^2	i	S_6^5	S_6		
A_g	1	1	1	1	1	1		R_z
A_u	1	1	1	-1	-1	-1	z	
E_g	$\left\{\begin{matrix}1 \\ 1\end{matrix}\right.$	$\begin{matrix}\varepsilon \\ \varepsilon^*\end{matrix}$	$\begin{matrix}\varepsilon^* \\ \varepsilon\end{matrix}$	$\begin{matrix}1 \\ 1\end{matrix}$	$\begin{matrix}\varepsilon \\ \varepsilon^*\end{matrix}$	$\left.\begin{matrix}\varepsilon^* \\ \varepsilon\end{matrix}\right\}$		R_x, R_y
E_u	$\left\{\begin{matrix}1 \\ 1\end{matrix}\right.$	$\begin{matrix}\varepsilon \\ \varepsilon^*\end{matrix}$	$\begin{matrix}\varepsilon^* \\ \varepsilon\end{matrix}$	$\begin{matrix}-1 \\ -1\end{matrix}$	$\begin{matrix}-\varepsilon \\ -\varepsilon^*\end{matrix}$	$\left.\begin{matrix}-\varepsilon^* \\ -\varepsilon\end{matrix}\right\}$	x, y	

$S_6 = C_3 \otimes S_2 \qquad \varepsilon = \exp(2\pi i/3)$

Table C.5 The groups D_n

D_2	E	C_2'	C_2''	C_2'''		
A	1	1	1	1		
B_1	1	1	-1	-1	x	R_x
B_2	1	-1	1	-1	y	R_y
B_3	1	-1	-1	1	z	R_z

In D_2 there is no principal axis as the three two-fold axes are indistinguishable. It is assumed that

C_2' is a two-fold rotation about x
C_2'' is a two-fold rotation about y
C_2''' is a two-fold rotation about z

D_3	E	$2C_3$	$3C_2$		
A_1	1	1	1		
A_2	1	1	-1	z	R_z
E	2	-1	0	x, y	R_x, R_y

D_4	E	$2C_4$	C_2	$2C_2'$	$2C_2''$		
A_1	1	1	1	1	1		
A_2	1	1	1	-1	-1	z	R_z
B_1	1	-1	1	1	-1		
B_2	1	-1	1	-1	1		
E	2	0	-2	0	0	x, y	R_x, R_y

Table C.5 (cont.)

D_6	E	$2C_6$	$2C_3$	C_2	$3C'_2$	$3C''_2$		
A_1	1	1	1	1	1	1		
A_2	1	1	1	1	-1	-1	z	R_z
B_1	1	-1	1	-1	1	-1		
B_2	1	-1	1	-1	-1	1		
E_1	2	1	-1	-2	0	0	x, y	R_x, R_y
E_2	2	-1	-1	2	0	0		

Table C.6 The groups D_{nd}

D_{2d}	E	$2S_4$	C_2	$2C'_2$	$2\sigma_d$		
A_1	1	1	1	1	1		
A_2	1	1	1	-1	-1		R_z
B_1	1	-1	1	1	-1		
B_2	1	-1	1	-1	1	z	
E	2	0	-2	0	0	x, y	R_x, R_y

D_{3d}	E	$2C_3$	$3C_2$	i	$2S_6$	$3\sigma_d$		
A_{1g}	1	1	1	1	1	1		
A_{2g}	1	1	-1	1	1	-1		R_z
A_{1u}	1	1	1	-1	-1	-1		
A_{2u}	1	1	-1	-1	-1	1	z	
E_g	2	-1	0	2	-1	0		R_x, R_y
E_u	2	-1	0	-2	1	0	x, y	

$$D_{3d} = D_3 \otimes S_2$$

Table C.7 The groups D_{nh}

D_{2h}	E	C'_2	C''_2	C'''_2	i	σ_v^{yz}	σ_v^{zx}	σ_h		
A_g	1	1	1	1	1	1	1	1		
B_{1g}	1	1	-1	-1	1	1	-1	-1		R_x
B_{2g}	1	-1	1	-1	1	-1	1	-1		R_y
B_{3g}	1	-1	-1	1	1	-1	-1	1		R_z
A_u	1	1	1	1	-1	-1	-1	-1		
B_{1u}	1	1	-1	-1	-1	-1	1	1	x	
B_{2u}	1	-1	1	-1	-1	1	-1	1	y	
B_{3u}	1	-1	-1	1	-1	1	1	-1	z	

$$D_{2h} = D_2 \otimes S_2$$

D_{3h}	E	$2C_3$	$3C_2$	σ_h	$2S_3$	$3\sigma_v$		
A'_1	1	1	1	1	1	1		
A'_2	1	1	-1	1	1	-1		R_z
A''_1	1	1	1	-1	-1	-1		
A''_2	1	1	-1	-1	-1	1	z	
E'	2	-1	0	2	-1	0	x, y	
E''	2	-1	0	-2	1	0		R_x, R_y

$$D_{3h} = D_3 \otimes C_{1h}$$

Table C.7 (cont.)

D_{4h}	E	$2C_4$	C_2	$2C'_2$	$2C''_2$	i	$2S_4$	σ_h	$2\sigma_v$	$2\sigma_d$		
A_{1g}	1	1	1	1	1	1	1	1	1	1		
A_{2g}	1	1	1	−1	−1	1	1	1	−1	−1		R_z
B_{1g}	1	−1	1	−1	1	1	−1	1	−1	1		
B_{2g}	1	−1	1	1	−1	1	−1	1	1	−1		
A_{1u}	1	1	1	1	1	−1	−1	−1	−1	−1		
A_{2u}	1	1	1	−1	−1	−1	−1	−1	1	1	z	
B_{1u}	1	−1	1	−1	1	−1	1	−1	1	−1		
B_{2u}	1	−1	1	1	−1	−1	1	−1	−1	1		
E_{1g}	2	0	−2	0	0	2	0	−2	0	0		R_x, R_y
E_{2u}	2	0	−2	0	0	−2	0	2	0	0	x, y	

$$D_{4h} = D_4 \otimes S_2$$

D_{6h}	E	$2C_6$	$2C_3$	C_2	$3C'_2$	$3C''_2$	i	$2S_3$	$2S_6$	σ_h	$3\sigma_d$	$3\sigma_v$		
A_{1g}	1	1	1	1	1	1	1	1	1	1	1	1		
A_{2g}	1	1	1	1	−1	−1	1	1	1	1	−1	−1		R_z
B_{1g}	1	−1	1	−1	1	−1	1	−1	1	−1	1	−1		
B_{2g}	1	−1	1	−1	−1	1	1	−1	1	−1	−1	1		
A_{1u}	1	1	1	1	1	1	−1	−1	−1	−1	−1	−1		
A_{2u}	1	1	1	1	−1	−1	−1	−1	−1	−1	1	1	z	
B_{1u}	1	−1	1	−1	1	−1	−1	1	−1	1	−1	1		
B_{2u}	1	−1	1	−1	−1	1	−1	1	−1	1	1	−1		
E_{1g}	2	1	−1	−2	0	0	2	1	−1	−2	0	0		R_x, R_y
E_{2g}	2	−1	−1	2	0	0	2	−1	−1	2	0	0		
E_{1u}	2	1	−1	−2	0	0	−2	−1	1	2	0	0	x, y	
E_{2u}	2	−1	−1	2	0	0	−2	1	1	−2	0	0		

$$D_{6h} = D_6 \otimes S_2$$

Table C.8 The cubic groups

T	E	$4C_3$	$4C_3^2$	$3C_2$		
A	1	1	1	1		
E	$\left\{\begin{matrix}1 \\ 1\end{matrix}\right.$	$\begin{matrix}\varepsilon \\ \varepsilon^*\end{matrix}$	$\begin{matrix}\varepsilon^* \\ \varepsilon\end{matrix}$	$\left.\begin{matrix}1 \\ 1\end{matrix}\right\}$		
T	3	0	0	−1	x, y, z	R_x, R_y, R_z

$$\varepsilon = \exp(2\pi i/3)$$

T_d	E	$8C_3$	$3C_2$	$6S_4$	$6\sigma_d$		
A_1	1	1	1	1	1		
A_2	1	1	1	−1	−1		
E	2	−1	2	0	0		
T_1	3	0	−1	1	−1		R_x, R_y, R_z
T_2	3	0	−1	−1	1	x, y, z	

Table C.8 (cont.)

T_h	E	$4C_3$	$4C_3^2$	$3C_2$	i	$4S_6$	$4S_6^2$	$3\sigma_d$		
A_g	1	1	1	1	1	1	1	1		
A_u	1	1	1	1	-1	-1	-1	-1		
E_g	$\begin{cases}1\\1\end{cases}$	ε ε^*	ε^* ε	1 1	1 1	ε ε^*	ε^* ε	1 1		
E_u	$\begin{cases}1\\1\end{cases}$	ε ε^*	ε^* ε	1 1	-1 -1	$-\varepsilon$ $-\varepsilon^*$	$-\varepsilon^*$ $-\varepsilon$	-1 -1		
T_g	3	0	0	-1	3	0	0	-1		R_x, R_y, R_z
T_u	3	0	0	-1	-3	0	0	1	x, y, z	

$T_h = T \otimes S_2 \qquad \varepsilon = \exp(2\pi i/3)$

O	E	$8C_3$	$3C_2$	$6C_4$	$6C_2'$		
A_1	1	1	1	1	1		
A_2	1	1	1	-1	-1		
E	2	-1	2	0	0		
T_1	3	0	-1	1	-1	x, y, z	R_x, R_y, R_z
T_2	3	0	-1	-1	1		

O_h	E	$8C_3$	$6C_2'$	$6C_4$	$3C_2$	i	$8S_6$	$6\sigma_d$	$6S_4$	$3\sigma_h$		
A_{1g}	1	1	1	1	1	1	1	1	1	1		
A_{2g}	1	1	-1	-1	1	1	1	-1	-1	1		
A_{1u}	1	1	1	1	1	-1	-1	-1	-1	-1		
A_{2u}	1	1	-1	-1	1	-1	-1	1	1	-1		
E_g	2	-1	0	0	2	2	-1	0	0	2		
E_u	2	-1	0	0	2	-2	1	0	0	-2		
T_{1g}	3	0	-1	1	-1	3	0	-1	1	-1		R_x, R_y, R_z
T_{2g}	3	0	1	-1	-1	3	0	1	-1	-1		
T_{1u}	3	0	-1	1	-1	-3	0	1	-1	1	x, y, z	
T_{2u}	3	0	1	-1	-1	-3	0	-1	1	1		

$O_h = O \otimes S_2$

Table C.9 The axial and spherical groups

C_∞	E	C_ϕ		
A	1	1	z	R_z
E_1	$\begin{cases}1\\1\end{cases}$	$e^{i\phi}$ $e^{-i\phi}$	x, y	R_x, R_y
E_2	$\begin{cases}1\\1\end{cases}$	$e^{2i\phi}$ $e^{-2i\phi}$		
$\vdots$	$\vdots$	$\vdots$		
E_m	$\begin{cases}1\\1\end{cases}$	$e^{im\phi}$ $e^{-im\phi}$		

Table C.9 (cont.)

R_3	E	R		
$D^0(R)$	1	1		
$D^1(R)$	3	$\frac{\sin(1+\frac{1}{2})\phi}{\sin\phi/2}$	x, y, z	R_x, R_y, R_z
$D^2(R)$	5	$\frac{\sin(2+\frac{1}{2})\phi}{\sin\phi/2}$		
	$\vdots$	$\vdots$		
$D^l(R)$	$2l+1$	$\frac{\sin(l+\frac{1}{2})\phi}{\sin\phi/2}$		

$R_3 \otimes S_2$	E	R	i	iR		
$D^{0\otimes g}(R)$	1	1	1	1		
$D^{0\otimes u}(R)$	1	1	-1	-1		
$D^{1\otimes g}(R)$	3	$\frac{\sin\frac{3}{2}\phi}{\sin\phi/2}$	3	$\frac{\sin\frac{3}{2}\phi}{\sin\phi/2}$		R_x, R_y, R_z
$D^{1\otimes u}(R)$	3	$\frac{\sin\frac{3}{2}\phi}{\sin\phi/2}$	-3	$-\frac{\sin\frac{3}{2}\phi}{\sin\phi/2}$	x, y, z	
	$\vdots$	$\vdots$	$\vdots$	$\vdots$		
$D^{l\otimes g}(R)$	$2l+1$	$\frac{\sin(l+\frac{1}{2})\phi}{\sin\phi/2}$	$2l+1$	$\frac{\sin(l+\frac{1}{2})\phi}{\sin\phi/2}$		
$D^{l\otimes u}(R)$	$2l+1$	$\frac{\sin(l+\frac{1}{2})\phi}{\sin\phi/2}$	$-(2l+1)$	$-\frac{\sin(l+\frac{1}{2})\phi}{\sin\phi/2}$		

Solutions to problems

Chapter 1

1.1 The appropriate combinations are

$$(1)\,(a);\ (2)\,(a);\ (2)\,(c);\ (3)\,(a);\ (4)\,(a)$$

The other combinations do not form groups for the following reasons:

(1) (b), (2) (b), (2) (d), (3) (b), (4) (b) non-associative
(1) (c) no inverse
(1) (d), (3) (e), (3) (f), (4) (e) not closed
(1) (e), (1) (f), (2) (e), (2) (f), (3) (c), (3) (d), (4) (c), (4) (d) no relationship.
(4) (f) no identity

1.3 $AB = F, FC = B, BD = A, AF = B$ therefore $ABCDF = B$.

1.4 The group is D_4 with elements

$C_4 = A,\ C_4^2 = B,\ C_4^3 = C,\ C_4^4 = E$

$C_2^{(y)} = D,\ C_2^{(x)} = F$ (rotations about perpendicular bisectors of the sides of the square).

$C_2' = G,\ C_2'' = H$ (rotations about the diagonals).

The multiplication table is

	E	A	B	C	D	F	G	H
E	E	A	B	C	D	F	G	H
A	A	B	C	E	G	H	F	D
B	B	C	E	A	F	D	H	G
C	C	E	A	B	H	G	D	F
D	D	H	F	G	E	B	C	A
F	F	G	D	H	B	E	A	C
G	G	D	H	F	A	C	E	B
H	H	F	G	D	C	A	B	E

The classes are

$$(E), (A, C), (B), (D, F), (G, H)$$

The subgroups are

$$(E), (EABC), (EB), (ED), (EF), (EG), (EH), (EBDF), (EBGH)$$

1.5 Hint: solve this problem by drawing diagrams such as those in Fig. 1.3.

1.6 (a) D_{3h} with elements $E, 2C_3, 2S_3, \sigma_h, 3\sigma_v, 3C_2'$
(b) C_{1h} with elements E, σ_h
(c) C_{3h} with elements $E, 2C_3, 2S_3, \sigma_h$
Let the operator which turns black to white be Θ then the new groups are
(b) $E, \sigma_h, \sigma_v\Theta, C_2\Theta$
(c) $E, 2C_3, 2S_3, \sigma_h, 3\sigma_v\Theta, 3C_2\Theta$

Chapter 2

2.1

$$\mathbf{T}^{-1} = \begin{bmatrix} \frac{1}{\sqrt{2}} & \frac{1}{\sqrt{2}} \\ -\frac{1}{\sqrt{2}} & \frac{1}{\sqrt{2}} \end{bmatrix}$$

e.g.,

$$\mathbf{T}^{-1}\mathbf{\Gamma}^3(B)\mathbf{T} = \begin{bmatrix} \frac{1}{\sqrt{2}} & \frac{1}{\sqrt{2}} \\ -\frac{1}{\sqrt{2}} & -\frac{1}{\sqrt{2}} \end{bmatrix} \begin{bmatrix} -\frac{1}{2} & -\frac{\sqrt{3}}{2} \\ -\frac{\sqrt{3}}{2} & \frac{1}{2} \end{bmatrix} \begin{bmatrix} \frac{1}{\sqrt{2}} & -\frac{1}{\sqrt{2}} \\ \frac{1}{\sqrt{2}} & \frac{1}{\sqrt{2}} \end{bmatrix}$$

$$= \tfrac{1}{4}\begin{bmatrix} 1 & 1 \\ -1 & 1 \end{bmatrix} \begin{bmatrix} -1 & -\sqrt{3} \\ -\sqrt{3} & 1 \end{bmatrix} \begin{bmatrix} 1 & -1 \\ 1 & 1 \end{bmatrix}$$

$$= \tfrac{1}{4}\begin{bmatrix} 1 & 1 \\ -1 & 1 \end{bmatrix} \begin{bmatrix} -(1+\sqrt{3}) & (1-\sqrt{3}) \\ (1-\sqrt{3}) & (1+\sqrt{3}) \end{bmatrix}$$

$$= \tfrac{1}{4}\begin{bmatrix} -2\sqrt{3} & 2 \\ 2 & 2\sqrt{3} \end{bmatrix} = \begin{bmatrix} -\frac{\sqrt{3}}{2} & \frac{1}{2} \\ \frac{1}{2} & \frac{\sqrt{3}}{2} \end{bmatrix} = \mathbf{\Gamma}^4(B)$$

2.2 The vectors are

$$[1, 1, 1, 1, 1, 1]$$
$$[1, -1, -1, -1, 1, 1]$$
$$\left[1, 0, -\frac{\sqrt{3}}{2}, \frac{\sqrt{3}}{2}, -\frac{1}{2}, -\frac{1}{2}\right]$$
$$\left[0, -1, \frac{1}{2}, \frac{1}{2}, \frac{\sqrt{3}}{2}, -\frac{\sqrt{3}}{2}\right]$$
$$\left[0, -1, \frac{1}{2}, \frac{1}{2}, -\frac{\sqrt{3}}{2}, \frac{\sqrt{3}}{2}\right]$$
$$\left[1, 0, \frac{\sqrt{3}}{2}, -\frac{\sqrt{3}}{2}, -\frac{1}{2}, -\frac{1}{2}\right]$$

These are mutually orthogonal, for example,

$$\left[1, 0, \frac{\sqrt{3}}{2}, -\frac{\sqrt{3}}{2}, -\frac{1}{2}, -\frac{1}{2}\right]\begin{bmatrix} 0 \\ -1 \\ \frac{1}{2} \\ \frac{1}{2} \\ \frac{\sqrt{3}}{2} \\ -\frac{\sqrt{3}}{2} \end{bmatrix} = 0 + 0 + \frac{\sqrt{3}}{4} - \frac{\sqrt{3}}{4} - \frac{\sqrt{3}}{4} + \frac{\sqrt{3}}{4} = 0$$

Hence since we have six mutually orthogonal vectors spanning a space whose dimension is equal to the number of group elements, $\mathbf{\Gamma}^1(R)$, $\mathbf{\Gamma}^2(R)$, $\mathbf{\Gamma}^4(R)$ form a set of inequivalent, irreducible representations of the group D_3.

2.3 Check that the matrices obey the multiplication table, e.g.,

$$\begin{bmatrix} -2 & 1 & 1 \\ -\frac{3}{2} & \frac{3}{2} & \frac{1}{2} \\ -\frac{3}{2} & \frac{1}{2} & \frac{3}{2} \end{bmatrix}\begin{bmatrix} -2 & 1 & 1 \\ -\frac{3}{2} & \frac{3}{2} & \frac{1}{2} \\ -\frac{3}{2} & \frac{1}{2} & \frac{3}{2} \end{bmatrix} = \begin{bmatrix} 1 & 0 & 0 \\ 0 & 1 & 0 \\ 0 & 0 & 1 \end{bmatrix}$$

i.e.,

$$\mathbf{\Gamma}(A)\mathbf{\Gamma}(B) = \mathbf{\Gamma}(F)$$

Since $\sum_i l_i^2 = 6$ there cannot be a three-dimensional irreducible representation of the group. The traces are

$$3 \quad 1 \quad 1 \quad 1 \quad 3 \quad 3$$

Hence using eq. (2.37) and the character table of D_3 it can be shown that

$$\mathbf{\Gamma}(R) = 2\mathbf{A}_1 \oplus \mathbf{A}_2$$

2.4 Example: the vectors for A_2 and T_1 are

$$[1, \sqrt{8}, \sqrt{3}, -\sqrt{6}, -\sqrt{6}]$$
$$[3, 0, -\sqrt{3}, \sqrt{6}, -\sqrt{6}]$$

which are obviously orthogonal.

2.5 $T_{1g} \rightarrow A_2 \oplus E$
$T_{2u} \rightarrow B_1 \oplus E$

Chapter 3

3.1 The representation $\mathbf{\Gamma}(R)$ is

$$\mathbf{\Gamma}(R); \quad \overset{E}{\begin{bmatrix} 1 & 0 \\ 0 & 1 \end{bmatrix}} \overset{A}{\begin{bmatrix} 1 & -1 \\ 0 & -1 \end{bmatrix}} \overset{B}{\begin{bmatrix} -1 & 0 \\ -1 & 1 \end{bmatrix}} \overset{C}{\begin{bmatrix} 0 & 1 \\ 1 & 0 \end{bmatrix}} \overset{F}{\begin{bmatrix} 0 & -1 \\ 1 & -1 \end{bmatrix}} \overset{D}{\begin{bmatrix} -1 & 1 \\ -1 & 0 \end{bmatrix}}$$

The representation is not unitary because in general $\tilde{\mathbf{\Gamma}}(R) \neq \mathbf{\Gamma}(R)^{-1}$, this has arisen because $\mathbf{e}_1$ and $\mathbf{e}_2$ are not orthogonal.

3.2 The group D_4 has eight elements

$$E, A = C_4, B = C_4^2, C = C_4^3$$

D and $E = C_2$ about axes bisecting the sides of the square

G and $H = C_2$ about diagonal axes

The representation of these elements using the (x, y, z) co-ordinates system is

$$\underset{E}{\begin{bmatrix} 1 & 0 & 0 \\ 0 & 1 & 0 \\ 0 & 0 & 1 \end{bmatrix}} \underset{A}{\begin{bmatrix} 0 & 1 & 0 \\ -1 & 0 & 0 \\ 0 & 0 & 1 \end{bmatrix}} \underset{B}{\begin{bmatrix} -1 & 0 & 0 \\ 0 & -1 & 0 \\ 0 & 0 & 1 \end{bmatrix}} \underset{C}{\begin{bmatrix} 0 & -1 & 0 \\ 1 & 0 & 0 \\ 0 & 0 & 1 \end{bmatrix}}$$

$$\underset{D}{\begin{bmatrix} -1 & 0 & 0 \\ 0 & 1 & 0 \\ 0 & 0 & -1 \end{bmatrix}} \underset{F}{\begin{bmatrix} 1 & 0 & 0 \\ 0 & -1 & 0 \\ 0 & 0 & -1 \end{bmatrix}} \underset{G}{\begin{bmatrix} 0 & 1 & 0 \\ 1 & 0 & 0 \\ 0 & 0 & -1 \end{bmatrix}} \underset{H}{\begin{bmatrix} 0 & -1 & 0 \\ -1 & 0 & 0 \\ 0 & 0 & -1 \end{bmatrix}}$$

now x, y, and z each form a basis for the representation $\mathbf{\Gamma}^u(S)$ of S_2, i.e.,

	E	i
$\mathbf{\Gamma}^u(S)$	1	-1

and using this representation we can obtain a representation of the group $D_{4h} = D_4 \otimes S_2$ using the equation

$$\mathbf{\Gamma}(R \,.\, S) = \mathbf{\Gamma}(R) \otimes \mathbf{\Gamma}^u(S)$$

so that

$$\mathbf{\Gamma}(R \,.\, E) = \mathbf{\Gamma}(R)$$

and

$$\mathbf{\Gamma}(Ri) = -\mathbf{\Gamma}(R)$$

3.3 (a) the group is D_{2d} with elements

$$E; A = S_4; B = S_4^2 = C_2; C = S_4^3;$$

D and $F = \sigma_v$ about vertical planes

G and $H = C_2$ about horizontal axes

(b) the representation is

$$\underset{E}{\begin{bmatrix} 1 & 0 \\ 0 & 1 \end{bmatrix}} \underset{A}{\begin{bmatrix} 0 & 1 \\ -1 & 0 \end{bmatrix}} \underset{B}{\begin{bmatrix} -1 & 0 \\ 0 & -1 \end{bmatrix}} \underset{C}{\begin{bmatrix} 0 & -1 \\ 1 & 0 \end{bmatrix}}$$

$$\underset{D}{\begin{bmatrix} 1 & 0 \\ 0 & -1 \end{bmatrix}} \underset{F}{\begin{bmatrix} -1 & 0 \\ 0 & 1 \end{bmatrix}} \underset{G}{\begin{bmatrix} 0 & 1 \\ 1 & 0 \end{bmatrix}} \underset{H}{\begin{bmatrix} 0 & -1 \\ -1 & 0 \end{bmatrix}}$$

(c) from the matrices derived in 3.3(b) we obtain

$$P_E f(x) = f(x); P_A f(x) = f(-y) \quad P_B x = f(-x) \quad P_C f(x) = f(y)$$

$$P_D f(x) = f(x) \quad P_F f(x) = f(-x) \quad P_G x = f(y) \quad P_H f(x) = f(-y)$$

we obtain four functions $f_1 = f(x)$; $f_2 = f(-y)$; $f_3 = f(-x)$; $f_4 = f(y)$. By using all these functions we obtain the representation

$$\overset{E}{\begin{bmatrix}1&0&0&0\\0&1&0&0\\0&0&1&0\\0&0&0&1\end{bmatrix}}\overset{A}{\begin{bmatrix}0&0&0&1\\1&0&0&0\\0&1&0&0\\0&0&1&0\end{bmatrix}}\overset{B}{\begin{bmatrix}0&0&1&0\\0&0&0&1\\1&0&0&0\\0&1&0&0\end{bmatrix}}\overset{C}{\begin{bmatrix}0&1&0&0\\0&0&1&0\\0&0&0&1\\1&0&0&0\end{bmatrix}}$$

$$\overset{D}{\begin{bmatrix}1&0&0&0\\0&0&0&1\\0&0&1&0\\0&1&0&0\end{bmatrix}}\overset{F}{\begin{bmatrix}0&0&1&0\\0&1&0&0\\1&0&0&0\\0&0&0&1\end{bmatrix}}\overset{G}{\begin{bmatrix}0&0&0&1\\0&0&1&0\\0&1&0&0\\1&0&0&0\end{bmatrix}}\overset{H}{\begin{bmatrix}0&1&0&0\\1&0&0&0\\0&0&0&1\\0&0&1&0\end{bmatrix}}$$

(d)

$$O^{A_1}F = \frac{c+d}{2}(x^2+y^2)$$

$$O^{A_2}F = 0$$

$$O^{B_1}F = \frac{c-d}{2}(x^2-y^2)$$

$$O^{B_2}F = exy$$

$$O^{E}_{11}F = ax, \qquad O^{E}_{21}ax = ay$$

$$O^{E}_{22}F = by, \qquad O^{E}_{12}by = bx$$

(e) For example

$$\begin{aligned}\langle\phi^{A_1}|\phi^{A_1}\rangle &= \langle x^2+y^2|x^2+y^2\rangle\\ &= \int_{-a}^{a}\int_{-a}^{a}(x^4+2x^2y^2+y^4)\,\mathrm{d}x\,\mathrm{d}y\\ &= \int_{-a}^{a}\left(\frac{2a^5}{5}+\frac{4a^3y^2}{3}+2y^4a\right)\mathrm{d}y\\ &= \frac{4a^6}{5}+\frac{8a^6}{3.3}+\frac{4a^6}{5}=\frac{112}{45}a^6\end{aligned}$$

it is clear that ϕ^{A_1} is not a normalized function,

$$\langle\phi^{A_1}|\phi^{B_2}\rangle = \int_{-a}^{a}\int_{-a}^{a}(x^3y+xy^3)\,\mathrm{d}x\,\mathrm{d}y = 0$$

3.4 For example

$$\boldsymbol{\Gamma}^3(A) \otimes \boldsymbol{\Gamma}^4(A) = \begin{bmatrix} 0 & -1 & 0 & 0 \\ -1 & 0 & 0 & 0 \\ 0 & 0 & 0 & 1 \\ 0 & 0 & 1 & 0 \end{bmatrix}$$

$$\boldsymbol{\Gamma}^3(B) \otimes \boldsymbol{\Gamma}^4(B) = \tfrac{1}{4}\begin{bmatrix} \sqrt{3} & -1 & 3 & -\sqrt{3} \\ -1 & -\sqrt{3} & -\sqrt{3} & -3 \\ 3 & -\sqrt{3} & -\sqrt{3} & 1 \\ -\sqrt{3} & -3 & 1 & \sqrt{3} \end{bmatrix}$$

$$\boldsymbol{\Gamma}^3(F) \times \boldsymbol{\Gamma}^4(F) = \tfrac{1}{4}\begin{bmatrix} 1 & \sqrt{3} & \sqrt{3} & 3 \\ -\sqrt{3} & 1 & -3 & \sqrt{3} \\ -\sqrt{3} & -3 & 1 & \sqrt{3} \\ 3 & -\sqrt{3} & -\sqrt{3} & 1 \end{bmatrix}$$

and $\boldsymbol{\Gamma}^3(A) \otimes \boldsymbol{\Gamma}^4(A) \,.\, \boldsymbol{\Gamma}^3(B) \otimes \boldsymbol{\Gamma}^4(B) = \boldsymbol{\Gamma}^3(F) \otimes \boldsymbol{\Gamma}^4(F)$.

3.5 z is perpendicular to the plane. Using the notation for the elements as described in solution to 3.2 we can see that

$$\begin{aligned} P_E z &= z & P_{iE} z &= -z \\ P_A z &= z & P_{iA} z &= -z \\ P_B z &= z & P_{iB} z &= -z \\ P_C z &= z & P_{iC} z &= -z \\ P_D z &= -z & P_{iD} z &= z \\ P_F z &= -z & P_{iF} z &= z \\ P_G z &= -z & P_{iG} z &= z \\ P_H z &= -z & P_{iH} z &= z \end{aligned}$$

then

$$O^{A_{1g}}\, e^{iz} = \tfrac{1}{16}\{8\, e^{iz} + 8\, e^{-iz}\} = \frac{e^{iz} + e^{-iz}}{2}$$

$$= \cos z$$

$$O^{B_{2g}}\, e^{iz} = \tfrac{1}{16}\{8\, e^{iz} - 8\, e^{-iz}\} = \frac{e^{iz} - e^{-iz}}{2}$$

$$= i \sin z$$

All other operators on e^{iz} produce zero.

3.6 C_4 has elements

$$A = C_4, \quad B = C_4^2, \quad C = C_4^3, \quad E = C_4^4$$

	x	y	z
P_E	x	y	z
P_A	y	$-x$	z
P_B	$-x$	$-y$	z
P_C	$-y$	x	z

(a) $O^A z = \frac{1}{4}\{P_E + P_A + P_B + P_C\}z = z$

$\therefore$ z belongs to A_1

(b)

$$O^{E_1} = \tfrac{1}{4}\{P_E - P_B + iP_A - iP_C\}$$

$$O^{E_2} = \tfrac{1}{4}\{P_E - P_B - iP_A + iP_C\}$$

$$O^{E_1}x = \frac{x + iy}{2}$$

$$O^{E_2}x = \frac{x - iy}{2}$$

(c)

$$O^E = \tfrac{1}{4}\{2P_E - 2P_B\}$$

$$O^E x = x; \qquad O^E y = y.$$

(d) z^2 belongs to A
xy belongs to B
xz belongs to E
yz belongs to E

Chapter 4

4.1 $\mathbf{a}_2 \,.\, \mathbf{b}_1 = 0$ and $\mathbf{a}_3 \,.\, \mathbf{b}_1 = 0$, hence $\mathbf{b}_1$ must be perpendicular to *both* $\mathbf{a}_2$ and $\mathbf{a}_3$. Thus

$$\mathbf{b}_1 = A(\mathbf{a}_2 \times \mathbf{a}_3)$$

where A is to be determined. Now $\mathbf{a}_1 \,.\, \mathbf{b}_1 = 2\pi$, therefore

$$A\mathbf{a}_1 \,.\, (\mathbf{a}_2 \times \mathbf{a}_3) = 2\pi$$

Set up right-angled (x, y, z) axes at the corner of the cube. Then letting unit vectors be $\mathbf{i}, \mathbf{j}, \mathbf{k}$, the lattice site in the centre of the $y - z$ plane passing through origin is situated at $\mathbf{a}_1 = (a/2)(\mathbf{k} + \mathbf{j})$ and so on.

$$\mathbf{b}_1 = \frac{2\pi(\mathbf{a}_2 \times \mathbf{a}_3)}{\mathbf{a}_1 \,.\, (\mathbf{a}_2 \times \mathbf{a}_3)};$$

$$(\mathbf{a}_2 \times \mathbf{a}_3) = \frac{a^2}{4}(\mathbf{k} + \mathbf{j}) \times (\mathbf{j} - \mathbf{i})$$

$$= \frac{a^2}{4}(-\mathbf{i} - \mathbf{j} + \mathbf{k})$$

Therefore,

$$\mathbf{a}_1 \, . \, (\mathbf{a}_2 \times \mathbf{a}_3) = \frac{a^3}{8}(\mathbf{k} - \mathbf{i}) \, . \, (-\mathbf{i} - \mathbf{j} + \mathbf{k})$$

$$= \frac{a^3}{8}(1 + 1) = \frac{a^3}{4}$$

4.2 $\mathbf{R} = m_1\mathbf{a}_1 + m_2\mathbf{a}_2 + m_3\mathbf{a}_3$
Hence

$$R_x = m_1 a_{1x} + m_2 a_{2x} + m_3 a_{3x} \equiv m_1 a_{11} + m_2 a_{12} + m_3 a_{13}$$

$$R_y = m_1 a_{1y} + m_2 a_{2y} + m_3 a_{3y} \equiv m_1 a_{21} + m_2 a_{22} + m_3 a_{23}$$

$$R_z = m_1 a_{1z} + m_3 a_{2z} + m_3 a_{3z} \equiv m_1 a_{31} + m_2 a_{32} + m_3 a_{33}$$

Similarly for the vector **k** but using a row vector $[s_1 s_2 s_3]$. Taking, now,

$$\mathbf{a}_1 = \frac{a}{2}(\mathbf{k} + \mathbf{j}), \qquad \mathbf{a}_2 = \frac{a}{2}(\mathbf{i} + \mathbf{k}), \qquad \mathbf{a}_3 = \frac{a}{2}(\mathbf{i} + \mathbf{j})$$

we obtain

$$a_{1x} = 0, \qquad a_{1y} = \frac{a}{2}, \qquad a_{1z} = \frac{a}{2}$$

$$a_{2x} = \frac{a}{2}, \qquad a_{2y} = 0, \qquad a_{2z} = \frac{a}{2}$$

$$a_{3x} = \frac{a}{2}, \qquad a_{3y} = \frac{a}{2}, \qquad a_{3z} = 0$$

$$\mathbf{B}_1 = 2\pi \mathbf{A}_1^{-1}$$

$$|\mathbf{A}_1| = a^3/4$$

$$\mathbf{A}_1^{-1} = \frac{4}{a^3}\begin{bmatrix} -a^2/4 & a^2/4 & a^2/4 \\ a^2/4 & -a^2/4 & a^2/4 \\ a^2/4 & a^2/4 & -a^2/4 \end{bmatrix}$$

4.3 The only symmetry elements in one dimension are identity and reflection through the origin, i.e., inversion. The elements of the point group are (E, i) and the elements of the translation group are (E, T, T^2, T^3). The space group elements are therefore

$$(E, T, T^2, T^3, i, iT, iT^2, iT^3)$$

4.4 The sets of symmetry operations which must of course be combined with $(E, T, T^2, T^3, \ldots)$ to form space groups are (N.B. $C_2 \equiv i$)

$$(E), \qquad (E, i), \qquad (E, \sigma_h), \qquad (E, \sigma_v), \qquad (E, i, \sigma_v, \sigma_h)$$

involving 1, 2, 2, 2, and 4 motif units, respectively, at each lattice site.
The groups involving glides use 1 and 2 motif units, respectively, at each lattice site.

4.5 The symmetry operations are

$$E, \sigma\tau, \sigma^2\tau^2, \sigma^3\tau^3, \ldots$$

and

$$\tau^2 = T, \qquad \sigma^2 = E.$$

4.6 Consider first the rotational symmetry. From Fig. 4.6 it can be seen that the base of the triclinic, i.e., an arbitrary parallelogram, can act as a framework for one- and two-fold rotational symmetry. Similarly the $2\pi/3$ rhombus of the hexagonal cell supports three- and six-fold rotational symmetry. The square base of the tetragonal and cubic cells supports four-fold rotational symmetry. Thus on the grounds of rotational symmetry we have monoclinic, hexagonal, and tetragonal lattices. The introduction of vertical mirror planes to the lattice with three-, four-, and six-fold symmetry produces no new lattices.

By combining vertical mirror planes and two-fold rotational symmetry we can produce two lattices, one with a rectangular cell and one whose cell is a diamond shaped rhombus. The point group symmetry elements (which need, of course, to be combined with the lattice translations) of the thirteen space-groups are, taking z as 0 vertical and x and y parallel to the primitive translations

Monoclinic: (E); $(E, C_2^{(z)})$;
Orthorhombic (rectangular cell): $(E, C_2^{(x)})$; $(E, C_2^{(x)}, C_2^{(y)}, C_2^{(z)})$;
Orthorhombic (diamond rhombus): $(E, C_2^{(x)})$; $(E, C_2^{(x)}, C_2^{(y)}, C_2^{(z)})$;

(Consider this as a centred orthorhombic lattice)

($C_2^{(x,y,z)}$ are two-fold rotations about x, y, or z)

$$\text{Tetragonal}: (E, 2C_4, C_2); (E, 2C_4, C_2, 2C_2', 2C_2'')$$

(C_4 is about the z axis, $C_2 = C_4^2$; C_2' is about x or y and C_2'' is about a diagonal axis in the plane of the lattice)

$$\begin{aligned}\text{Hexagonal}: &(E, 2C_3); (E, 2C_3, 3C_2')\\ &(E, 2C_6, 2C_3, C_2); (E, 2C_3, 3C_2'')\\ &(E, 2C_6, 3C_2', 2C_3, C_2, 3C_2'')\end{aligned}$$

(C_6 is about the z axis; $C_3 = C_6^2$; $C_2 = C_6^3$; C_2' and C_2'' are rotations about axes in the plane of the lattice.)

4.7 The answer for the first part is the same as for 4.6. We may use the following result. Set up Cartesian axes (x, y) and consider a line passing through the origin O making an angle θ with the y axis. Rotate the line through an angle 2θ and then translate it along the x axis through a distance b. The rotation of 2θ about O is then equivalent to a rotation of 2θ about the point of the intersection of the original line and the line in its final position. Consider a rhombus of side b which, of course, is two equilateral triangles joined together. Examine the lines of length b which make an angle of 120° to each other. Next consider a line making an angle of 30° to $\mathbf{a}_1$ (which is horizontal say). Then rotate this selected line through 120° and translate it through a distance b. The intersection point (i.e., the position of an equivalent three-fold rotation axis) is then at the centre of the equilateral triangle of side b. The other position can be found in a similar way. The point of intersection is on the perpendicular bisector of b at a height $b/2 \cot \theta$ above the x axis. If a three-fold rotation axis passes through one corner then the additional three-fold rotation axes can be found as follows:

$$2\theta = \frac{2\pi}{3}$$

Therefore $\theta = 60°$, $\cot 60° = 1/\sqrt{3}$ so that the rotation axes pass through points on the bisectors of the triangles at heights $b/2\sqrt{3}$ above their bases.

Hence, for this non-orthogonal basis, the additional rotation axes pass through

$$\left(\frac{\mathbf{a}_1}{3}, \frac{2}{3}\mathbf{a}_2\right) \quad \text{and} \quad \left(\frac{2}{3}\mathbf{a}_1, \frac{\mathbf{a}_2}{3}\right)$$

A similar procedure to the above can be applied to the six-fold rotation axis.

Chapter 5

5.1 The vector product $\mathbf{C} = \mathbf{A} \times \mathbf{B}$ has components

$$C_1 = A_2B_3 - A_3B_2$$
$$C_2 = A_3B_1 - A_1B_3$$
$$C_3 = A_1B_2 - A_2B_1$$

C_1 for example can be written as

$$\begin{aligned} C_1 &= \varepsilon_{1jk}A_jB_k \\ &= \varepsilon_{111}A_1B_1 + \varepsilon_{112}A_1B_2 + \varepsilon_{113}A_1B_3 + \varepsilon_{121}A_2B_1 \\ &\quad + \varepsilon_{122}A_2B_2 + \varepsilon_{123}A_2B_3 + \varepsilon_{131}A_3B_1 + \varepsilon_{132}A_3B_2 + \varepsilon_{133}A_3B_3 \end{aligned}$$

where

$$\varepsilon_{111} = \varepsilon_{112} = \varepsilon_{113} = \varepsilon_{121} = \varepsilon_{122} = \varepsilon_{131} = \varepsilon_{133} = 0, \qquad \varepsilon_{123} = 1;$$
$$\varepsilon_{132} = -1$$

The determinant of a is

$$\begin{aligned} |a| &= a_{11}\{a_{22}a_{33} - a_{32}a_{23}\} - a_{21}\{a_{12}a_{33} - a_{32}a_{13}\} + a_{31}\{a_{12}a_{23} - a_{22}a_{13}\} \\ &= \varepsilon_{ijk}a_{i1}a_{j2}a_{k3} \end{aligned}$$

Thus,

$$|a|\varepsilon_{pqr} = \varepsilon_{ijk}a_{pi}a_{qj}a_{rk}$$

Now

$$C'_i = \varepsilon_{ijk}A'_jB'_k$$
$$A'_j = a_{js}A_s$$
$$B'_k = a_{kt}B_t$$

So that

$$C'_i = \varepsilon_{ijk}a_{js}A_sa_{kt}B_t$$

and

$$\begin{aligned} \varepsilon_{ijk}a_{js}a_{kt} &= |a|\,(a_{pi}^{-1})\varepsilon_{pst} \\ &= |a|a_{ip}\varepsilon_{pst} \end{aligned}$$

Therefore

$$C'_i = |a|a_{ip}\varepsilon_{pst}A_sB_t = |a|a_{ip}C_p$$

5.2 Consider the vector p_i. If the crystal possesses a centre of symmetry then

$$\begin{bmatrix} p_1 \\ p_2 \\ p_3 \end{bmatrix} = \begin{bmatrix} -1 & 0 & 0 \\ 0 & -1 & 0 \\ 0 & 0 & -1 \end{bmatrix} \begin{bmatrix} p_1 \\ p_2 \\ p_3 \end{bmatrix} \qquad \text{i.e., } p_i = a_{ik}p_k$$

Hence

$$p_1 = -p_1, \qquad p_2 = -p_2, \qquad p_3 = -p_3$$

Therefore

$$\mathbf{p} = 0$$

For cubic symmetry, **A** for a four-fold rotation about an x axis is

$$\begin{bmatrix} 1 & 0 & 0 \\ 0 & 0 & 1 \\ 0 & -1 & 0 \end{bmatrix}$$

so that

$$p_1 = p_1, \qquad p_2 = p_3, \qquad p_3 = -p_2$$

Also a three-fold rotation about a [111] axis gives

$$\begin{bmatrix} p_1 \\ p_2 \\ p_3 \end{bmatrix} = \begin{bmatrix} 0 & 1 & 0 \\ 0 & 0 & 1 \\ 1 & 0 & 0 \end{bmatrix} \begin{bmatrix} p_1 \\ p_2 \\ p_3 \end{bmatrix}$$

giving $p_1 = p_2 = p_3$. Hence $\mathbf{p} = 0$. The crystal system for which $p_1 \neq 0$, $p_2 \neq 0$, $p_3 \neq 0$ is triclinic.

5.3 Example: for a three-fold rotation about the z axis $\psi = 120°$, $l = 0$, $m = 0$, $n = 1$, $\cos\psi = -\frac{1}{2}$ and $\sin\psi = \sqrt{3}/2$ so that

$$\mathbf{A} = \begin{bmatrix} -1/2 & \sqrt{3}/2 & 0 \\ -\sqrt{3}/2 & -1/2 & 0 \\ 0 & 0 & 1 \end{bmatrix}$$

5.4 Using the symmetry operations of C_{3v} we obtain three-fold rotation about the z axis:

$$\mathbf{A}_1 = \begin{bmatrix} -1/2 & \sqrt{3}/2 & 0 \\ -\sqrt{3}/2 & -1/2 & 0 \\ 0 & 0 & 1 \end{bmatrix}$$

Reflection in the $x - z$ plane

$$\mathbf{A}_2 = \begin{bmatrix} 1 & 0 & 0 \\ 0 & -1 & 0 \\ 0 & 0 & 1 \end{bmatrix}$$

let B_{im} be the second rank tensor and use

$$B_{im} = a_{ij} a_{mk} B_{jk}$$

Then, using $\mathbf{A}_2$

$$B_{21} = a_{2j} a_{1k} B_{jk}$$

i.e., $B_{21} = 0$

$$B_{12} = a_{1j} a_{2k} B_{jk} = -B_{12}$$

i.e., $B_{12} = 0$. Using $\mathbf{A}_1$ we obtain

$$\begin{aligned} B_{11} &= a_{1j}a_{1k}A_{jk} \\ &= \frac{1}{4}B_{11} - \frac{\sqrt{3}}{4}B_{12} - \frac{\sqrt{3}}{4}B_{21} + \frac{3}{4}B_{22} \end{aligned}$$

i.e., $A_{11} = A_{22}$. Thus in this way it can be shown that

$$\mathbf{B} = \begin{bmatrix} B_{11} & 0 & 0 \\ 0 & B_{11} & 0 \\ 0 & 0 & B_{33} \end{bmatrix}$$

5.5 Consider a tensor of arbitrary rank, i.e., $T_{ijkl}\ldots$ then if the crystal possesses a centre of symmetry $a_{ij} = -\delta_{ij}$ so that

$$T_{ijkl}\ldots = (-1)^n \delta_{il}\delta_{jp}\delta_{ks}\delta_{lt}\ldots T_{lpst}$$

where n is the rank of the tensor, i.e., the number of suffices. Obviously

$$T_{ijkl}\ldots = (-1)^n T_{ijkl}\ldots.$$

Hence if n is odd all components of the tensor vanish. We imagine an experimental set-up in which a stress is being applied to a crystal. We now perform an inversion by actually reconstructing the apparatus. Now the second-rank stress tensor is symmetrical so even after re-building the apparatus, physically, we are still applying the same stress. However, the sign of the polarization must now be in the opposite direction. This is physically impossible unless the polarization is zero.

Chapter 6

6.1 We use eq. (4.17), i.e.,

$$\{\boldsymbol{\alpha}|\mathbf{t}\}^{-1} = \{\boldsymbol{\alpha}^{-1}| -\boldsymbol{\alpha}^{-1}\mathbf{t}\}$$

so that

$$\begin{aligned} \{\boldsymbol{\alpha}|\mathbf{t}\}^{-1}\{E|\mathbf{R}\}\{\boldsymbol{\alpha}|\mathbf{t}\} &= \{\boldsymbol{\alpha}^{-1}|\boldsymbol{\alpha}^{-1}\mathbf{R} - \boldsymbol{\alpha}^{-1}\mathbf{t}\}\{\boldsymbol{\alpha}|\mathbf{t}\} \\ &= \{\boldsymbol{\alpha}^{-1}\boldsymbol{\alpha}|\boldsymbol{\alpha}^{-1}\mathbf{t} + \boldsymbol{\alpha}^{-1}\mathbf{R} - \boldsymbol{\alpha}^{-1}\mathbf{t}\} \\ &= \{E|\boldsymbol{\alpha}^{-1}\mathbf{R}\} = \{E|\mathbf{R}'\} \end{aligned}$$

Hence

$$\langle\kappa|\mathbf{R}'\rangle = \langle\kappa|\boldsymbol{\alpha}^{-1}|\mathbf{R}\rangle = \langle\kappa'|\mathbf{R}\rangle$$

6.2 If H is the Hamiltonian then

$$H\{\boldsymbol{\alpha}|\mathbf{t}\}\psi_{n,\kappa} = E(\kappa)\{\boldsymbol{\alpha}|\mathbf{t}\}\psi_{n,\kappa}$$

Also by Bloch's theorem and the result of 6.1

$$\{E|\mathbf{R}\}\{\boldsymbol{\alpha}|\mathbf{t}\}\psi_{n,\kappa} = e^{i\kappa'.\mathbf{R}}\{\boldsymbol{\alpha}|\mathbf{t}\}\psi_{n,\kappa}$$

so that

$$(\boldsymbol{\alpha}|\mathbf{t}\}\psi_{n,\kappa} = \psi_{n,\kappa'}$$

Hence, since

$$H\psi_{n,\kappa'} = E(\kappa')\psi_{n,\kappa'}$$

we have

$$E(\kappa') = E(\kappa)$$

6.3 We observe first that from Bloch's theorem

$$\{E|\mathbf{R}\}\psi_{n,\kappa} = e^{i\kappa.\mathbf{R}}\psi_{n,\kappa}$$
$$\{E|\mathbf{R}\}\psi^*_{n,\kappa} = e^{-i\kappa.\mathbf{R}}\psi^*_{n,\kappa}$$

Hence

$$\psi^*_{n,\kappa} = \psi_{n,-\kappa}$$

Now since

$$H\psi_{n,\kappa} = E_n(\kappa)\psi_{n,\kappa}$$
$$H\psi^*_{n,\kappa} = E_n(\kappa)\psi^*_{n,\kappa}$$

we have

$$H\psi_{n,-\kappa} = E_n(-\kappa)\psi_{n,-\kappa}$$

Therefore

$$E_n(\kappa) = E_n(-\kappa)$$

However, from

$$H\psi = -ih\frac{\partial\psi}{\partial t}$$

we can see that ψ^* satisfies

$$H\psi^* = ih\frac{\partial\psi^*}{\partial t}$$

This last step is equivalent to reversing t. It is therefore correct to say that if $E_n(\kappa) = E_n(-\kappa)$ the bands possess time-inversion symmetry.

6.4 Consider the (001) vector and use

$$\psi = e^{(2\pi i/a)(\alpha x + z)}$$

$$\Delta_1: \quad \phi_1 = \tfrac{1}{8}\sum_R \chi^i P_R(e^{2\pi i z/a})\, e^{(2\pi i/a)\alpha x}$$
$$= \tfrac{1}{8}\{e^{2\pi i z/a} + e^{-2\pi i z/a} + e^{2\pi i y/a} + e^{-2\pi i y/a}$$
$$+ e^{2\pi i z/a} + e^{-2\pi i z/a} + e^{2\pi i y/a} + e^{-2\pi i y/a}\}$$
$$= \frac{1}{2}\left\{\cos\left(\frac{2\pi z}{a}\right) + \cos\left(\frac{2\pi y}{a}\right)\right\} e^{2\pi i\alpha x/a}$$

$$\Delta_5: \quad \phi_5 = \tfrac{1}{8} e^{2\pi i\alpha x/a}\{2\,e^{2\pi i z/a} - 2\,e^{-2\pi i z/a}\}$$
$$= i\, e^{2\pi i\alpha x/a}\sin\left(\frac{2\pi z}{a}\right)$$

Try also

$$F = e^{2\pi i\alpha x/a}\, e^{2\pi i y/a}\, 2e^{-2\pi i y/a}\}$$

leading to

$$\phi_5 = \tfrac{1}{4} e^{2\pi i\alpha x/a}\{2\,e^{2\pi i y/a} - 2e^{-2\pi i y/a}\}$$

For Γ_1 try $F = e^{2\pi i x/a}$ and use table 6.2.

6.5 The first two normalized reciprocal lattice vectors are $(p, q, r) = (0, 0, 0)$ and $(p, q, r) = (\pm 1, \pm 1, \pm 1)$. Hence we arrive at (for the direction α)

$$\xi_1 = \alpha^2$$
$$\xi_2 = (\alpha + 1)^2 + 2$$
$$\xi_3 = (\alpha - 1)^2 + 2$$

and so on. The state functions are

$$\psi = e^{2\pi i/a((\alpha \pm 1)x \pm y \pm z)}$$

so that the state $\Gamma = e^{2\pi i/a(\pm x \pm y \pm z)}$ at Γ is eight-fold degenerate. These eight functions can be used as basis functions to construct a reducible representation $\Gamma(R)$. The character of $\Gamma(R)$ can be determined as explained in section 6.6. We use the basis functions

$$e^{B(x+y+z)}, \quad e^{B(-x+y+z)}, \quad e^{B(x-y+z)}, \quad e^{B(-x-y+z)},$$
$$e^{B(x+y-z)}, \quad e^{B(x-y-z)}, \quad e^{B(-x-y-z)}, \quad e^{B(-x+y-z)}$$

where $B = 2\pi i/a$. All we have to do to find the traces of the representation matrices $\Gamma(R)$ is to look for non-zero diagonal elements appearing for a selected group operation (remember that the representation matrices of all elements in a class have the *same* trace). Using table 6.2 we find that the trace of $\Gamma(E)$ is 8. For C_4^2 we select the rotation of π about the z axis and we find that *none* of the eight basis functions remains unchanged. Hence $\Gamma(C_4^2)$ cannot possess other than zero diagonal elements. Its trace is therefore zero. For C_3 we select the first operation in table 6.2 which sends (xyz) to (zxy) and we find that

$$e^{B(x+y+z)} \quad \text{and} \quad e^{B(-x-y-z)}$$

remain unchanged. The trace of $\Gamma(C_3)$ is therefore 2. The character of $\Gamma(R)$ is therefore the set

$$8, 0, 0, 0, 2, 0, 0, 0, 4, 0$$

arranged in order of the classes in table 6.2. Now use

$$a_j = \frac{1}{g} \sum_R \chi^j(R) \chi^{\text{red}}(R)$$

to show that

$$a_1 = \tfrac{1}{48}(8 + 0 + 0 + 0 + 16 + 0 + 0 + 0 + 24 + 0) = 1$$
$$a_2' = \tfrac{1}{48}(8 + 0 + 0 + 0 + 16 + 0 + 0 + 0 + 24 + 0) = 1$$
$$a_{15} = \tfrac{1}{48}(24 + 0 + 0 + 0 + 0 + 0 + 0 + 0 + 24 + 0) = 1$$
$$a_{25'} = \tfrac{1}{48}(24 + 0 + 0 + 0 + 0 + 0 + 0 + 0 + 24 + 0) = 1$$

The point $(p, q, r) = (0, 1, \frac{1}{2})$ corresponds to a vector $(0, 2\pi/a, \pi/a)$. The Brillouin zone has the shape of Fig. 6.1. If we call the point $(0, 1, \frac{1}{2})$ W then there are four *equivalent* W points namely $W_1 = (0, 1, \frac{1}{2})$, $W_2 = (0, -1, \frac{1}{2})$, $W_3 = (-1, 0, -\frac{1}{2})$, $W_4 = (1, 0, -\frac{1}{2})$: all these points are equivalent because they are separated by a reciprocal lattice vector. Set up co-ordinate axes (k_x, k_y, k_z) and show that the operations E, C_4^2, C_2, JC_4^2, JC_4 transform W points into each other.

There are five irreducible representations because there are only five classes of eq. (2.31). From eq. (2.23) the dimensions of the irreducible representations are 1, 1, 1, 1, and 2.

6.6 Set up co-ordinate axes (k_x, k_y, k_z). The Σ axis is the (110) axis. Using table 6.2 we find the operations which leave the Σ axis invariant are

$$E: xyz \rightarrow xyz$$

$$C_2: xyz \rightarrow yx-z$$

$$JC_4^2: xyz \rightarrow -x-yz$$

followed by

$$-x-yz \rightarrow xy-z$$

$$JC_2: xyz \rightarrow -y-x-z$$

followed by

$$-y-x-z \rightarrow yxz.$$

Check from character tables in Appendix C that these symmetry operations are $E, C_2, \sigma'_v, \sigma''_v$. The energy relationship for the Γ–M direction is, setting $\alpha = \beta$,

$$\xi = (\alpha + p)^2 + (\alpha + q)^2 + r^2$$

Hence we obtain:

$$(pqr) = (000); \qquad \xi = 2\alpha^2$$

(pqr)	ξ
(100) (010)	$(\alpha + 1)^2 + \alpha^2$
(001) (00 – 1)	$2\alpha^2 + 1$
(– 100) (0 – 10)	$(\alpha - 1)^2 + \alpha^2$

Therefore from Γ, for $\xi = 1$, three bands emerge. At M the band $\xi = 2\alpha^2$ gives $\xi = 0{\cdot}5$. This merges with the band $\xi = (\alpha - 1)^2 + \alpha^2$ and the band $\xi = 2(\alpha - 1)^2$ formed with $(pqr) = (-1-10)$. The group of $\mathbf{k}$ at M contains the symmetry operations, E, $2C_4^2$, C_4^2, $2C_4$, $2C_2$, J, $2JC_4^2$, JC_4^2, $2JC_4$, $2JC_2$ which is identical to the group of $\mathbf{k}$ at X. The group of $\mathbf{k}$ at M therefore has the character table of the group D_{4h}. The nature of the levels can now be determined.

Chapter 7

7.1

$$T = \tfrac{1}{2}\sum_i \dot{q}_i^2; \qquad V = \tfrac{1}{2}\sum_i \lambda_i q_i^2$$

$$V = \frac{k}{2}(x_2 - x_1)^2$$

$$T = \frac{m_1}{2}(\dot{x}_1^2 + \dot{y}_1^2 + \dot{z}_1^2) + \frac{m_2}{2}(\dot{x}_2^2 + \dot{y}_2^2 + \dot{z}_2^2)$$

Therefore,

$$V = \frac{1}{2}\left(\frac{k}{\mu}\right)\frac{m_1 m_2}{m_1 + m_2}(x_2 - x_1)^2$$
$$= \frac{k}{2}(x_2 - x_1)^2$$

and

$$T = \frac{1}{2}\left\{\frac{m_1 m_2}{m_1 + m_2}(\dot{x}_1 - \dot{x}_2)^2 + \frac{1}{m_1 + m_2}(m_1\dot{x}_1 + m_2\dot{x}_2)^2 + m_1\dot{y}_1^2 + m_2\dot{y}_2^2 + m_1\dot{z}_1^2 + m_2\dot{z}_2^2\right\}$$
$$= \tfrac{1}{2}(m_1\dot{x}_1^2 + m_2\dot{x}_2^2 + m_1\dot{y}_1^2 + m_2\dot{y}_2^2 + m_1\dot{z}_1^2 + m_2\dot{z}_2^2)$$

7.2 For an equilateral triangle $\phi = 30°$

$$s = \tfrac{1}{2} \qquad c = \sqrt{3}/2 \qquad \kappa = 1$$

so that, ignoring the x-coordinates, we have

$$\mathbf{K} = \frac{k}{4}\begin{bmatrix} 2 & -1 & -1 & 0 & -\sqrt{3} & \sqrt{3} \\ -1 & 5 & -4 & -\sqrt{3} & \sqrt{3} & 0 \\ -1 & -4 & 5 & \sqrt{3} & 0 & -\sqrt{3} \\ 0 & -\sqrt{3} & \sqrt{3} & 6 & -3 & -3 \\ -\sqrt{3} & \sqrt{3} & 0 & -3 & 3 & 0 \\ \sqrt{3} & 0 & -\sqrt{3} & -3 & 0 & 3 \end{bmatrix}$$

and

$$\mathbf{M} = m\mathbf{I} \quad \text{(where } \mathbf{I} \text{ is of order 6)}$$

Apart from the factor $k/4m$ the solution of

$$|K - \lambda M| = 0$$

is given by

$$\begin{vmatrix} 2-\lambda & -1 & -1 & 0 & -\sqrt{3} & \sqrt{3} \\ -1 & 5-\lambda & -4 & -\sqrt{3} & \sqrt{3} & 0 \\ -1 & -4 & 5-\lambda & \sqrt{3} & 0 & -\sqrt{3} \\ 0 & -\sqrt{3} & \sqrt{3} & 6-\lambda & -3 & -3 \\ -\sqrt{3} & \sqrt{3} & 0 & -3 & 3-\lambda & 0 \\ \sqrt{3} & 0 & -\sqrt{3} & -3 & 0 & 3-\lambda \end{vmatrix} = 0$$

Let us denote the ith row by r_i and the ith column by c_i. Then if we perform

$$r_1 + r_2 + r_3 \quad \text{and} \quad r_4 + r_5 + r_6$$

we obtain

$$\lambda^2 \begin{vmatrix} 1 & 1 & 1 & 0 & 0 & 0 \\ -1 & 5-\lambda & -4 & -\sqrt{3} & \sqrt{3} & 0 \\ -1 & -4 & 5-\lambda & \sqrt{3} & 0 & -\sqrt{3} \\ 0 & 0 & 0 & 1 & 1 & 1 \\ -\sqrt{3} & \sqrt{3} & 0 & -3 & 3-\lambda & 0 \\ \sqrt{3} & 0 & -\sqrt{3} & -3 & 0 & 3-\lambda \end{vmatrix} = 0$$

Now perform $c_2 - c_1, c_3 - c_1, c_5 - c_4, c_6 - c_4$ and expand the determinant about the first row and the fourth row to obtain:

$$\lambda^2 \begin{vmatrix} 6-\lambda & -3 & 2\sqrt{3} & \sqrt{3} \\ -3 & 6-\lambda & -\sqrt{3} & -2\sqrt{3} \\ 2\sqrt{3} & \sqrt{3} & 6-\lambda & 3 \\ -\sqrt{3} & -2\sqrt{3} & 3 & 6-\lambda \end{vmatrix} = 0$$

Now perform $r_3/\sqrt{3}$, $r_4/\sqrt{3}$, $c_3/\sqrt{3}$, and $c_4/\sqrt{3}$ in succession to obtain

$$\lambda^2 \begin{vmatrix} 6-\lambda & -3 & 2 & 1 \\ -3 & 6-\lambda & -1 & -2 \\ 2 & 1 & 2-\lambda/3 & 1 \\ -1 & -2 & 1 & 2-\lambda/3 \end{vmatrix} = 0$$

Follow this by $r_1 + r_2 - r_3 + r_4$:

$$\lambda^3 \begin{vmatrix} 1 & 1 & -\frac{1}{3} & \frac{1}{3} \\ -3 & 6-\lambda & -1 & -2 \\ 2 & 1 & 2-\lambda/3 & 1 \\ -1 & -2 & 1 & 2-\lambda/3 \end{vmatrix} = 0$$

and by $r_2 + 3r_1, r_3 - 2r_1, r_4 + r_1$

$$\lambda^3 \begin{vmatrix} 9-\lambda & -2 & -1 \\ -1 & \frac{1}{3}(8-\lambda) & 1/3 \\ -1 & 2/3 & \frac{1}{3}(7-\lambda) \end{vmatrix} = 0$$

$r_3 - r_2$ then gives

$$\lambda^3 \frac{(6-\lambda)}{3} \begin{vmatrix} 9-\lambda & -2 & -1 \\ -1 & \frac{1}{3}(8-\lambda) & 1/3 \\ 0 & -1 & 1 \end{vmatrix} = 0$$

and finally $c_2 + c_3$ gives

$$\lambda^3 \frac{(6-\lambda)}{3} \begin{vmatrix} 9-\lambda & -3 & -1 \\ -1 & \frac{1}{3}(9-\lambda) & 1/3 \\ 0 & 0 & 1 \end{vmatrix}$$

which when expanded about the third row gives

$$\lambda^3 \frac{(6-\lambda)}{3}\left\{(9-\lambda)\frac{(9-\lambda)}{3} - 3\right\} = 0$$

hence $\lambda = 0, 0, 0, 6$ or the roots of

$$(9-\lambda)^2 = 9$$

$$\lambda = 6, 12$$

The actual roots, allowing for the factor $k/4m$ are

$$\lambda = 0, 0, 0, \frac{3k}{2m}, \frac{3k}{2m}, \frac{3k}{m},$$

in agreement with eqs. (7.45) and (7.46) when we substitute

$$a = \frac{1}{2}, c = \frac{\sqrt{3}}{2}, M = m, \kappa = 1, k_1 = k.$$

7.3 *The symmetric basis*
The group character table is

	E	i
A_1	1	1
A_2	1	−1

and if we consider motion in the x-direction only, the characters of $\Gamma^c\Gamma^t\Gamma^r\Gamma^v$ are

Γ^c	2	0
Γ^t	1	−1
Γ^r	0	0
Γ^v	1	1

Hence

$$\Gamma^{v(\text{red})}(R) = A_1$$

$$\Gamma^{t(\text{red})}(R) = A_2$$

The projection operators yield the following independent vectors:

$$O^{A_1}\mathbf{x}_1 = \mathbf{x}_1 - \mathbf{x}_2 = \mathbf{s}_1$$

$$O^{A_2}\mathbf{x}_1 = \mathbf{x}_1 + \mathbf{x}_2 = \mathbf{s}_2$$

Hence

$$\boldsymbol{\beta}^\dagger = \begin{bmatrix} 1 & -1 \\ 1 & 1 \end{bmatrix}$$

The eigenvalues
Now

$$T = \tfrac{1}{2}m_1\dot{x}_1^2 + \tfrac{1}{2}m_2\dot{x}_2^2$$

$$V = \tfrac{1}{2}k(x_1 - x_2)^2$$

$$\therefore \quad \mathbf{M} = \begin{bmatrix} m_1 & 0 \\ 0 & m_2 \end{bmatrix}, \quad \mathbf{K} = k\begin{bmatrix} 1 & -1 \\ -1 & 1 \end{bmatrix}$$

We then find

$$\boldsymbol{\beta}^{\dagger}\mathbf{M}\boldsymbol{\beta} = \begin{bmatrix} m_1 + m_2 & m_1 - m_2 \\ m_1 - m_2 & m_1 + m_2 \end{bmatrix}$$

$$\boldsymbol{\beta}^{\dagger}\mathbf{K}\boldsymbol{\beta} = \begin{bmatrix} 4k & 0 \\ 0 & 0 \end{bmatrix}$$

The solution of $|K - \lambda M| = 0$ is then the solution of

$$\begin{vmatrix} 4k - \lambda(m_1 + m_2) & -\lambda(m_1 - m_2) \\ -\lambda(m_1 - m_2) & -\lambda(m_1 + m_2) \end{vmatrix} = 0$$

The solution of the above determinant is

$$\lambda = 0, \qquad \lambda = k\frac{(m_1 + m_2)}{m_1 m_2}$$

in agreement with eq. (7.34).

The normal basis

To show that if $m_1 = m_2$ then the symmetrized basis is also the normal basis we can carry out the exact diagonalization of **M** and **K**, as shown in appendix A.20, to obtain

$$\tilde{\boldsymbol{\alpha}}\mathbf{M}\boldsymbol{\alpha} = \mathbf{I}$$

$$\tilde{\boldsymbol{\alpha}}\mathbf{K}\boldsymbol{\alpha} = \text{diag}\left\{0, k\left(\frac{1}{m_1} + \frac{1}{m_2}\right)\right\}$$

where

$$\boldsymbol{\alpha} = \frac{1}{\sqrt{m_1 + m_2}}\begin{bmatrix} 1 & \sqrt{m_2/m_1} \\ 1 & -\sqrt{m_1/m_2} \end{bmatrix}$$

If $m_1 = m_2 = m$

$$\boldsymbol{\alpha} = \frac{1}{\sqrt{2m}}\begin{bmatrix} 1 & 1 \\ 1 & -1 \end{bmatrix}$$

and the normal basis, given by eq. (7.9), is

$$\begin{bmatrix} \mathbf{q}_1 \\ \mathbf{q}_2 \end{bmatrix} = \boldsymbol{\alpha}\begin{bmatrix} \mathbf{x}_1 \\ \mathbf{x}_2 \end{bmatrix} = \frac{1}{\sqrt{2m}}\begin{bmatrix} 1 & 1 \\ 1 & -1 \end{bmatrix}\begin{bmatrix} \mathbf{x}_1 \\ \mathbf{x}_2 \end{bmatrix}$$

Hence

$$\mathbf{q}_1 = \frac{1}{\sqrt{2m}}(\mathbf{x}_1 + \mathbf{x}_2) = \mathbf{s}_2/\sqrt{2m}$$

$$\mathbf{q}_2 = \frac{1}{\sqrt{2m}}(\mathbf{x}_1 - \mathbf{x}_2) = \mathbf{s}_1/\sqrt{2m}$$

7.4 The given solutions can be obtained by the following operations:

$$\mathbf{s}_1 = O^{A_1}\mathbf{y}_1$$
$$\mathbf{s}_2 = O^{A_2}\mathbf{y}_2$$
$$\mathbf{s}_3 = O^{A_2}\mathbf{z}_1$$
$$\mathbf{s}_4 = O^{E}(\mathbf{x}_1 + \mathbf{x}_2 + \mathbf{x}_3)$$
$$\mathbf{s}_5 = O^{E}(\mathbf{y}_1 + \mathbf{y}_2 + \mathbf{y}_3)$$
$$\mathbf{s}_6 = O^{E}\mathbf{z}_1$$
$$\mathbf{s}_7 = O^{E}(\mathbf{z}_3 - \mathbf{z}_2)$$
$$\mathbf{s}_8 = O^{E}\mathbf{y}_1 - \mathbf{s}_5$$
$$\mathbf{s}_9 = O^{E}\mathbf{x}_1 - \mathbf{s}_4$$

The symmetric basis can be written in terms of the Cartesian basis as

$$\mathbf{s} = (\boldsymbol{\beta}^{+})\mathbf{x}$$

where

$$\boldsymbol{\beta}^{\dagger} = \frac{1}{2}\begin{bmatrix} 0 & -\sqrt{3} & \sqrt{3} & 2 & -1 & -1 & 0 & 0 & 0 \\ 2 & -1 & -1 & 0 & \sqrt{3} & -\sqrt{3} & 0 & 0 & 0 \\ 0 & 0 & 0 & 0 & 0 & 0 & 2 & 2 & 2 \\ 2 & 2 & 2 & 0 & 0 & 0 & 0 & 0 & 0 \\ 0 & 0 & 0 & 2 & 2 & 2 & 0 & 0 & 0 \\ 0 & 0 & 0 & 0 & 0 & 0 & 4 & -2 & -2 \\ 0 & 0 & 0 & 0 & 0 & 0 & 0 & -2 & 2 \\ 0 & \sqrt{3} & -\sqrt{3} & 2 & -1 & -1 & 0 & 0 & 0 \\ 2 & -1 & -1 & 0 & -\sqrt{3} & \sqrt{3} & 0 & 0 & 0 \end{bmatrix}$$

The **K**-matrix of eq. (7.42) must be rearranged, (because in this problem the principal axis z is perpendicular to the plane of the atoms), by changing x to z, y to x, and z to y. We then find that

$$\boldsymbol{\beta}^{+}\mathbf{K}\boldsymbol{\beta} = \text{diag}\,\{9k, 0, 0, 0, 0, 0, 0, 9k/2, 9k/2\}$$

and

$$\boldsymbol{\beta}^{+}\mathbf{M}\boldsymbol{\beta} = \text{diag}\,\{3m, 3m, 3m, 3m, 3m, 6m, 2m, 3m, 3m\}$$

7.5

$$2T = m(2\dot{x}_1^2 + \dot{x}_2^2 + \dot{x}_3^2)$$

$$\begin{aligned} 2V &= k(x_2 - x_1)^2 + k(x_3 - x_1)^2 \\ &= k(2x_1^2 + x_2^2 + x_3^2 - 2x_1x_2 - 2x_3x_1) \end{aligned}$$

Hence

$$\mathbf{M} = m\begin{bmatrix} 2 & 0 & 0 \\ 0 & 1 & 0 \\ 0 & 0 & 1 \end{bmatrix}, \qquad \mathbf{K} = k\begin{bmatrix} 2 & -1 & -1 \\ -1 & 1 & 0 \\ -1 & 0 & 1 \end{bmatrix}$$

From the methods of appendix A.20 we have

$$Q = \sqrt{\frac{1}{m}}\begin{bmatrix} 1/\sqrt{2} & 0 & 0 \\ 0 & 1 & 0 \\ 0 & 0 & 1 \end{bmatrix}$$

and $\tilde{\mathbf{Q}}\mathbf{M}\mathbf{Q} = \mathbf{I}$. Now

$$\mathbf{K}' = \tilde{\mathbf{Q}}\mathbf{K}\mathbf{Q} = \frac{k}{m}\begin{bmatrix} 1 & -1/\sqrt{2} & -1/\sqrt{2} \\ -1/\sqrt{2} & 1 & 0 \\ -1/\sqrt{2} & 0 & 1 \end{bmatrix}$$

The eigenvalues of $\mathbf{K}'$ are $\lambda = 0, k/m, 2k/m$ with normalized eigenvectors $\frac{1}{2}(\sqrt{2} \quad 1 \quad 1), \sqrt{2}(0 \quad 1 \quad -1), \frac{1}{2}(\sqrt{2} \quad -1 \quad -1)$ respectively. Hence if

$$\mathbf{R} = \tfrac{1}{2}\begin{bmatrix} \sqrt{2} & 0 & \sqrt{2} \\ 1 & \sqrt{2} & -1 \\ 1 & -\sqrt{2} & -1 \end{bmatrix}$$

$\tilde{\mathbf{R}}\mathbf{K}'\mathbf{R} = \mathrm{diag}\,(0 \quad 1 \quad 2)$. The matrix that diagonalizes $\mathbf{K}$ and reduces $\mathbf{M}$ to the unit matrix is thus

$$\boldsymbol{\alpha} = \mathbf{QR} = \frac{1}{2}\sqrt{\frac{1}{m}}\begin{bmatrix} 1 & 0 & 1 \\ 1 & \sqrt{2} & -1 \\ 1 & -\sqrt{2} & -1 \end{bmatrix}$$

and

$$\tilde{\boldsymbol{\alpha}}\mathbf{M}\boldsymbol{\alpha} = \mathbf{I}$$

$$\tilde{\boldsymbol{\alpha}}\mathbf{K}\boldsymbol{\alpha} = \mathrm{diag}\left(0 \quad \frac{k}{m} \quad \frac{2k}{m}\right)$$

Now from eq. (7.10), the normal coordinates are given by

$$|q\rangle = \boldsymbol{\alpha}^{-1}|x\rangle$$

that is

$$\begin{bmatrix} q_1 \\ q_2 \\ q_3 \end{bmatrix} = \sqrt{m}\begin{bmatrix} 1 & 1/2 & 1/2 \\ 0 & 1/\sqrt{2} & -1/\sqrt{2} \\ 1 & -1/2 & -1/2 \end{bmatrix}\begin{bmatrix} x_1 \\ x_2 \\ x_3 \end{bmatrix}$$

or

$$q_1 = \sqrt{m}\left(x_1 + \frac{x_2}{2} + \frac{x_3}{2}\right)$$

$$q_2 = \sqrt{\frac{m}{2}}(x_2 - x_3)$$

$$q_3 = \sqrt{m}\left(x_1 - \frac{x_2}{2} - \frac{x_3}{2}\right)$$

We can easily check that

$$\sum \dot{q}_i^2 = T$$

and

$$\sum \lambda_i q_i^2 = V$$

The symmetric basis is obtained from use of the character table for S_2, given in the solution to problem 7.6 above. By applying the projection operators we can find

$$O^{A_1}\mathbf{x}_1 = 0$$

$$O^{A_1}\mathbf{x}_2 = \mathbf{x}_2 - \mathbf{x}_3 = \mathbf{s}_1$$

$$O^{A_2}\mathbf{x}_1 = 2\mathbf{x}_1 = \mathbf{s}_2$$

$$O^{A_2}\mathbf{x}_2 = \mathbf{x}_2 + \mathbf{x}_3 = \mathbf{s}_3$$

Hence

$$\begin{bmatrix} \mathbf{s}_1 \\ \mathbf{s}_2 \\ \mathbf{s}_3 \end{bmatrix} = \begin{bmatrix} 0 & 1 & -1 \\ 2 & 0 & 0 \\ 0 & 1 & 1 \end{bmatrix} \begin{bmatrix} \mathbf{x}_1 \\ \mathbf{x}_2 \\ \mathbf{x}_3 \end{bmatrix}$$

and

$$(\mathbf{s}) = \boldsymbol{\beta}^\dagger(\mathbf{x})$$

We then find that

$$\boldsymbol{\beta}^\dagger \mathbf{M} \boldsymbol{\beta} = \frac{m}{2} \begin{bmatrix} 0 & 1 & -1 \\ 2 & 0 & 0 \\ 0 & 1 & 1 \end{bmatrix} \begin{bmatrix} 2 & 0 & 0 \\ 0 & 1 & 0 \\ 0 & 0 & 1 \end{bmatrix} \begin{bmatrix} 0 & 2 & 0 \\ 1 & 0 & 1 \\ -1 & 0 & 1 \end{bmatrix}$$

$$= \frac{m}{2} \begin{bmatrix} 2 & 0 & 0 \\ 0 & 8 & 0 \\ 0 & 0 & 2 \end{bmatrix}$$

Similarly

$$\boldsymbol{\beta}^\dagger \mathbf{K} \boldsymbol{\beta} = k \left[\begin{array}{c|cc} 2 & 0 & 0 \\ \hline 0 & 8 & -4 \\ 0 & -4 & 2 \end{array}\right] \begin{matrix} A_1 \\ A_2 \\ \end{matrix}$$

7.6 If $\sum m_i x_i = 0$ then $x_1 = -\frac{1}{2}(x_2 + x_3)$. Hence on substituting into the expression

$$T = \frac{m}{2}(2\dot{x}_i^2 + \dot{x}_2^2 + \dot{x}_3^2)$$

we find

$$T = \frac{m}{2}(3\dot{x}_2^2 + 3\dot{x}_3^2 + 2\dot{x}_2\dot{x}_3)$$

and on substituting into the expression

$$V = k(x_2 - x_1)^2 + k(x_3 - x_1)^2$$

we find

$$V = \frac{k}{2}(5x_2^2 + 5x_3^2 + 6x_2x_3)$$

Hence

$$M = \frac{m}{4}\begin{bmatrix} 3 & 1 \\ 1 & 3 \end{bmatrix}, \qquad K = \frac{k}{2}\begin{bmatrix} 5 & 3 \\ 3 & 5 \end{bmatrix}$$

The secular equation is:

$$|K - \lambda M| = \begin{vmatrix} \frac{5k}{2} - \frac{3\lambda m}{4} & \frac{3k}{2} - \frac{\lambda m}{4} \\ \frac{3k}{2} - \frac{\lambda m}{4} & \frac{5k}{2} - \frac{3\lambda m}{4} \end{vmatrix} = 0$$

$$\therefore \quad \left(\frac{5k}{2} - \frac{3\lambda m}{4}\right) = \pm\left(\frac{3k}{2} - \frac{\lambda m}{4}\right)$$

$$\lambda = \frac{2k}{m}, \quad \frac{4k}{m}$$

From the method of section A.20, as solved in problem A.15, we have that

$$\boldsymbol{\alpha} = \frac{1}{\sqrt{2m}}\begin{bmatrix} 1 & \sqrt{2} \\ 1 & -\sqrt{2} \end{bmatrix}$$

The normal basis is given by eq. (7.9), that is,

$$\begin{bmatrix} \mathbf{q}_2 \\ \mathbf{q}_3 \end{bmatrix} = \frac{1}{\sqrt{2m}}\begin{bmatrix} 1 & 1 \\ \sqrt{2} & -\sqrt{2} \end{bmatrix}\begin{bmatrix} \mathbf{x}_2 \\ \mathbf{x}_3 \end{bmatrix}$$

$$\mathbf{q}_2 = \frac{1}{\sqrt{2m}}(\mathbf{x}_2 + \mathbf{x}_3)$$

$$\mathbf{q}_3 = \frac{1}{\sqrt{m}}(\mathbf{x}_2 - \mathbf{x}_3)$$

and the normal coordinates by eq. (7.10)

$$|q\rangle = \boldsymbol{\alpha}^{-1}|x\rangle$$

$$\begin{bmatrix} q_2 \\ q_3 \end{bmatrix} = \frac{\sqrt{m}}{2}\begin{bmatrix} \sqrt{2} & \sqrt{2} \\ 1 & -1 \end{bmatrix}\begin{bmatrix} x_1 \\ x_2 \end{bmatrix}$$

$$q_2 = \frac{\sqrt{m}}{2}(x_1 + x_2)$$

$$q_3 = \frac{\sqrt{m}}{2}(x - x_2)$$

Chapter 8

8.1 The symmetry elements are E, $C_2^{(z)}$, $C_2^{(y)}$, $C_2^{(x)}$, i, σ_h, $\sigma^{(zx)}$, $\sigma^{(zy)}$. By considering the movement of the ten p_z-orbitals under each of these operators one can find the character of the representation which is

	E	$C_2^{(z)}$	$C_2^{(y)}$	$C_2^{(x)}$	i	σ_h	$\sigma^{(zx)}$	$\sigma^{(zy)}$
$\chi(R)$	10	0	−2	0	0	−10	0	2

Remember that under a horizontal reflection a p_z-orbital changes sign. The character under $C_2^{(y)}$ for example is obtained by remembering that under this operation two orbitals remain at the same atom but change sign whereas all the other orbitals change places.

By use of the character table of D_{2h} and eq. (2.37) we obtain

$$\Gamma^{\text{red}}(R) = 2\mathbf{B}_{2g} \oplus 3\mathbf{B}_{1g} \oplus 2\mathbf{A}_u \oplus 3\mathbf{B}_{3u}$$

All these representations are one-dimensional, hence all the energy levels will be highly degenerate.

8.2 Number the carbon atoms: for example let one of the carbon atoms linking the two benzene rings be number one and then number the remaining atoms in a clockwise manner. Set up a table indicating the action of the operation P_R on all these functions. For example,

$$P_E\phi_1 = \phi_1; \quad P_{C_2^{(z)}}\phi_1 = \phi_6; \quad P_{C_2^{(y)}}\phi_1 = -\phi_1; \quad P_{C_2^{(x)}}\phi_1 = -\phi_6$$
$$P_i\phi_1 = -\phi_6; \quad P_{\sigma_h}\phi_1 = -\phi_1; \quad P_\sigma^{(zx)}\phi_1 = \phi_6; \quad P_\sigma^{(yz)}\phi_1 = \phi_1$$

By using the operators $O^{B_{2g}}$, $O^{B_{3g}}$, O^{A_u}, $O^{B_{1u}}$ on the functions ϕ_1, ϕ_2, and ϕ_3 we obtain the following symmetrized functions:

$$f_1 = \tfrac{1}{4}(\phi_2 + \phi_5 - \phi_7 - \phi_{10}) = O^{B_{2g}}\phi_2$$
$$f_2 = \tfrac{1}{4}(\phi_3 + \phi_4 - \phi_8 - \phi_9) = O^{B_{2g}}\phi_3$$
$$f_3 = \tfrac{1}{4}(2\phi_1 - 2\phi_6) = O^{B_{1g}}\phi_1$$
$$f_4 = \tfrac{1}{4}(\phi_2 - \phi_5 - \phi_7 + \phi_{10}) = O^{B_{1g}}\phi_2$$
$$f_5 = \tfrac{1}{4}(\phi_3 - \phi_4 - \phi_8 + \phi_9) = O^{B_{1g}}\phi_3$$
$$f_6 = \tfrac{1}{4}(\phi_2 - \phi_5 + \phi_7 - \phi_{10}) = O^{A_u}\phi_2$$
$$f_7 = \tfrac{1}{4}(\phi_3 - \phi_4 + \phi_8 - \phi_9) = O^{A_u}\phi_3$$
$$f_8 = \tfrac{1}{4}(2\phi_1 + 2\phi_6) = O^{B_{3u}}\phi_1$$
$$f_9 = \tfrac{1}{4}(\phi_2 + \phi_5 + \phi_7 + \phi_{10}) = O^{B_{3u}}\phi_2$$
$$f_{10} = \tfrac{1}{4}(\phi_3 + \phi_4 + \phi_8 + \phi_9) = O^{B_{3u}}\phi_3$$

Use the Hückel approximation with

$$\langle\phi_i|H|\phi_i\rangle = \alpha$$
$$\langle\phi_i|H|\phi_j\rangle = \beta \quad \text{for adjacent atoms}$$
$$\langle\phi_i|\phi_i\rangle = 1$$
$$\langle\phi_i|\phi_j\rangle = 0 \qquad i \neq j$$

We can now find the matrix elements

$$H_{ij} = \langle f_i|H|f_j\rangle$$

and

$$S_{ij} = \langle f_i|f_j\rangle,$$

for example,

$$\begin{aligned}
H_{22} &= \langle f_2|H|f_2\rangle \\
&= \tfrac{1}{16}\{\langle\phi_3|H|\phi_3\rangle + \langle\phi_3|H|\phi_4\rangle + 0 + 0 + \langle\phi_4|H|\phi_3\rangle \\
&\quad + \langle\phi_4|H|\phi_4\rangle + 0 + 0 + 0 + 0 + \langle\phi_8|H|\phi_8\rangle \\
&\quad + \langle\phi_8|H|\phi_9\rangle + 0 + 0 + \langle\phi_9|H|\phi_8\rangle + \langle\phi_9|H|\phi_9\rangle\} \\
&= \tfrac{1}{4}\{\alpha + \beta\} \\
H_{35} &= \tfrac{1}{16}\{0 + 0 + 0 + 0 + 0 + 0 + 0 + 0\}
\end{aligned}$$

The matrices **H** and **S** are both in block form because we have used symmetrized functions and we obtain four secular equations one for each of the irreducible components.
The secular equations are:

$$B_{2g} \quad \begin{vmatrix} \alpha - \varepsilon & \beta \\ \beta & \alpha - \varepsilon + \beta \end{vmatrix} = 0$$

$$B_{1g} \quad \begin{vmatrix} 2(\alpha - \varepsilon - \beta) & 2\beta & 0 \\ 2\beta & \alpha - \varepsilon & \beta \\ 0 & \beta & \alpha - \varepsilon - \beta \end{vmatrix} = 0$$

$$A_u \quad \begin{vmatrix} \alpha - \varepsilon & \beta \\ \beta & \alpha - \varepsilon - \beta \end{vmatrix} = 0$$

$$B_{3u} \quad \begin{vmatrix} 2(\alpha - \varepsilon + \beta) & 2\beta & 0 \\ 2\beta & \alpha - \varepsilon & \beta \\ 0 & \beta & \alpha - \varepsilon + \beta \end{vmatrix} = 0$$

Each of these equations can be solved by dividing each line by β and substituting $(\alpha - \varepsilon)/\beta = x$. The equations then become

$$B_{2g} \quad \begin{vmatrix} x & 1 \\ 1 & x + 1 \end{vmatrix} = 0$$

which gives solutions

$$x = -\frac{1 \pm \sqrt{5}}{2}$$

$$B_{1g} \quad \begin{vmatrix} 2(x - 1) & 2 & 0 \\ 2 & x & 1 \\ 0 & 1 & x - 1 \end{vmatrix} = 0$$

which yields

$$(x - 1)(x^2 - x - 3) = 0$$

so that

$$x = 1, \qquad \frac{1 \pm \sqrt{13}}{2}$$

$$A_u \quad \begin{vmatrix} x & 1 \\ 1 & x - 1 \end{vmatrix} = 0$$

which yields

$$x = -\frac{1 \pm \sqrt{5}}{2}$$

$$B_{1u} \begin{vmatrix} 2(x+1) & 2 & 0 \\ 2 & x & 1 \\ 0 & 1 & x+1 \end{vmatrix} = 0$$

which yields

$$(x+1)(x^2 + x - 3) = 0$$

so that

$$x = -1, \qquad -\frac{1 \pm \sqrt{13}}{2}$$

Hence the energies are

$$B_{2g}:\ \varepsilon_1 = \alpha + 1{\cdot}618\beta; \qquad \varepsilon_2 = \alpha - 0{\cdot}618\beta$$

$$B_{1g}:\ \varepsilon_3 = \alpha - \beta; \qquad \varepsilon_4 = \alpha - 2{\cdot}302\beta; \qquad \varepsilon_5 = \alpha + 1{\cdot}302\beta$$

$$A_u:\ \varepsilon_6 = \alpha - 1{\cdot}618\beta; \qquad \varepsilon_7 = \alpha + 0{\cdot}618\beta.$$

$$B_{3u}:\ \varepsilon_8 = \alpha + \beta; \quad \varepsilon_9 = \alpha + 2{\cdot}302\beta; \quad \varepsilon_{10} = \alpha - 1{\cdot}302\beta.$$

8.3 x, y, and z belong to B_{3u}, B_{2u}, and B_{1u} respectively in the group D_{2h}. As all representations are one-dimensional we require

$$\mathbf{A}_{1g} \otimes \mathbf{\Gamma}^{x,y,z} \otimes \mathbf{\Gamma}^{f} = \mathbf{A}_{1g}$$

therefore $\mathbf{\Gamma}^{x,y,z} \otimes \mathbf{\Gamma}^{f} = \mathbf{A}_{1g}$ and hence $\mathbf{\Gamma}^{f} = \mathbf{\Gamma}^{x}, \mathbf{\Gamma}^{y}$, or $\mathbf{\Gamma}^{z}$.
The transitions are therefore

$$A_{1g} \leftrightarrow B_{3u}:\ x$$

$$A_{1g} \leftrightarrow B_{2u}:\ y$$

$$A_{1g} \leftrightarrow B_{1u}:\ z$$

8.4 x, y, and z belong to B_1, B_2, and A_1 respectively in the group C_{2v}, and again all representations are one-dimensional. By use of the character table we can see that the following transitions are allowed:

$$x:\ A_1 \leftrightarrow B_1; \qquad A_2 \leftrightarrow B_2$$

$$y:\ A_1 \leftrightarrow B_2; \qquad A_2 \leftrightarrow B_1$$

$$z:\ A_1 \leftrightarrow A_1; \qquad A_2 \leftrightarrow A_2; \qquad B_1 \leftrightarrow B_1; \qquad B_2 \leftrightarrow B_2$$

8.5 The group is D_{2h}. By considering the operations of this group on the 18-dimensional Cartesian co-ordinates we see that only E_1, $C_2^{(x)}$, σ_h, and $\sigma^{(xz)}$ leave some atoms unchanged, and thus the character under all other operations is zero. Now under the operation σ_h, for example, each atom remains fixed, two axes remain fixed, and one axis is reversed at each atom hence under σ_h the character equals 6; under the operation $C_2^{(x)}$ two atoms remain fixed, two axes are reversed and one axis remains unchanged, hence the character is -2. The characters are therefore:

E	$C_2^{(z)}$	$C_2^{(y)}$	$C_2^{(x)}$	i	σ_h	$\sigma^{(xz)}$	$\sigma^{(yz)}$
18	0	0	-2	0	6	2	0

from which we find

$$\Gamma^{c}(R) = 3\mathbf{A}_g \oplus 3\mathbf{B}_{3g} \oplus 2\mathbf{B}_{2g} \oplus \mathbf{B}_{1g} \oplus \mathbf{A}_u \oplus 2\mathbf{B}_{3u} \oplus 3\mathbf{B}_{2u} \oplus 3\mathbf{B}_{1u}$$

By subtracting the translations and vibrations we have:

$$\Gamma^{\text{vib}}(R) = 3\mathbf{A}_g \oplus 2\mathbf{B}_{3g} \oplus \mathbf{B}_{2g} \oplus \mathbf{A}_u \oplus \mathbf{B}_{3u} \oplus 2\mathbf{B}_{2u} \oplus 2\mathbf{B}_{1u}$$

As all representations are one-dimensional all the two quantum state overtones are of the type

$$\mathbf{B}_{1g} \otimes \mathbf{B}_{1g} = \mathbf{A}_g$$

which is identical for all the vibrations, and hence all these states belong to A_g. For the two quantum combination states we have for example

$$\begin{aligned} &A_g \quad \text{and} \quad B_{1g} \leftrightarrow B_{1g} \\ &B_{1g} \quad \text{and} \quad B_{2g} \leftrightarrow B_{3g} \\ &B_{2u} \quad \text{and} \quad B_{3u} \leftrightarrow B_{1g} \\ &B_{1u} \quad \text{and} \quad B_{2g} \leftrightarrow B_{3u} \end{aligned}$$

Chapter 9

9.1 The elements of the group T_d are

$$E \quad 8C_3 \quad 3C_2 \quad 6S_4 \quad 6\sigma_d$$

where $\sigma_d = S_2C_2$ and $S_4 = S_2C_4$. The character of a rotation operator of the group $R_3 \otimes S_2$ is

$$\chi^2(R) = \frac{\sin[(2+\frac{1}{2})\phi]}{\sin(\phi/2)}$$

where ϕ is the angle of rotation, therefore in the group T_d the characters of the reducible representations are

$$\begin{aligned} &E \quad \phi = 0 && \chi = 5 \\ &C_3 \quad \phi = \frac{2\pi}{3} && \chi = -1 \\ &C_2 \quad \phi = \pi && \chi = 1 \\ &S_4 \quad \phi = \pi/2 && \chi = -1 \\ &\sigma_d \quad \phi = \pi && \chi = 1 \end{aligned}$$

using the character table of the group T_d we obtain:

$$\chi^2(R) = T_2 + E$$

and so the atomic D state splits into two states, one doubly degenerate and the other triply degenerate.

9.2

$$\mathbf{h} = \mathbf{r} \cdot \boldsymbol{\sigma} = \begin{bmatrix} z & y - ix \\ y + ix & -z \end{bmatrix}$$

$$\mathbf{h}' = \mathbf{u}^{-1}\mathbf{h}\mathbf{u} = \begin{bmatrix} z' & y' - ix' \\ y' + ix' & -z' \end{bmatrix}$$

where

$$z' = i(ab - a^*b^*)x + (ab + a^*b^*)y + (aa^* - bb^*)z$$
$$y' - ix' = -i(a^{*2} + b^2)x + (a^{*2} - b^2)y - 2a^*bz$$
$$y' + ix' = i(a^2 + b^{*2})x + (a^2 - b^{*2})y - 2ab^*z$$

Hence

$$y' = \frac{i}{2}\{a^2 - a^{*2} + b^{*2} - b^2\}x + \tfrac{1}{2}\{a^2 + a^{*2} - b^{*2} - b^2\}y - (ab^* + a^*b)z$$

$$x' = \tfrac{1}{2}\{a^2 + a^{*2} + b^{*2} + b^2\}x - \frac{i}{2}\{a^2 - a^{*2} + b^2 - b^{*2}\}y + i(ab^* - a^*b)z$$

so that we can write that $\mathbf{r}' = \mathbf{R}\mathbf{r}$ where $\mathbf{R}$ is a three by three matrix. We can see that all the elements are real. By looking at the determinant of $\mathbf{h}$ we see that

$$|\mathbf{h}'| = |\mathbf{u}^{-1}||\mathbf{h}||\mathbf{u}| = |\mathbf{h}|$$

and also

$$|\mathbf{h}| = -(x^2 + y^2 + z^2) = -\mathbf{r}\cdot\mathbf{r}$$
$$|\mathbf{h}'| = -(x'^2 + y'^2 + z'^2) = -\mathbf{r}'\cdot\mathbf{r}'$$

Hence $\langle\mathbf{r}'|\mathbf{r}'\rangle = \langle\mathbf{r}|\mathbf{R}^\dagger\mathbf{R}|\mathbf{r}\rangle = \langle\mathbf{r}|\mathbf{r}\rangle$ which implies that the matrix $\mathbf{R}$ is unitary and it must therefore represent a real rotation.
Note that for every matrix $\mathbf{u}$ of the group SU_2 one can set up a matrix $\mathbf{R}$ belonging to the group R_3. If we choose $\mathbf{u}$ to be

$$\mathbf{u} = \begin{bmatrix} e^{i\alpha/2} & 0 \\ 0 & e^{-i\alpha/2} \end{bmatrix} \qquad \text{i.e., } a = e^{i\alpha/2}, \quad b = 0$$

it is easy to see that $\mathbf{R}$ becomes

$$\mathbf{R} = \begin{bmatrix} \cos\alpha & \sin\alpha & 0 \\ -\sin\alpha & \cos\alpha & 0 \\ 0 & 0 & 1 \end{bmatrix}$$

9.3 The group C_3 has the three elements

$$C_3, C_3^2 \text{ and } C_3^3 = E$$

If we introduce a rotation of $2\pi = \theta \neq E$ such that $\theta^2 = E$ we obtain the double group

$$C_3, C_3^2, C_3^3 = \theta, \quad C_3^4 = \theta C_3, \quad C_3^5 = \theta C_3^2 \text{ and } C_3^6 = E$$

This is a cyclic group of order 6 and therefore will have the same properties and hence character table as the ordinary cyclic group of order 6, i.e., the group C_6.

9.4

$$\chi^1(\alpha) = e^{i\alpha} + 1 + e^{-i\alpha}$$
$$\chi^1(\alpha)\chi^1(\alpha) = e^{2i\alpha} + 2e^{i\alpha} + 3 + 2e^{-i\alpha} + e^{-2i\alpha}$$

which by inspection we can rewrite as:

$$e^{2i\alpha} + e^{i\alpha} + 1 + e^{-i\alpha} + e^{-2i\alpha} + e^{i\alpha} + 1 + e^{-i\alpha} + 1$$

which is $\chi^2(\alpha) + \chi^1(\alpha) + \chi^0(\alpha)$.

9.5 The six operators of the group P_3 are

$$\overset{E}{\begin{bmatrix}1&2&3\\1&2&3\end{bmatrix}}\overset{A}{\begin{bmatrix}1&2&3\\1&3&2\end{bmatrix}}\overset{B}{\begin{bmatrix}1&2&3\\3&2&1\end{bmatrix}}\overset{C}{\begin{bmatrix}1&2&3\\2&1&3\end{bmatrix}}\overset{D}{\begin{bmatrix}1&2&3\\3&1&2\end{bmatrix}}\overset{F}{\begin{bmatrix}1&2&3\\2&3&1\end{bmatrix}}$$

A, *B*, and *C* are simple operators and are therefore odd operators. *E*, *D*, and *F* are even operators. The multiplication table is:

E	*A*	B	*C*	*D*	*F*
A	E	*D*	*F*	B	*C*
B	*F*	E	*D*	*C*	*A*
C	*D*	*F*	E	*A*	B
D	*C*	*A*	B	*F*	E
F	B	*C*	*A*	E	*D*

There are three classes

$$E; D, \text{ and } F; A, B \text{ and } C$$

so that

$$\sum_{i=1}^{3} l_i^2 = 6$$

therefore $l_1 = 1$, $l_2 = 1$, $l_3 = 2$, so that there are two one-dimensional representations, one of which is symmetric and the other antisymmetric.

9.6 Let the four functions be

$$\Phi_{2s}(r_1)Y_0^0(\theta_1\phi_1)u^+(\zeta_1) = f_2^+(1)$$
$$\Phi_{2s}(r_1)Y_0^0(\theta_1\phi_1)u^-(\zeta_1) = f_2^-(1)$$
$$\Phi_{3s}(r_2)Y_0^0(\theta_2\phi_2)u^+(\zeta_2) = f_3^+(2)$$

and

$$\Phi_{3s}(r_2)Y_0^0(\theta_2\phi_2)u^-(\zeta_2) = f_3^-(2)$$

The four product functions are

$$F_1 = f_2^+(1)f_3^+(2)$$
$$F_2 = f_2^+(1)f_3^-(2)$$
$$F_3 = f_2^-(1)f_3^+(2)$$
$$F_4 = f_2^-(1)f_3^-(2)$$

The projection operator of the group P_2 for the antisymmetric representation is $\frac{1}{2}(P_{12} - P_{21})$ so that the symmetrized functions are

$$S_1 = \tfrac{1}{2}\{f_2^+(1)f_3^+(2) - f_2^+(2)f_3^+(1)\}$$
$$S_2 = \tfrac{1}{2}\{f_2^+(1)f_3^-(2) - f_2^+(2)f_3^-(1)\}$$
$$S_3 = \tfrac{1}{2}\{f_2^-(1)f_3^+(2) - f_2^-(2)f_3^+(1)\}$$
$$S_4 = \tfrac{1}{2}\{f_2^-(1)f_3^-(2) - f_2^-(2)f_3^-(1)\}$$

These four functions will form a basis for the reducible representation of the group R_3

$$\mathbf{D}^{1/2}(R) \otimes \mathbf{D}^{1/2}(R) = \mathbf{D}^1(R) \oplus \mathbf{D}^0(R)$$

9.7 In the case of two equivalent electrons we find that

$$S_1 = S_4 = 0$$
$$S_2 = -S_3$$

so there is only one symmetrized product function which then forms a basis for $\mathbf{D}^0(R)$.

We conclude from the results of problems 9.6 and this problem that a $(2s3s)$ configuration becomes a P and an S state whereas a $(2s, 2s) = (2s^2)$ configuration becomes an S state only.

Appendix A

A.1 Use the property of eq. (A.65) and let $\mathbf{B} = \mathbf{A}^{-1}$. Then

$$\mathbf{AA}^{-1} = \mathbf{I}$$

hence

$$(\mathbf{AA}^{-1})^\dagger = \mathbf{I}$$
$$(\mathbf{A}^{-1})^\dagger \mathbf{A}^\dagger = \mathbf{I}$$

and thus

$$(\mathbf{A}^{-1})^\dagger = (\mathbf{A}^\dagger)^{-1}$$

A.2 (a) Let $|\mathbf{A}| = \sum a_{ij}\alpha_{ij}$ (eq. (A.50)), then

$$|\mathbf{A}|^* = \sum a_{ij}^*\alpha_{ij}^*$$
$$= |\mathbf{A}^*|$$

(b) $|\mathbf{A}| = \sum a_{ji\,ji} = \sum a_{ij\,ij} = |\mathbf{A}|$

(c) From eq. (A.53)

$$|\mathbf{AA}^{-1}| = |\mathbf{A}||\mathbf{A}^{-1}| = |\mathbf{I}| = 1$$
$$\therefore \quad |\mathbf{A}^{-1}| = 1/|\mathbf{A}| = |\mathbf{A}|^{-1}$$

(d)

$$|k\mathbf{A}| = |k\mathbf{IA}| = |k\mathbf{I}||\mathbf{A}| \quad \text{by eq. (A.53)}$$
$$= k^n|\mathbf{A}|$$

A.3 Let $|x\rangle$ be an eigenvector, then

(a)

$$\mathbf{A}|x\rangle = \lambda|x\rangle$$

hence

$$\langle x|\mathbf{A}^\dagger = \lambda^*\langle x|$$

thus λ^* is an eigenvalue of $\mathbf{A}^\dagger$.

(b)

$$\mathbf{A}|x\rangle = \lambda|x\rangle$$

therefore

$$|x\rangle = \lambda\mathbf{A}^{-1}|x\rangle$$

$$\lambda^{-1}|x\rangle = \mathbf{A}^{-1}|x\rangle$$

hence λ^{-1} is an eigenvalue of $\mathbf{A}^{-1}$.

(c)

$$\mathbf{A}|x\rangle = \lambda|x\rangle$$

hence

$$\text{adj}\,\mathbf{A}\,.\,\mathbf{A}|x\rangle = \lambda\,\text{adj}\,\mathbf{A}|x\rangle$$

or

$$|\mathbf{A}|\mathbf{I}|x\rangle = \lambda\,\text{adj}\,\mathbf{A}|x\rangle$$

$$\lambda^{-1}|\mathbf{A}||x\rangle = \text{adj}\,\mathbf{A}|x\rangle$$

hence $\lambda^{-1}|\mathbf{A}|$ is an eigenvalue of adj $\mathbf{A}$.

(d)

$$\mathbf{A}|x\rangle = \lambda|x\rangle$$

therefore

$$k\mathbf{A}|x\rangle = k\lambda|\rangle$$

$$(k\mathbf{A})|x\rangle = k\lambda|x\rangle$$

hence an eigenvalue of $k\mathbf{A}$ is $k\lambda$.

A.4 (a)

$$\mathbf{AB} = \begin{bmatrix} 0 & 0 & 0 \\ 0 & 0 & 0 \\ 0 & 0 & 0 \end{bmatrix} \qquad \mathbf{BA} = 4\begin{bmatrix} 2 & 0 & -1 \\ 6 & 0 & -3 \\ 4 & 0 & -2 \end{bmatrix}$$

(b) From eqs. (A.43) and (A.49) we find

$$|\mathbf{A}| = 1(-1) - 1(3) - 2(-2) = 0$$

$$|\mathbf{B}| = 1(12 - 12) - 2(6 - 6) + 1(12 - 12) = 0$$

(c) From eq. (A.196) the eigenvalues λ of $\mathbf{A}$ are given by the equation

$$|\mathbf{A} - \lambda\mathbf{I}| = 0$$

but

$$|\mathbf{A}| = 0$$

hence $\lambda = 0$ is a solution. The same result is obtained for $\mathbf{B}$.

A.5 (a) Let

$$\mathbf{A}|x_i\rangle = \lambda_i|x_i\rangle$$

then

$$\mathbf{BA}|x_i\rangle = \lambda_i\mathbf{B}|x_i\rangle$$

or

$$\mathbf{AB}|x_i\rangle = \lambda_i \mathbf{B}|x_i\rangle \quad \text{as } \mathbf{AB} = \mathbf{BA}$$

hence $\mathbf{B}|x_i\rangle$ is an eigenvector of $\mathbf{A}$ which is possible if

$$\mathbf{B}|x_i\rangle = \mu_i|x_i\rangle$$

and hence $|x_i\rangle$ is also an eigenvector of $\mathbf{B}$ with a generally different eigenvalue μ_i.
(b) From eqs. (A.204)–(A.207) we see that if we form the matrix $\mathbf{R}$ of the eigen-columns then

$$\mathbf{R}^{-1}\mathbf{AR} = \mathbf{D}_A$$

$$\mathbf{R}^{-1}\mathbf{BR} = \mathbf{D}_B$$

where $\mathbf{D}_A$, $\mathbf{D}_B$ are the diagonal forms of $\mathbf{A}$ and $\mathbf{B}$ respectively and $\mathbf{R}$ is the same matrix because the eigenvectors are the same.
(c) Let

$$\mathbf{A}|x_i\rangle = \lambda_A|x_i\rangle$$

$$\mathbf{B}|x_i\rangle = \lambda_B|x_i\rangle$$

then

$$\mathbf{AB}|x_i\rangle = \lambda_B \mathbf{A}|x_i\rangle = \lambda_B\lambda_A|x_i\rangle$$

hence $|x_i\rangle$ is an eigenvector of $\mathbf{AB}$ with eigenvalue $\lambda_A\lambda_B$.

A.6 (a) From eqs. (A.197), (A.202), (A.203) we know that $\mathbf{A}$ has eigenvectors

$$|x_1\rangle = \begin{bmatrix} 1 \\ -1 \end{bmatrix}, \qquad |x_2\rangle = \begin{bmatrix} 2 \\ 5 \end{bmatrix}$$

and that

$$\mathbf{A}|x_1\rangle = -1|x_1\rangle$$

$$\mathbf{A}|x_2\rangle = 6|x_2\rangle$$

By substitution we find that

$$\mathbf{B}|x_1\rangle = -3|x_1\rangle$$

$$\mathbf{B}|x_2\rangle = 11|x_2\rangle$$

Now

$$\mathbf{AB} = \mathbf{BA} = \begin{bmatrix} 21 & 18 \\ 45 & 48 \end{bmatrix}$$

and by substitution we find

$$\mathbf{AB}|x_1\rangle = 3|x_1\rangle$$

$$\mathbf{AB}|x_2\rangle = 66|x_2\rangle$$

in agreement with $\lambda_{AB} = \lambda_A\lambda_B$.
(b) Similarly we can find $\mathbf{AB}$ and $\mathbf{AB} = \mathbf{BA}$ have eigenvectors

$$|x_1\rangle = \begin{bmatrix} 1 \\ -1 \end{bmatrix}, \qquad |x_2\rangle = \begin{bmatrix} 2 \\ 5 \end{bmatrix}$$

with eigenvalues

A	−4	3
B	−1	6
AB	4	18

A.7 (a) Let

$$\mathbf{A}|x_A\rangle = \lambda_A|x_A\rangle$$
$$\mathbf{B}|x_B\rangle = \lambda_B|x_B\rangle$$

Then from eq. (A.35) we have

$$\begin{aligned}(\mathbf{A} \otimes \mathbf{B})(|x_A\rangle \otimes |x_B\rangle) &= \mathbf{A}|x_A\rangle \otimes \mathbf{B}|x_B\rangle \\ &= \lambda_A|x_A\rangle \otimes \lambda_B|x_B\rangle \\ &= \lambda_A\lambda_B(|x_A\rangle \otimes |x_B\rangle)\end{aligned}$$

(b) From eq. (A.229) we have

$$\begin{aligned}\text{trace}\,(\mathbf{A} \otimes \mathbf{B}) &= \sum_i \sum_j (\lambda_A)_i(\lambda_B)_j \\ &= \sum_i (\lambda_A)_i \sum_j (\lambda_B)_j \\ &= \text{trace}\,\mathbf{A}\,.\,\text{trace}\,\mathbf{B}\end{aligned}$$

(c) From eq. (A.35) we have

$$\begin{aligned}(\mathbf{A} \otimes \mathbf{B})(\mathbf{A}^{-1} \otimes \mathbf{B}^{-1}) &= (\mathbf{A}\mathbf{A}^{-1}) \otimes (\mathbf{B}\mathbf{B}^{-1}) \\ &= \mathbf{I}\end{aligned}$$

A.8 (a) Detailed solution for **A** is as follows. The eigenvector satisfies the equation

$$\begin{vmatrix} 5-\lambda & -1 & -1 \\ -1 & 5-\lambda & -1 \\ -1 & -1 & 5-\lambda \end{vmatrix} = 0$$

which can be expanded as

$$\lambda^3 - 15\lambda^2 + 72\lambda - 108 = 0$$

or

$$(\lambda - 3)(\lambda - 6)(\lambda - 6) = 0$$

so that the roots are

$$\lambda = 3, 6, 6$$

and

$$\sum \lambda_i = 15 = \text{trace}\,\mathbf{A}$$

When $\lambda = 3$

$$[\mathbf{A} - \lambda\mathbf{I}]|x\rangle = \begin{bmatrix} 2 & -1 & -1 \\ -1 & 2 & -1 \\ -1 & -1 & 2 \end{bmatrix}\begin{bmatrix} x_1 \\ x_2 \\ x_3 \end{bmatrix} = 0$$

If we add the first two rows to the last then

$$\begin{bmatrix} 2 & -1 & -1 \\ -1 & -1 & 2 \\ 0 & 0 & 0 \end{bmatrix} \begin{bmatrix} x_1 \\ x_2 \\ x_3 \end{bmatrix} = 0$$

Add twice the second row to the first

$$\begin{bmatrix} 0 & -3 & 3 \\ -1 & -1 & 2 \\ 0 & 0 & 0 \end{bmatrix} \begin{bmatrix} x_1 \\ x_2 \\ x_3 \end{bmatrix}$$

Divide the top row by 3 and subtract from second row

$$\begin{bmatrix} 0 & -1 & 1 \\ -1 & 0 & 1 \\ 0 & 0 & 0 \end{bmatrix} \begin{bmatrix} x_1 \\ x_2 \\ x_3 \end{bmatrix}$$

hence $x_1 = x_2 = x_3$
So that a solution, in integers, is

$$\begin{bmatrix} 1 \\ 1 \\ 1 \end{bmatrix}$$

When $\lambda = 6$ then

$$[\mathbf{A} - \lambda\mathbf{I}] = \begin{bmatrix} -1 & -1 & -1 \\ -1 & -1 & -1 \\ -1 & -1 & -1 \end{bmatrix}$$

$$\equiv \begin{bmatrix} 1 & 1 & 1 \\ 0 & 0 & 0 \\ 0 & 0 & 0 \end{bmatrix}$$

hence $x_1 = -(x_2 + x_3)$ of which two mutually orthogonal solutions are

$$\begin{bmatrix} 2 \\ -1 \\ -1 \end{bmatrix} \quad \text{and} \quad \begin{bmatrix} 0 \\ 1 \\ -1 \end{bmatrix}$$

both of which are orthogonal to the eigenvalue corresponding to $\lambda = 3$.
If the solutions are normalized then a matrix **R** can be constructed

$$\mathbf{R} = \begin{bmatrix} 1/\sqrt{3} & 2/\sqrt{6} & 0 \\ 1/\sqrt{3} & -1/\sqrt{6} & 1/\sqrt{2} \\ 1/\sqrt{3} & -1/\sqrt{6} & -1/\sqrt{2} \end{bmatrix}$$

and it can be checked that

$$\mathbf{R}^{-1}\mathbf{A}\mathbf{R} = \text{diag}\,\{3 \quad 6 \quad 6\}$$

where $\mathbf{R}^{-1} = \tilde{\mathbf{R}}$.

The eigenrows are the same as the eigencolumns because the matrix is symmetric. The solutions to the others are as follows. Unless separately shown, the eigencolumn is the complex conjugate of the transpose of the eigenrow.

D

$$\lambda^3 - \lambda^2 + 5\lambda - 3 = 0$$

$$-1 \quad [1 \quad 0 \quad -1], \qquad 3 \quad 3 \quad [1 \quad 0 \quad 1]$$

$$\begin{bmatrix} 4 & 0 \\ 1 & 1 \\ -4 & 0 \end{bmatrix}$$

E

$$\lambda^2 - \lambda - 2 = 0$$

$$-1 \quad [1 \quad -i \quad -2], \qquad 2 \quad [2 \quad 1 + i]$$

G

$$\lambda^2 - 2i\lambda - 1 = 0$$

$$i \quad i \quad [1 \quad -i]$$

$$\begin{bmatrix} 1 \\ -i \end{bmatrix}$$

J

$$\lambda^3 - \lambda^2 - \lambda + 1 = 0$$

$$-1 \quad [3 \quad -1 \quad -1], \qquad 1 \quad 1 \quad [0 \quad 1 \quad -1] \qquad [1 \quad 0 \quad -2]$$

$$\begin{bmatrix} 2 \\ 1 \\ 1 \end{bmatrix} \quad \begin{bmatrix} 0 \\ 1 \\ -1 \end{bmatrix} \quad \begin{bmatrix} 1 \\ 0 \\ 3 \end{bmatrix}$$

K

$$16\sqrt{2}\lambda^3 - 8(3 + \sqrt{2})\lambda^2 + 8(3 + \sqrt{2})\lambda - 16\sqrt{2} = 0$$

$$1 \quad [1 \quad \sqrt{2} + 1 \quad \sqrt{3}], \qquad e^{\pm i\theta}$$

$$\theta = \cos^{-1}\left(\frac{3\sqrt{2} - 2}{8}\right)$$

L

$$\lambda^2 - 5\lambda - 6$$

$$6 \quad [1 \quad 1], \qquad -1 \quad [5 \quad 2]$$

$$\begin{bmatrix} 2 \\ 5 \end{bmatrix} \quad \begin{bmatrix} 1 \\ -1 \end{bmatrix}$$

(b) A diagonalizing matrix **R** can be constructed from the eigencolumns, which may be normalized.

The solution for **A** has already been given. Solution for **E** is:

$$\mathbf{R} = \frac{1}{\sqrt{6}}\begin{bmatrix} 1 + i & 2 \\ -2 & 1 - i \end{bmatrix}, \qquad \mathbf{R}^{-1}\mathbf{E}\mathbf{R} = \text{diag}\,(-1 \quad 2)$$

(ii) **G** cannot be diagonalized because although there are two equal eigenvalues there is only one eigenvector.

(iii) Similarly **D** cannot be diagonalized because corresponding to the double eigenvalue 3 there is only one eigenvector.

A.9 The procedure is tedious as it uses a step by step method and produces no general formula. The first few can be obtained as follows; let

$$|x_0\rangle = 1$$
$$|x_1\rangle = x$$
$$|x_2\rangle = x^2$$

etc. Let $|y_0\rangle = |x_0\rangle = 1$, then, by eqs. (A.117) and (A.163), where we write eq. (A.117) in Dirac notation as

$$|y_i\rangle = |x_i\rangle - \sum_{j=1}^{i-1} \frac{|y_i\rangle\langle y_j|x_i\rangle}{\langle y_j|y_j\rangle}$$

and

$$\sum_j \frac{|y_j\rangle\langle y_j|}{\langle y_j|y_j\rangle}$$

is the projection operator of problem A.13.

$$|y_1\rangle = |x_1\rangle - \frac{|y_0\rangle\langle y_0|x_1\rangle}{\langle y_0|y_0\rangle}$$

$$= x - \frac{\int_{-1}^{+1} x\,dx}{\int_{-1}^{+1} dx}$$

$$|y_1\rangle = x.$$

We can then find

$$|y_2\rangle = |x_2\rangle - \frac{|y_1\rangle\langle y_1|x_2\rangle}{\langle y_1|y_1\rangle} - \frac{\langle y_0\rangle\langle y_0|x_2\rangle}{\langle y_0|y_0\rangle}$$

$$= x^2 - x\frac{\int_{-1}^{+1} x^3\,dx}{\int_{-1}^{+1} x^2\,dx} - 1\frac{\int_{-1}^{+1} x^2\,dx}{\int_{-1}^{+1} dx}$$

$$= x^2 - \tfrac{1}{3}$$

$$|y_3\rangle = |x_3\rangle - \frac{|y_2\rangle\langle y_2|x_3\rangle}{\langle y_2|y_2\rangle} - \frac{|y_1\rangle\langle y_1|x_3\rangle}{\langle y_1|y_1\rangle} - \frac{|y_0\rangle\langle y_0|x_3\rangle}{\langle y_0|y_0\rangle}$$

$$= x^3 - \frac{(x^2 - \frac{1}{3})\int_{-1}^{+1}(x^5 - x^3/3)\,dx}{\int_{-1}^{+1}(x^2 - \frac{1}{3})^2\,dx} - \frac{x\int_{-1}^{+1} x^4\,dx}{\int_{-1}^{+1} x^2\,dx} - \frac{\int x^3\,dx}{\int dx}$$

$$|y_3\rangle = x^3 - \frac{3x}{5}$$

$$|y_4\rangle = |x_4\rangle - \frac{|y_3\rangle\langle y_3|x_4\rangle}{\langle y_3|y_3\rangle} - \frac{|y_2\rangle\langle y_2|x_4\rangle}{\langle y_2|y_2\rangle} - \frac{|y_1\rangle\langle y_1|x_4\rangle}{\langle y_1|y_1\rangle} - \frac{|y_0\rangle\langle y_0|x_4\rangle}{\langle y_0|y_0\rangle}$$

$$= x^4 - \frac{(x^3 - 3x/5)\int(x^7 - x)\,dx}{\int(x^3 - x/5)^2\,dx} - \frac{(x^2 - \frac{1}{3})\int(x^6 - x^4/3)\,dx}{\int(x^2 - \frac{1}{3})^2\,dx}$$

$$- \frac{x\int x^5\,dx}{\int x^2\,dx} - \frac{\int x^4\,dx}{\int dx}$$

$$= x^4 - \tfrac{6}{7}(x^2 - \tfrac{1}{3}) - \tfrac{1}{5}$$

$$= x^4 - \tfrac{6}{7}x^2 + \tfrac{3}{35}$$

Further functions can obviously be found. The series may be normalized by dividing each $|y_n\rangle$ by the root of scalar product $(\sqrt{\langle y_n|y_n\rangle})$. Appropriate values are

$$\sqrt{\langle y_0|y_0\rangle} = \sqrt{2}$$
$$\sqrt{\langle y_1|y_1\rangle} = \sqrt{\tfrac{2}{3}}$$
$$\sqrt{\langle y_2|y_2\rangle} = \tfrac{2}{3}\sqrt{\tfrac{2}{5}}$$
$$\sqrt{\langle y_3|y_3\rangle} = \tfrac{2}{5}\sqrt{\tfrac{2}{7}}$$
$$\sqrt{\langle y_4|y_4\rangle} = \frac{8\sqrt{2}}{7 \times 5 \times 3}$$

A.10 We have, eq. (A.133)

$$\mathbf{e}_1' = \mathbf{e}_1 + 2\mathbf{e}_2$$
$$\mathbf{e}_2' = 5\mathbf{e}_1 + 4\mathbf{e}_2$$

where $\mathbf{e}_1$ and $\mathbf{e}_2$ are orthonormal. These can be written as

$$\begin{bmatrix} \mathbf{e}_1' \\ \mathbf{e}_2' \end{bmatrix} = \begin{bmatrix} 1 & 2 \\ 5 & 4 \end{bmatrix} \begin{bmatrix} \mathbf{e}_1 \\ \mathbf{e}_2 \end{bmatrix}$$

so that, from eq. (A.267)

$$\tilde{\mathbf{T}} = \begin{bmatrix} 1 & 2 \\ 5 & 4 \end{bmatrix}$$

(a) To find the reciprocal basis $\bar{\mathbf{e}}$, let

$$\bar{\mathbf{e}}_1' = a_{11}\mathbf{e}_1 + a_{12}\mathbf{e}_2$$
$$\bar{\mathbf{e}}_2' = a_{21}\mathbf{e}_1 + a_{22}\mathbf{e}_2$$

Then, from eq. (A.108) we must have,

$$a_{11} + 2a_{12} = 1 \qquad 5a_{21} + 4a_{22} = 1$$
$$5a_{11} + 4a_{12} = 0 \qquad a_{21} + 2a_{22} = 0$$

of which the solutions are

$$a_{11} = -\tfrac{2}{3} \qquad a_{21} = \tfrac{1}{3}$$
$$a_{12} = \tfrac{5}{6} \qquad a_{22} = -\tfrac{1}{6}$$

so that

$$\bar{\mathbf{e}}_1' = -\tfrac{2}{3}\mathbf{e}_1 + \tfrac{5}{6}\mathbf{e}_2$$
$$\bar{\mathbf{e}}_2' = \tfrac{1}{3}\mathbf{e}_1 + \tfrac{1}{6}\mathbf{e}_2$$

or

$$\begin{bmatrix} \bar{\mathbf{e}}_1' \\ \bar{\mathbf{e}}_2' \end{bmatrix} = \frac{1}{6}\begin{bmatrix} -4 & 5 \\ 2 & -1 \end{bmatrix} \begin{bmatrix} \mathbf{e}_1 \\ \mathbf{e}_2 \end{bmatrix}$$

(b) In the basis $\mathbf{e}$ the vector $\mathbf{z}$ is given by eq. (A.89) as

$$\mathbf{z} = 4\mathbf{e}_1 + 2\mathbf{e}_2$$

Now if we solve for the basis $\mathbf{e}$ in terms of the reciprocal basis $\bar{\mathbf{e}}'$ we find

$$\mathbf{e}_1 = \bar{\mathbf{e}}_1' + 5\bar{\mathbf{e}}_2'$$
$$\mathbf{e}_2 = 2\bar{\mathbf{e}}_1' + 4\bar{\mathbf{e}}_2'$$

Hence

$$\mathbf{z} = 4(\bar{\mathbf{e}}_1' + 5\bar{\mathbf{e}}_2') + 2(\bar{\mathbf{e}}_1' + 4\bar{\mathbf{e}}_2')$$
$$= 8\bar{\mathbf{e}}_1' + 28\bar{\mathbf{e}}_2'$$

(c) By eq. (A.132) we have

$$|\bar{z}'\rangle = \tilde{\mathbf{T}}|z\rangle$$

$$\begin{bmatrix} \bar{z}_1' \\ \bar{z}_2' \end{bmatrix} = \begin{bmatrix} 1 & 2 \\ 5 & 4 \end{bmatrix} \begin{bmatrix} 4 \\ 2 \end{bmatrix} = \begin{bmatrix} 8 \\ 28 \end{bmatrix}$$

By eq. (A.142) we have

$$(\bar{\mathbf{e}}') = \tilde{\mathbf{T}}^{-1}(\mathbf{e})$$

But

$$\tilde{\mathbf{T}} = \begin{bmatrix} 1 & 2 \\ 5 & 4 \end{bmatrix}$$

$$\therefore \quad \tilde{\mathbf{T}}^{-1} = \tfrac{1}{6} \begin{bmatrix} -4 & 2 \\ 5 & -1 \end{bmatrix}$$

$$\therefore \quad \begin{bmatrix} \bar{\mathbf{e}}_1' \\ \bar{\mathbf{e}}_2' \end{bmatrix} = \tfrac{1}{6} \begin{bmatrix} 4 & 2 \\ 5 & -1 \end{bmatrix} \begin{bmatrix} \mathbf{e}_1 \\ \mathbf{e}_2 \end{bmatrix}$$

(d) As $\mathbf{e}_1$ and $\mathbf{e}_2$ are orthonormal then

$$\mathbf{g} = \begin{bmatrix} \mathbf{e}_1 . \mathbf{e}_1 & \mathbf{e}_1 . \mathbf{e}_2 \\ \mathbf{e}_2 . \mathbf{e}_1 & \mathbf{e}_2 . \mathbf{e}_2 \end{bmatrix} = \begin{bmatrix} 1 & 0 \\ 0 & 1 \end{bmatrix}$$

$$\mathbf{g}' = \begin{bmatrix} \mathbf{e}_1' . \mathbf{e}_1' & \mathbf{e}_1' . \mathbf{e}_2' \\ \mathbf{e}_2' . \mathbf{e}_1' & \mathbf{e}_2' . \mathbf{e}_2' \end{bmatrix} = \begin{bmatrix} 5 & 13 \\ 13 & 41 \end{bmatrix}$$
$$= \tilde{\mathbf{T}}\mathbf{g}\mathbf{T}$$

$$\bar{\mathbf{g}}' = \begin{bmatrix} \bar{\mathbf{e}}_1' . \bar{\mathbf{e}}_1' & \bar{\mathbf{e}}_1' . \bar{\mathbf{e}}_2' \\ \bar{\mathbf{e}}_2' . \bar{\mathbf{e}}_1' & \bar{\mathbf{e}}_2' . \bar{\mathbf{e}}_2' \end{bmatrix} = \tfrac{1}{36} \begin{bmatrix} 41 & -13 \\ -13 & 5 \end{bmatrix}$$
$$= (\mathbf{g}')^{-1}$$

(e) By eq. (A.102) we have

$$\langle z|z\rangle = \sum_{ij} z_i^* z_j g_{ij}$$

The components of z in the different bases are

$\mathbf{e}$	$z_1 = 4$	$z_2 = 2$
$\mathbf{e}'$	$z_1' = -1$	$z_2' = 1$
$\bar{\mathbf{e}}'$	$\bar{z}_1' = 8$	$\bar{z}_2' = 28$

Substituting in eq. (A.102). In the basis $\mathbf{e}$

$$\langle z|z\rangle = z_1^2 g_{11} + 2z_1 z_2 g_{12} + z_2^2 g_{22}$$
$$= (16)1 + (16)0 + (4)1$$
$$= 20$$

In the basis $\mathbf{e}'$

$$\langle z'|z'\rangle = z_1'^2 g_{11} + 2z_1'z_2'g_{12} + z_2'^2 g_{22}$$
$$= (1)5 + 2(-1)(13) + (1)41$$
$$= 20$$

In the basis $\bar{\mathbf{e}}'$

$$\langle \bar{z}'|\bar{z}'\rangle = \bar{z}_1'^2 g_{11} + 2\bar{z}_1'\bar{z}_2'g_{12} + \bar{z}_2'^2 g_{22}$$
$$= 64(\tfrac{41}{36}) - 2(\tfrac{224}{36})13 + (\tfrac{784}{36})5$$
$$= 20$$

(f) In terms of a basis and its reciprocal

$$\langle z|z\rangle = z_1'\bar{z}_1'^* + z_2'\bar{z}_2'^*$$
$$= (-1)(8) + (1)(28)$$
$$= 20$$

A.11 (a) The eigenvalues are 1, -1. The corresponding eigenvectors are

$$\boldsymbol{\sigma}_1 \quad [1 \quad 1] \quad [1 \quad -1]$$
$$\boldsymbol{\sigma}_2 \quad [1 \quad -i] \quad [1 \quad i]$$
$$\boldsymbol{\sigma}_3 \quad [1 \quad 0] \quad [0 \quad 1]$$

(b) These results follow immediately by matrix multiplication.

A.12 Let $\mathbf{U}$ be a unitary matrix with eigenvalue λ and eigenvector $|u\rangle$. Then

$$\mathbf{U}|u\rangle = \lambda|u\rangle$$

therefore

$$\langle u|\mathbf{U}^\dagger = \lambda^*\langle u|$$
$$\langle u|\mathbf{U}^\dagger|u\rangle = \lambda^*\langle u|u\rangle$$

Also

$$\mathbf{U}^\dagger\mathbf{U}|u\rangle = \lambda\mathbf{U}^\dagger|u\rangle$$

but

$$\mathbf{U}^\dagger = \mathbf{U}^{-1}$$
$$\therefore \quad \langle u|u\rangle = \lambda\langle u|\mathbf{U}^\dagger|u\rangle$$

hence

$$\lambda^* = \lambda^{-1}$$

and thus

$$\lambda = e^{i\theta}$$

A.13 (a) Let $|c_i\rangle$ be an eigencolumn of a real symmetric matrix $\mathbf{A}$ belonging to an eigenvalue λ_i, and $\langle r_i|$ be the corresponding eigenrow. Then

$$\mathbf{A}|c_i\rangle = \lambda_i|c_i\rangle$$

therefore

$$\langle c_i|\tilde{\mathbf{A}} = \lambda_i\langle c_i|$$

(because, by section A.19 λ_i real). But because $\mathbf{A}$ is symmetric, $\tilde{\mathbf{A}} = \mathbf{A}$ therefore

$$\langle c_i|\mathbf{A} = \lambda_i\langle c_i|$$

but

$$\langle r_i|\mathbf{A} = \lambda_i\langle r_i|$$

Hence

$$\langle r_i| = \langle c_i|$$

(b)

$$\langle r_i|\mathbf{A} = \lambda_i\langle r_i|$$

therefore

$$\langle r_i|\mathbf{A}|c_j\rangle = \lambda_i\langle r_i|c_j\rangle$$

also

$$\mathbf{A}|c_j\rangle = \lambda_j|c_j\rangle$$

therefore

$$\langle r_i|\mathbf{A}|c_j\rangle = \lambda_j\langle r_i|c_j\rangle$$

Hence

$$\langle r_i|\mathbf{A}|c_j\rangle = \lambda_i\langle r_i|c_j\rangle = \lambda_j\langle r_i|c_j\rangle$$

and as $\lambda_i \neq \lambda_j$ we must have

$$\langle r_i|c_j\rangle = 0$$

(c) (i)

$$\mathbf{P}^j \sum_i a_i|c_i\rangle = \sum_i a_i \frac{|c_j\rangle\langle r_j|c_i\rangle}{\langle r_j|c_j\rangle}$$
$$= a_i|c_j\rangle\delta_{ij}$$

because $\langle r_j|c_i\rangle = 0$ when $i \neq j$

$$\mathbf{P}^j \sum_i a_i|c_i\rangle = a_j|c_j\rangle$$

(ii)

$$\mathbf{P}^j\mathbf{P}^j = \frac{|c_j\rangle\langle r_j|c_j\rangle\langle r_j|}{\langle r_j|c_j\rangle\langle r_j|c_j\rangle}$$
$$= \frac{|c_j\rangle\langle r_j|}{\langle r_j|c_j\rangle}$$
$$= \mathbf{P}^j$$

(iii) Let $|c_i\rangle$ be the eigencolumns and let $|c\rangle = \sum_i |c_i\rangle$ then

$$\sum_j \mathbf{P}^j|c\rangle = \sum_j \sum_i \frac{|c_j\rangle\langle r_j|c_i\rangle}{\langle r_j|c_j\rangle}$$
$$= \sum_j |c_j\rangle$$
$$= \mathbf{I}|c\rangle$$

so that

$$\sum_j \mathbf{P}^j = \mathbf{I}$$

(iv)

$$\sum_j \lambda_j \mathbf{P}^j|c_i\rangle = \lambda_i|c_i\rangle$$

$$= \mathbf{A}|c_i\rangle$$

hence

$$\sum_j \lambda_j \mathbf{P}^j = \mathbf{A}$$

A.14

$$\lambda_i = \quad 3 \qquad 6 \qquad 6$$

$$\mathbf{A} = \begin{bmatrix} 5 & -1 & -1 \\ -1 & 5 & -1 \\ -1 & -1 & 5 \end{bmatrix} \qquad |c_i\rangle = \begin{bmatrix} 1 \\ 1 \\ 1 \end{bmatrix}\begin{bmatrix} 2 \\ -1 \\ -1 \end{bmatrix}\begin{bmatrix} 0 \\ 1 \\ -1 \end{bmatrix}$$

$$\lambda_1 = 3 \qquad \mathbf{P}^1 = \tfrac{1}{3}\begin{bmatrix} 1 \\ 1 \\ 1 \end{bmatrix}[1 \quad 1 \quad 1] = \tfrac{1}{3}\begin{bmatrix} 1 & 1 & 1 \\ 1 & 1 & 1 \\ 1 & 1 & 1 \end{bmatrix}$$

For $\lambda = 6$ there are two projection operators that we can designate as $\mathbf{P}_1^2, \mathbf{P}_2^2$.

$$\mathbf{P}_1^2 = \tfrac{1}{6}\begin{bmatrix} 2 \\ -1 \\ -1 \end{bmatrix}[2 \quad -1 \quad -1] = \tfrac{1}{6}\begin{bmatrix} 4 & -2 & -2 \\ -2 & 1 & 1 \\ -2 & 1 & 1 \end{bmatrix}$$

$$\mathbf{P}_2^2 = \tfrac{1}{2}\begin{bmatrix} 0 \\ 1 \\ -1 \end{bmatrix}[0 \quad 1 \quad -1] = \tfrac{1}{2}\begin{bmatrix} 0 & 0 & 0 \\ 0 & 1 & -1 \\ 0 & -1 & 1 \end{bmatrix}$$

It is easily checked that

$$\sum \mathbf{P}^j = \mathbf{I}$$

and that

$$\sum \lambda_j \mathbf{P}^j = 3\mathbf{P}^1 + 6(\mathbf{P}_1^2 + \mathbf{P}_2^2) = \mathbf{A}$$

An example of $\mathbf{P}^j\mathbf{P}^j$ is provided by

$$\mathbf{P}_1^2\mathbf{P}_1^2 = \tfrac{1}{36}\begin{bmatrix} 4 & -2 & -2 \\ -2 & 1 & 1 \\ -2 & 1 & 1 \end{bmatrix}\begin{bmatrix} 4 & -2 & -2 \\ -2 & 1 & 1 \\ -2 & 1 & 1 \end{bmatrix}$$

$$= \tfrac{1}{36}\begin{bmatrix} 24 & -12 & -12 \\ -12 & 6 & 6 \\ -12 & 6 & 6 \end{bmatrix} = \mathbf{P}_1^2$$

In the case of **L** we have different eigenrows and eigencolumns

$$\lambda_1 = 6 \qquad \mathbf{P}^1 = \tfrac{1}{7}\begin{bmatrix} 2 \\ 5 \end{bmatrix}[1 \quad 1] = \tfrac{1}{7}\begin{bmatrix} 2 & 2 \\ 5 & 5 \end{bmatrix}$$

$$\lambda_2 = -1 \qquad \mathbf{P}^2 = \tfrac{1}{7}\begin{bmatrix} 1 \\ -1 \end{bmatrix}[5 \quad -2] = \tfrac{1}{7}\begin{bmatrix} 5 & -2 \\ -5 & 2 \end{bmatrix}$$

The matrix **D** cannot be expanded in this way because for $\lambda = 3$ we find that $\langle r|c\rangle = 0$

A.15 First diagonalize **M** which is found to have eigenvalues 4 and 2 with corresponding normalized eigenrows

$$\frac{1}{\sqrt{2}}[1 \quad 1] \qquad \frac{1}{\sqrt{2}}[1 \quad -1]$$

Hence the matrix

$$\mathbf{P} = \frac{1}{\sqrt{2}}\begin{bmatrix} 1 & 1 \\ 1 & -1 \end{bmatrix},$$

for which $\tilde{\mathbf{P}} = \mathbf{P}^{-1}$ is such that

$$\tilde{\mathbf{P}}\mathbf{M}\mathbf{P} = \text{diag}\,[4, 2]$$

therefore

$$\mathbf{Q} = \text{diag}\,[1/2, 1/\sqrt{2}]$$

and

$$\tilde{\mathbf{Q}}\tilde{\mathbf{P}}\mathbf{M}\mathbf{P}\mathbf{Q} = \mathbf{I}$$

Now form

$$\begin{aligned}\mathbf{K}' &= \tilde{\mathbf{Q}}\tilde{\mathbf{P}}\mathbf{K}\mathbf{P}\mathbf{Q} \\ &= \begin{bmatrix} 2 & 0 \\ 0 & 1 \end{bmatrix}\end{aligned}$$

which is already diagonal. Hence the transformation matrix

$$\mathbf{S} = \mathbf{P}\mathbf{Q} = \frac{1}{2\sqrt{2}}\begin{bmatrix} 1 & \sqrt{2} \\ 1 & -\sqrt{2} \end{bmatrix}$$

is such that

$$\tilde{\mathbf{S}}\mathbf{M}\mathbf{S} = \mathbf{I}$$

$$\tilde{\mathbf{S}}\mathbf{K}\mathbf{S} = \text{diag}\,(2, 1)$$

and the roots λ of $|\mathbf{K} - \lambda\mathbf{M}| = 0$ are given by

$$\begin{vmatrix} 5 - 3\lambda & 3 - \lambda \\ 3 - \lambda & 5 - 3\lambda \end{vmatrix} = 0$$

$$(5 - 3\lambda)^2 = (3 - \lambda)^2$$

$$5 - 3\lambda = \pm(3 - \lambda)$$

$$\lambda = 2, 1$$

(b) **M** is already diagonal. Hence $\mathbf{P} = \mathbf{I}$

$$\mathbf{Q} = \text{diag}\,[1/\sqrt{2}, 1, 1]$$

and

$$\tilde{\mathbf{Q}}\mathbf{M}\mathbf{Q} = \mathbf{I}$$

$$\mathbf{K}' = \tilde{\mathbf{Q}}\mathbf{K}\mathbf{Q} = \begin{bmatrix} 1 & -1/\sqrt{2} & -1/\sqrt{2} \\ -1/\sqrt{2} & 1 & 0 \\ -1/\sqrt{2} & 0 & 1 \end{bmatrix}$$

The eigenvalues and normalized eigenvectors of $\mathbf{K}'$ are

$$\lambda = \quad 0 \qquad 1 \qquad 2$$

$$\begin{bmatrix} \sqrt{2}/2 \\ \frac{1}{2} \\ \frac{1}{2} \end{bmatrix} \begin{bmatrix} 0 \\ 1/\sqrt{2} \\ -1/\sqrt{2} \end{bmatrix} \begin{bmatrix} \sqrt{2}/2 \\ -\frac{1}{2} \\ -\frac{1}{2} \end{bmatrix}$$

Hence if

$$\mathbf{R} = \begin{bmatrix} 1/\sqrt{2} & 0 & 1/\sqrt{2} \\ \frac{1}{2} & 1/\sqrt{2} & -\frac{1}{2} \\ \frac{1}{2} & -1/\sqrt{2} & -\frac{1}{2} \end{bmatrix}$$

then

$$\mathbf{R}^{-1} = \tilde{\mathbf{R}}$$

and

$$\tilde{\mathbf{R}}\mathbf{K}'\mathbf{R} = \text{diag}\,[0, 1, 2]$$

The transformation matrix

$$\mathbf{QR} = \tfrac{1}{2}\begin{bmatrix} 1 & 0 & 1 \\ 1 & \sqrt{2} & -1 \\ 1 & -\sqrt{2} & -1 \end{bmatrix}$$

is such that

$$\tilde{\mathbf{R}}\tilde{\mathbf{Q}}\mathbf{MQR} = \mathbf{I}$$

$$\tilde{\mathbf{R}}\mathbf{QKQR} = \text{diag}\,[0, 1, 2]$$

where 0 1 2 are the roots of $|\mathbf{K} - \lambda\mathbf{M}| = 0$.

(c) The eigenvalues of $\mathbf{M}$ are $\frac{4}{3}$, 3, and 3, corresponding to the root $\frac{4}{3}$ there is a normalized eigenvector $1/\sqrt{5}[\sqrt{3} \quad 1 \quad -1]$. For the root 3 a general solution is $[a, b, a\sqrt{3} + b]$ from which an orthogonal pair must be selected. A convenient choice, when normalized, is

$$\frac{1}{\sqrt{2}}[0 \quad 1 \quad 1] \quad \text{and} \quad \frac{1}{\sqrt{10}}[2 \quad -\sqrt{3} \quad \sqrt{3}]$$

The matrix $\mathbf{P}$ is then

$$\mathbf{P} = \frac{1}{\sqrt{10}}\begin{bmatrix} 0 & 2 & \sqrt{6} \\ \sqrt{5} & -\sqrt{3} & \sqrt{2} \\ \sqrt{5} & \sqrt{3} & -\sqrt{2} \end{bmatrix}$$

and

$$\tilde{\mathbf{P}}\mathbf{MP} = \text{diag}\,[3, 3, \tfrac{4}{3}]$$

The matrix $\mathbf{Q}$ is therefore

$$\mathbf{Q} = \text{diag}\,[1/\sqrt{3}, 1/\sqrt{3}, \sqrt{3}/2]$$

and

$$\tilde{\mathbf{Q}}\tilde{\mathbf{P}}\mathbf{MPQ} = \mathbf{I}$$

We can then form

$$\mathbf{K}' = \tilde{\mathbf{Q}}\tilde{\mathbf{P}}\mathbf{KPQ}$$
$$= \frac{3}{20}\begin{bmatrix} 15 & -\sqrt{15} & -\sqrt{10} \\ -\sqrt{15} & 13 & \sqrt{6} \\ -\sqrt{10} & \sqrt{6} & 12 \end{bmatrix}$$

We must now diagonalize $\mathbf{K}'$ of which the eigenvalues are $\frac{3}{2}, \frac{3}{2}, 3$. For $\lambda = 3$ the normalized eigenvector is $1/\sqrt{10}[-\sqrt{5} \quad \sqrt{3} \quad \sqrt{2}]$ whilst for $\lambda = \frac{3}{2}$ the components x_1, x_2, x_3 of the eigenvector must satisfy $\sqrt{5}x_1 - \sqrt{3}x_2 - \sqrt{2}x_3 = 0$ and a suitably normalized pair are $1/\sqrt{10}(0 \quad -2 \quad \sqrt{6})$, $1/\sqrt{10}$ $(\sqrt{5} \quad \sqrt{3} \quad \sqrt{2})$. The matrix

$$\mathbf{R} = \frac{1}{\sqrt{10}}\begin{bmatrix} -\sqrt{5} & 0 & \sqrt{5} \\ \sqrt{3} & -2 & \sqrt{3} \\ \sqrt{2} & \sqrt{6} & \sqrt{2} \end{bmatrix}$$

is then such that

$$\tilde{\mathbf{R}}\mathbf{K}'\mathbf{R} = \text{diag}\,[3, \tfrac{3}{2}, \tfrac{3}{2}]$$

and the matrix

$$\mathbf{S} = \mathbf{PQR} = \frac{1}{2\sqrt{3}}\begin{bmatrix} \sqrt{3} & 1 & \sqrt{3} \\ -1 & \sqrt{3} & 1 \\ -1 & -\sqrt{3} & 1 \end{bmatrix}$$

is such that

$$\tilde{\mathbf{S}}\mathbf{MS} = \mathbf{I}$$

and

$$\tilde{\mathbf{S}}\mathbf{KS} = \text{diag}\,[3, \tfrac{3}{2}, \tfrac{3}{2}]$$

where $3, \frac{3}{2}, \frac{3}{2}$ are the roots of $|\mathbf{K} - \lambda\mathbf{M}| = 0$.

A.16 Let $|e_i\rangle$ be the covariant basis. Then the contravariant basis is the reciprocal basis $|\bar{e}_i\rangle$. In terms of the contravariant basis the contravariant components of a vector $|a\rangle$ are $\bar{a}_i$

$$|a\rangle = \sum_i \bar{a}_i|\bar{e}_i\rangle$$

Hence

$$\langle e_j|a\rangle = \sum \bar{a}_i\langle e_j|\bar{e}_i\rangle$$
$$= \bar{a}_i\delta_{ij}$$

by eq. (A.108). Therefore

$$\bar{a}_j = \langle e_j|a\rangle$$

Now the projection of

$$|a\rangle \text{ on } |e_j\rangle \text{ is } |e_j\rangle\frac{\langle e_j|a\rangle}{\langle e_j|e_j\rangle}$$

(compare the projection operator of question A.13) so that the contravariant component $\bar{a}_j$ of $|a\rangle$ is proportional to the projection of $|a\rangle$ onto the basis $|e_j\rangle$.

A.17

$$T_{ii'}T_{ij'} = T^{\dagger}_{i'i}T_{ij'} = T^{-1}_{i'i}T_{ij'} = \delta_{i'j'}$$

because T is unitary.

A.18

$$B_{lk} = (R^{-1})_{li}A_{ij}R_{jk}$$

The trace of **B** is B_{ll}

$$\begin{aligned} \therefore \quad B_{ll} &= (R^{-1})_{li}A_{ij}R_{jl} \\ &= (R_{jl})(R^{-1})_{li}A_{ij} \\ &= \delta_{ij}A_{ij} \\ &= A_{ii} \end{aligned}$$

Therefore Trace **B** = Trace **A**.

Further reading

We give, below, a list of books which will be of value to students who wish to develop their interest in symmetry and group theory, or their knowledge of the background material to individual chapters. In order to make the list more useful we have divided it into categories.

Complementary texts

Bhagavantam S. and Venkatarayudu T., *Theory of Groups and its Application to Physical Problems*, Academic Press, 1969.

Atkins P. E., Child M. S. and Philipps C. S. G., *Tables for Group Theory*, Oxford University Press, 1970.

Leech J. W. and Newman D., *How to Use Groups*, Science Paperbacks; Chapman & Hall, Methuen, and Spon, 1969.

Cracknell A. P., *Applied Group Theory*, Pergamon, 1968.

Nussbaum A., *Applied Group Theory for Chemists, Physicists and Engineers*, Prentice-Hall, 1971.

Salthouse J. A. and Ware M. J., *Point Group Character Tables and Related Data*, Cambridge UP, 1972.

Advanced texts

Hollingsworth C. A., *Vectors, Matrices and Group Theory for Scientists and Engineers*, McGraw-Hill, New York, 1967.

Heine V., *Group Theory and Quantum Mechanics*, Pergamon, 1960.

Tinkham M., *Group Theory and Quantum Mechanics*, McGraw-Hill, London, 1964.

McWeeny R., *Symmetry: An Introduction to Group Theory and its Applications*, Pergamon, 1963.

Hamermesh M., *Group Theory and its Application to Physical Problems*, Pergamon, 1962.

Hall G. G., *Applied Group Theory*, Longmans, 1967.

Nye J. F., *Physical Properties of Crystals*, Oxford UP, 1956.

Knox R. S. and Gold A., *Symmetry in the Solid State*, Benjamin, New York, 1964.

Nussbaum A., *Crystal Symmetry, Group Theory and Band Structure Calculations Solid State Physics: Advances in Research and Applications*, Volume 18, pp. 165–272, 1966.

Falicov L. M., *Group Theory and its Physical Applications*, The University of Chicago Press, 1966.

Hochstrasser R. M., *Molecular Aspects of Symmetry*, Benjamin, New York, 1966.

Background reading

Ayres F., *Matrices* (Schaum's Outline Series), Schaum, New York, 1962.

Spiegel M. R. *Vector Analysis* (Schaum's Outline Series), Schaum, New York, 1959.

Dennery P. and Krzywicki A., *Mathematics for Physicists*, Harper & Row, New York; Weatherhill, Tokyo, 1967.

Landau L. D. and Lifshitz E. M., *Quantum Mechanics* (especially pp. 332 *et seq.*), Pergamon, 1958.

Jones, H., *The Theory of Brillouin Zones and Electronic States in Crystals*, North-Holland Publishing, Amsterdam, 1962.

Smith A. C., Janak J. F., and Adler R. B., *Electronic Conduction in Solids*, McGraw-Hill, London, 1967.

Buerger M. J., *Elementary Crystallography*, Wiley, New York, 1963.
Wilson E. B., Decius J. G., and Cross P. C., *Molecular Vibrations*, McGraw-Hill, New York, 1955.
King G. W., *Spectroscopy and Molecular Structure*, Holt, Rinehart & Winston, 1964.
Ballhausen C. J. and Gray H. B., *Molecular Orbital Theory*, Benjamin, New York, 1965.
Hall G. G., *Matrices and Tensors*, Pergamon, 1963.
Cracknell A. P., *Crystals and Their Structures*, Pergamon, 1969.
Atkin R. H., *Mathematics and Wave Mechanics*, Heinemann, 1956.
Pauling L. and Wilson E. B., *Introduction to Quantum Mechanics*, McGraw-Hill, New York, 1935.
Schiff L. I., *Quantum Mechanics*, McGraw-Hill, New York, 1968.
Messiah A., *Quantum Mechanics*, Volumes I and II, North Holland Publishing, Amsterdam, 1965.
Leighton R. B., *Principles of Modern Physics*, McGraw-Hill, New York, 1959.

Index

Printed by J. W. Arrowsmith Ltd., Bristol, England.